Defense Applications of Nanomaterials

ACS SYMPOSIUM SERIES **891**

Defense Applications of Nanomaterials

Andrzej W. Miziolek, Editor
U.S. Army Research Laboratory

Shashi P. Karna, Editor
U.S. Army Research Laboratory

J. Matthew Mauro, Editor
U.S. Naval Research Laboratory

Richard A. Vaia, Editor
Air Force Research Laboratory

Sponsored by the ACS Division of Industrial and Engineering Chemistry, Inc.

American Chemical Society, Washington, DC

Library of Congress Cataloging-in-Publication Data

Defense applications of Nanomaterials / Andrzej W. Miziolek, editor ...[et al.].

p. cm.—(ACS symposium series ; 891)

Includes bibliographical references and index.

ISBN 0–8412–3806–5 (alk. paper)

1. Nanostructured materials—Military Applications.

I. Miziolek, Andrzej W., 1950- II. Series.

UG489.5.D44 2004
623—dc22 2004053126

The paper used in this publication meets the minimum requirements of American National Standard for Information Sciences—Permanence of Paper for Printed Library Materials, ANSI Z39.48–1984.

Distributed by Oxford University Press

PRINTED IN THE UNITED STATES OF AMERICA

Foreword

The ACS Symposium Series was first published in 1974 to provide a mechanism for publishing symposia quickly in book form. The purpose of the series is to publish timely, comprehensive books developed from ACS sponsored symposia based on current scientific research. Occasionally, books are developed from symposia sponsored by other organizations when the topic is of keen interest to the chemistry audience.

Before agreeing to publish a book, the proposed table of contents is reviewed for appropriate and comprehensive coverage and for interest to the audience. Some papers may be excluded to better focus the book; others may be added to provide comprehensiveness. When appropriate, overview or introductory chapters are added. Drafts of chapters are peer-reviewed prior to final acceptance or rejection, and manuscripts are prepared in camera-ready format.

As a rule, only original research papers and original review papers are included in the volumes. Verbatim reproductions of previously published papers are not accepted.

ACS Books Department

Contents

Novel Nanomaterials, Processes, and Nanomaterial Applications

Nanoenergetics

Nanoelectronics and Photonics

Indexes

Preface

Technology and innovation define who we are and the society in which we live. We need only to consider materials, where the crucial innovations are synonymous with broad cultural ages (Stone, Bronze, Iron, and Silicon) named for the vital discoveries in shaping the substances that made the driving force of technology of their day possible. Our ability to transform these fundamental understandings into the machines that define the world around us continues to accelerate as the tools available multiply with the breakthroughs of each generation. The world is truly getting smaller as the distance between the farthest locales on earth becomes mere seconds as communication becomes ubiquitous.

Incredible challenges and opportunities loom on the horizon, including exploration into the world of atomic-level, human-engineered materials and devices. Nanoscience and technology have been growing at a phenomenal rate in recent years and offer a diverse range of potential applications within communications, manufacturing, medicine, biotechnology, coatings, sensors, and processing. Worldwide, increasing numbers of researchers from a wide variety of disciplines enter the field each year, spawning novel ideas and creating unforetold opportunities.

This book highlights recent accomplishments in nanostructured materials and their applications to critical technologies for national defense. The contributions examine the unique chemical and physical aspects of nanostructured materials and discuss the latest fundamental and applied research in this area. Special emphasis was given to the necessary directions for the development of high-performance nanomaterials critical to future defense technologies, including sensors, data storage, information transfer, structural materials, and energetics. The chapters reflect the multidisciplinary nature of this research, including contributions from chemists, materials scientists, physicists, chemical engineers, and processing specialists. Perspectives from both the scientists working on the basic issues surrounding pushing back the frontiers of nanomaterials, as well as those working on applied problems associated with bringing to fruition the promise of nanomaterials for defense applications, create a unique contribution to this rapidly changing field.

The editors are especially grateful for the dedication and effort of our colleagues who participated and contributed to the symposium and this volume. Additionally, without the tireless effort of the American Chemical Society (ACS) Books Department, this book would not have been possible. Finally, Miziolek thanks Nora Beck-Tan, formerly of the U.S. Army Research Laboratory, for her insight and key suggestions that led to the successful ACS Symposium upon which this book is based.

Andrzej W. Miziolek
U.S. Army Research Laboratory
Weapons and Materials Research Directorate
AMSRD-ARL-WM-BD
Aberdeen Proving Ground, MD 21005–5067
miziolek@arl.army.mil

Shashi P. Karna
U.S. Army Research Laboratory
Weapons and Materials Research Directorate
AMSRD-ARL-WM-BD
Aberdeen Proving Ground, MD 21005–5067
karnas@plk.af.mil

J. Matthew Mauro*
U.S. Naval Research Laboratory
Department of the Navy
4554 Overlook Drive, SW
Washington, DC 20375
matt.mauro@probes.com

Richard A. Vaia
Air Force Research Laboratory
AFRL/MLBP, Building 654
2941 Hopson Way
Wright-Patterson Air Force Base, OH 45433–7750
richard.vaia@wpafb.af.mil

*Current address: Molecular Probes/Invitrogen Corporation, 29851 Willow Creek Road, Eugene, OR 97402

Defense Applications of Nanomaterials

Overview

Chapter 1

Overview of the Nanoscale Science and Technology Program in the Department of Defense

J. S. Murday[1], B. D. Guenther[2], C. G. Lau[3], C. R. K. Marrian[4], J. C. Pazik[5], and G. S. Pomrenke[6]

[1]Department of the Navy, Naval Research Laboratory, 4554 Overlook Avenue, SW, Washington, DC 20375–5320
[2]Department of Physics, Box 90305, Duke University, Durham, NC 27708
[3]DUSD (S&T/BR) Ballston Centre Tower 3, 4015 Wilson Boulevard, Arlington, VA 22203
[4]Defense Advanced Research Projects Agency and Department of the Navy, Naval Research Laboratory, 4554 Overlook Avenue, SW, Washington, DC 20375–5320
[5]Office of Naval Research, Ballston Centre Tower 1, 800 North Quincy Street, Arlington, VA 22217
[6]Air Force Office of Scientific Research, Ballston Centre Tower 3, 4015 Wilson Boulevard, Arlington, VA 22203–1954

The U.S. National Nanotechnology Initiative (NNI) is a coordinated multiagency/department program with the NSF leading the fundamental science investment and the other agencies/departments focused on science investment that promises revolutionary new technologies relevant to their missions. The Department of Defense has identified three nanoscience challenges for focused S&T investments – nanoelectronics, magnetics and photonics; nanomaterials; and nanobiodevices. This article will succinctly sketch the reasons for excitement over nanoscience and the DoD program.

Introduction

Nanoscience involves materials where some critical property is attributable to a structure with at least one dimension limited to the nanometer size scale, 1 – 100 nanometers[1,2]. Below that size the disciplines of Chemistry and Atomic/Molecular Physics have already provided detailed scientific understanding. Above that size scale, in the last 50 years Condensed Matter Physics and Materials Science have provided detailed scientific understanding of microstructures. So the nanoscale is the last "size" frontier for materials science.

The interest in nanostructures extends beyond their individual properties. We have learned to exploit the natural self-assembly of atoms/molecules into crystals. The directed self-assembly of nanostructures into more complex, hierarchical systems is also an important goal. Without direct self-assembly, manufacturing costs will severely limit the nanotechnology impact.

While the scientific understanding of nanostructures is deficient, their use in technology is at least two thousand years old. The Lycurgis cup, a Roman artifact pictured in the lower left of Figure 1, utilizes nanosized Au clusters to provide different colors depending on front or back lighting. The Roman artisans knew how to achieve the effect; they didn't know its nanocluster basis.

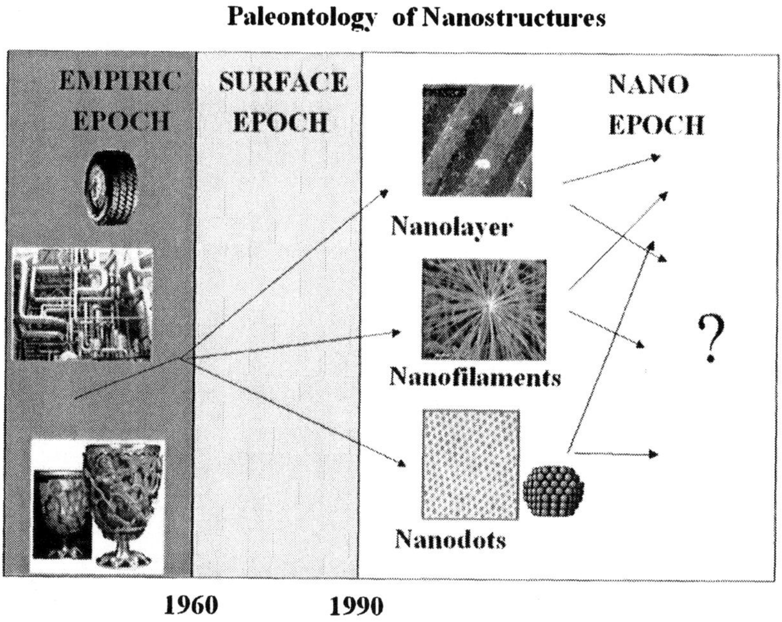

Figure 1. Paleontology of Nanostructures (See page 1 of color insert.)

In the last century nanostructures have contributed to <u>many</u> significant technologies - examples include the addition of nanosized carbon particles to rubber for improved mechanical properties (tires) and the use of nanosized particles for catalysis in the petrochemical industries. These technologies were all developed empirically. As depicted in Figure 1, one might assign these examples to an empiric epoch in the continuing evolution of nanotechnology.

Empirically based technology, without greater scientific understanding, is usually difficult to extend or control. Since the Department of Defense demands the highest performance from its systems, it is always looking to enhance material properties and to minimize their failure mechanisms. So there is a strong DOD interest in nanoscience. The scientific foundation of nanostructures received a quantum advance when surface science enjoyed a renaissance starting in the 1960s. Surface science constrained one material dimension into the nanometer size scale. Events catalyzing that renaissance were the development of new surface-sensitive analytical tools, the ready availability of ultra-high vacuum (a by-product of the space race), and the maturity of solid-state physics (surfaces representing a controlled lattice defect – termination of repeating unit cells). The principal economic driving force was the electronics industry. From 1960 to now, surface science has progressed from "clean, flat and cold" into thin films (two or more nanoscale interfaces) and film processing. Superlattice devices (see Figure 2) are a growing technological manifestation where the individual layers are nanometer in thickness and the interfaces are controlled with atomic precision.

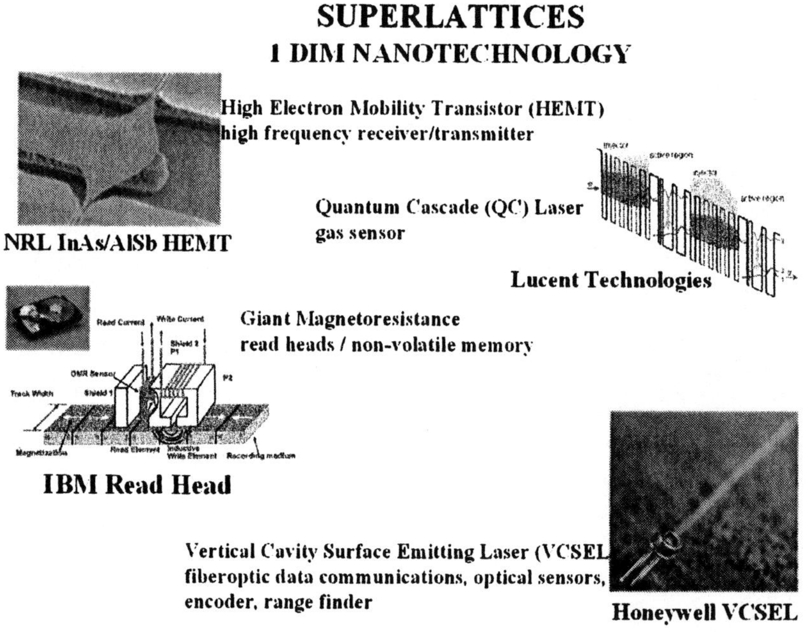

Figure 2. Superlattice-based Commercial Devices (See page 1 of color insert.)

If nanotechnology has been around so long, why is it the current rage? The 1990s nanoscience renaissance has close parallels to the 1960 surface science renaissance. First, beginning in 1980, the discovery and development of proximal probes – scanning tunneling microscopy/spectroscopy, atomic force microscopy/spectroscopy, near-field microscopy/spectroscopy – have provided tools for measurement and manipulation of individual nanosized structures[3]. Those tools needed 10-15 years for reliable commercial instruments to come onto the markets. Note the rapid increase in refereed nanoscience publications beginning in the early 1990s (see Figure 3). The properties of the individual nanostructures can now be observed, rather than the ensemble averaged values. In turn, those properties can be understood in terms of composition / structure, with that understanding comes the possibility for control, and with control comes the possibility for accelerated progress toward new technology.

Second, in addition to the new experimental measurement capabilities, computer hardware is now sufficiently advanced (speed and memory capacity) such that accurate predictions, even first principle, are enabled for the number of atoms incorporated in a nanostructure (see Table I). Accurate predictions for carbon nanotubes were made years before experimental measurements could confirm their validity. Modeling and simulation will play a leading role in the race toward nanotechnology.

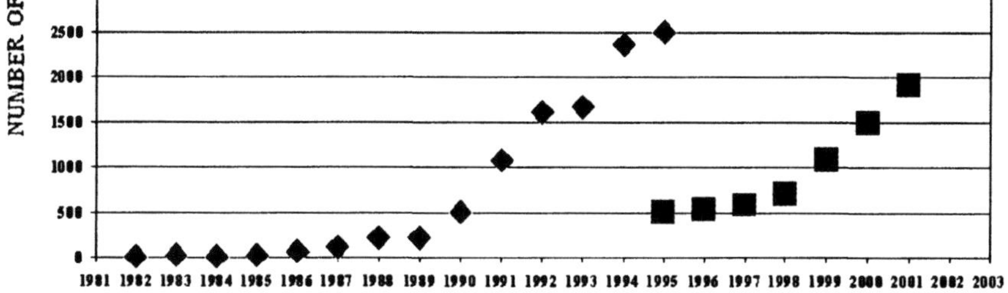

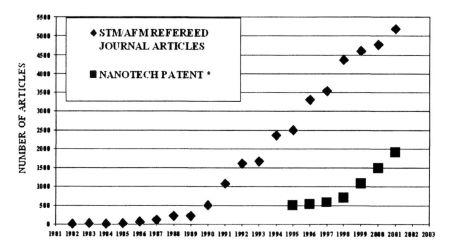

* "Nanotechnology – Size Matters", white paper, 10 July 2002, Institute of Nanotechnology

Figure 3. Annual Publication Count

Table I. Carbon Nanotubes: Prediction and Subsequent Confirmation

Prediction	*Experimental Confirmation*
Armchair SWNTs metallic.[4]	Individual single-wall carbon nanotubes as quantum wires.[5]
Elastic modulus should be similar to high stiff-ness in-plane modulus of graphite.[6]	Nanobeam Mechanics: Elasticity, Strength, and toughness of Nanorods and graphite Nanotubes.[7]
Stability of metallic SWNT conduction channels in presence of disorder, leading to exceptional ballistic transport properties.[8]	Fabry-Perot interference in a nanotube electron wave-guide.[9]

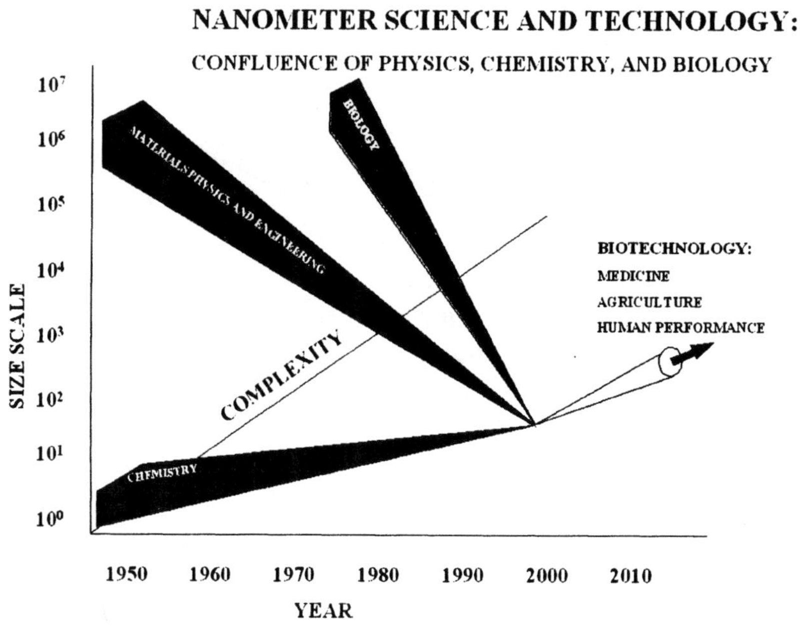

Figure 4. Convergence of biology, chemistry, and physics at the nanoscale.

Third, the disciplines of biology, chemistry, materials, and physics and have all reached a point where nanostructures are of interest – chemistry building up from simpler molecules, physics/materials working down from microstructures, and biology sorting out from very complex systems into simpler subsystems (see Figure 4).

If one expected to simply extrapolate the properties of nanostructures from the size scales above or below, then there would be little reason for the current interest in nanoscience/nanotechnology. There are three reasons for nanostructured materials to behave very differently: large surface/interface to volume ratios, size effects (where cooperative phenomena like ferromagnetism is compromised by the limited number of atoms/molecules) and quantum effects. Many of the models for materials properties at the micron and larger sizes have characteristic length scales of nanometers (see Table II). When the size of the structure is nanometer, those parameters will no longer be adequate to model/predict the property. One can expect "surprises" – new materials behavior that may be technologically exploitable.

Table II. Characteristic Lengths in Solid State Science Models

FIELD	PROPERTY	SCALE LENGTH
ELECTRONIC	ELECTRON WAVELENGTH	10 -- 100nm
	INELASTIC MEAN FREE PATH	1 -- 100nm
	TUNNELING	1 -- 10nm
MAGNETIC	DOMAIN WALL	10 -- 100nm
	EXCHANGE ENERGY	0.1 -- 1nm
	SPIN-FLIP SCATTERING LENGTH	1 -- 100nm
OPTIC	QUANTUM WELL	1 -- 100nm
	EVANESCENT WAVE DECAY LENGTH	10 -- 100nm
	METALLIC SKIN DEPTHS	10 -- 100nm
SUPER-CONDUCTIVITY	COOPER PAIR COHERENCE LENGTH	0.1 -- 100nm
	MEISSNER PENETRATION DEPTH	1 -- 100nm
MECHANICS	DISLOCATION INTERACTION	1 -- 1000nm
	GRAIN BOUNDARIES	1 -- 10nm
NUCLEATION/ GROWTH	DEFECT	0.1 -- 10nm
	SURFACE CORRUGATION	1 -- 10nm
CATALYSIS	LOCALIZED BONDING ORBITALS	0.01 -- 0.1nm
	SURFACE TOPOLOGY	1 -- 10nm
SUPRA-MOLECULES	PRIMARY STRUCTURE	0.1 -- 1nm
	SECONDARY STRUCTURE	1 -- 10nm
	TERTIARY STRUCTURE	10 -- 1000nm
IMMUNOLOGY	MOLECULAR RECOGNITION	1 -- 10nm

Finally, there are several economic engines driving the interest with information technology (electronics), biotechnology (pharmaceuticals, healthcare, and agriculture), and high performance materials the more certain beneficiaries. Several estimates have been made for the economic impact of nanotechnology. They all cite a worldwide commercial market on the order of $1T per year by 2020 for systems whose function is enabled by the properties of nanostructures – "Nano Inside."

With these substantial scientific and economic opportunities, it not surprising to find strong global interest in fostering nanoscience, with the intent of accelerating scientific discovery into innovative commercial product. The increasing nanotechnology patent literature shown in Figure 3 gives evidence for that acceleration. Table III has an estimate of governmental funding in the nanosciences around the globe. From estimates of FY03 budget projections, it is anticipated that well over $3B will be invested in nanotechnology S&T in 2003, with the U.S. contribution around $850M, and the DOD contribution around $300M. One should be careful in the comparison of funding levels between different countries. For instance, one estimate of nanotechnology investment in China is around $25M/year; however, the manpower costs in China are significantly less than in the U.S. and that investment is not so small as it might seem.

US National Nanotechnology Initiative

The U.S. response to the nanoscale scientific and economic opportunities is the National Nanotechnology Initiative (NNI), recently enacted into law by the 108th Congress as the "21st Century Nanotechnology research and Development Act." An outline of the funding categories in the NNI is provided below (for more details see National Nanotechnology Initiative, detailed technical report

Table III. "Nanotechnology" Research Program Investment ($M)

	1997	*2000*	*2001*	*2003*
USA	115	270	420	~800
Japan	120	245	410	~800
Western Europe	125	200	225	~700
Other Countries (FSU, China, Australia, others)	70	110	200	~800
Total	430	749	1230	>3000

associated with the supplemental report to the President's FY2004 Budget[10]. Table IV shows the annual investment levels in nanoscience:

Table IV. U.S. National Nanotechnology Initiative Program ($M)

Category	*Lead*	*FY00*	*FY01*	*FY03*
Knowledge Generation		87	140	230
Grand Challenges		71	125	400
Nanostructured Materials by Design	NSF			~120
Nanoelectronics, Photonics, Magnetics	DOD			~110
Advanced Healthcare/Therapeutics	NIH			~60
Environmental Improvements	EPA			~10
Energy Conversion/Storage	DOE			~10
Microcraft and Robotics	NASA			~ 5
CBRE Protection/Detection	DOD			~20
Instrumentation & Metrology	NIST			~50
Nanoscale Manufacturing	NSF			~10
Centers/Networks		47	66	130
Infrastructure		50	77	90
Ethical/Social Implications		15	21	16
Totals		270	422	~860

The Knowledge Generation investment is deliberately broad based. Science discovery frequently occurs in unexpected ways; it is important to have part of the investment portfolio available to give breadth and depth to the range of nanoscience projects. In the Grand Challenges, the research investments are selected with technological ramifications in mind. The lead agency has the responsibility for surveying the investment in the designated Grand Challenge and for identifying investment opportunities/shortfalls. Most of the Grand Challenge topics have been the subject of a recent workshop; copies of the workshop reports can be obtained from the NNI website (www.nano.gov).

DoD contributions to Nanoscale Science and Engineering

The DoD has been investing in fundamental nanoscience research for over 20 years. As examples: a) one of the early programs, dating into the early 1980s was Ultra-submicron Electronics Research (USER); b) ONR initiated an accelerated research initiative (ARI) in Properties of Interfacial Nanostructures in 1990; c) ARO initiated a nanoscience University Research Initiative (URI) in 1991; and d) DARPA initiated the ULTRA (ultra dense, ultra fast computing) program in 1991. In 1997 the DoD identified several S&T topics with the potential for significant impact on military technology; nanoscience was selected as one of those special research area (SRA) topics. Selected examples of anticipated impact are listed in Table V. A website[11] and coordinating committees have been established, and each Service has its own laboratory nanoscience programs.

Table V. Nanoscale Opportunities with Major DoD Impact

Nanoelectronics/Optoelectronics/Magnetics
- Network Centric Warfare
- Information Warfare
- Uninhabited Combat Vehicles
- Automation/Robotics for Reduced Manning
- Effective training Through Virtual Reality
- Digital Signal Processing and LPI

Nanomaterials "by Design"
- High Performance, Affordable Materials
- Energetic Materials and Controlled Release of Energy
- Multifunction Adaptive (Smart) Materials
- Nanoengineered Functional Materials (Metamaterials)
- Reduced Maintenance (halt nanoscale failure initiation)

BioNanotechnology – Warfighter Protection
- Chemical/Biological Agent detection/Destruction
- Human Performance/Health Monitor/Prophylaxis

While the NNI is an initiative focused on fundamental science (i.e., DoD 6.1 funding category), one of the principal NNI goals is to transition science discovery into new technology. The DoD structures its S&T investment into basic research (6.1), applied research (6.2) and exploratory development (6.3); the latter two focus on transitioning science discovery into innovative technology. MANTECH, SBIR and STTR programs are also available for transition efforts. There are nanoscience discoveries for which technology potentials are clear and transition funding is appropriate. Beginning in FY02, the DoD is tracking and encouraging the transitions into these applied programs. They are included under the label "6.2/6.3" in Table VI.

Table VI. DoD Investment in Nanoscience

	1998	*2000*	*2001*		*2002*		*2003*	
	6.1	6.1	6.1	6.2/6.3	6.1	6.2/6.3	6.1	6.2/6.3
Total	73	90	121	32	141	84	179	143

The Air Force is increasing its investment in nanoscience with a NAS panel report as guidance[12]. The Air Force basic research activities address topics in:

- *Nanocomposites*- hybrid polymer-inorganic nanocomposites for dramatic improvement over the properties of traditional polymers without sacrificing density, processability or toughness as in conventional composite/blend approaches; dispersed carbon nanotubes in polymer fiber to provide improvement in tensile and compressive properties of high performance polymer fibers.
- *Self-assembly and Nanoprocessing* - advances in organic and organic/inorganic nanoparticles and nanoscale materials and nanoscale materials processing techniques to create the opportunity for development of new paradigms for the realization of 3-D optical and electronic circuitry.
- *Highly Efficient Space Solar Cells* - if U.S. satellites could be developed which more efficiently convert light to electrical power, then lower launch costs and heavier payloads could be realized. Nanotechnology may enable space solar cells that can operate at an efficiency of 60%.
- *Nanoenergetics* - understanding the factors that control reactivity and energy release in nanostructured systems, and developing the structures and architectures to optimize them.
- *Nanostructures for Highly Selective Sensors and Catalysts* - nanostructures for the remote sensing and identification of chemical and physical species, and catalysts to selectively react with compounds to release energy (monopropellants) or destroy them (undesirable chemicals).
- *Nanoelectronic, Nanomagnetics* and *Nanophotonics* – ultra-dense magnetic memory, ultra-fast digital signal processing, nanosensing, spatial and temporal dispersion characteristics of photonic crystals, and a computer architecture based on one-dimensional cellular automata which offers an important solution to molecular scale limitations.
- *Nanostructured coatings, ceramics and metals* - tailor the structure of coatings at the nano-scale to provide unique and revolutionary properties, specifically nano-structured adaptive tribological (low friction) coatings for MEMS devices; economical, multifunctional ceramic materials to which operate in extreme environments; new material concepts and design tools to impact mechanics issues important to airframes.

The Army is augmenting its program with a University Affiliated Research Center (UARC) - Institute for Soldier Nanotechnologies. The purpose of this center of excellence is to develop unclassified nanometer-scale science and technology solutions for the soldier. MIT hosts this center, which will emphasize revolutionary materials research toward advanced soldier protection and survivability capabilities. The center works in close collaboration with industry, the Army's Natick Soldier Center (NSC), the Army Research Laboratory (ARL) and the other Army Research Development and Engineering Centers (RDECs) in pursuit of the Army's goals. The research will integrate a wide range of functionalities, including multi-threat protection against ballistics, sensory attack, chemical and biological agents; climate control (cooling, heating, and insulating), possible chameleon-like garments; biomedical monitoring; and load management. In addition, the Army has just created a second UARC entitled Institute for Collaborative Biotechnology at the University of California, Santa Barbara; this center will also seek to exploit nanoscale phenomena.

The Navy investment in nanotechnology has an emphasis in the area of nanoelectronics. Reprogramming $10M of its core funds, the Naval Research Laboratory has initiated an Institute for Nanoscience to enhance multidisciplinary thinking and critical infrastructure. The mission of the institute is to conduct highly innovative, interdisciplinary research at the intersection of the fields of materials, electronics and biology in the nanometer domain. A new Nanoscience Building has been constructed at NRL to provide modern, state-of-the art facilities dedicated to nanoscience research and nanotechnology development. The building has been specially designed and built to minimize sources of measurement noise (acoustic, vibration, electromagnectic, contamination, temperature/pressure fluctuation, electrical power).

Conclusion

Nanoscience shows great promise for arrays of inexpensive, integrated, miniaturized sensors for chemical / biological / radiological / explosive (CBRE) agents, for nanostructures enabling protection against agent, and for nanostructures that neutralize agents[13]. The recent terrorist events motivate accelerated insertion of innovative technologies to improve the national security posture relative to CBRE. The NNI has redefined a Grand Challenge to address this important topic; DoD expects to play a major role in this multiagency effort.

DoD programs in nanoscience are major contributors to the NNI with the expectation of significant technology innovations for national security and homeland defense. The articles in this book illustrate the breadth and quality of those programs. An Advanced Materials and Processing Technology Information Center (AMPTIAC) Newsletter[14] provides a succinct summary of many DOD programs, and their accomplishments and expectations.

References:

1. IWGN Workshop Report: Nanotechnology Research Directions, M.C. Roco, R.S. Williams and P. Alivisatos Eds., (Kluwer Academic Publishers, ISBN 0-7923-62220-9, 2000).
2. Nanotechnology: A Gentle Introduction to the Next Big Idea, Mark A. Ratner and Daniel Ratner, (Prentice Hall PTR; ISBN 0131014005, 2002).
3. Scanning Probe Microscopy and Spectroscopy: Methods and Applications, R. Wiesendanger (Cambridge University Press, ISBN 0-521-41810-0, 1994).
4. Mintmire, Dunlap, White, PRL 68, 631 (1992)
5. Tans, et.al. Nature 386, 474 (1997)
6. Robertson, Brenner, Mintmire, PRB 45, 12592 (1992)
7. Wong, Sheehan, and Lieber, Science 277, 1971 (1997)
8. White, Todorov, Nature 393, 240 (1998)
9. Liang et. al., Nature 411, 665 (2001)
10. http://nano.gov/nsetrpts.htm
11. http://www.nanosra.nrl.navy.mil
12. Implications of Emerging Micro and Nanotechnologies (National Academy Press, ISBN 0-309-08623-X, 2002).
13. Nanotechnology and Homeland Security, Daniel Ratner and Mark A. Ratner, (Prentice Hall PTR; ISBN 1-13-145307-6, 2002)
14. "A Look Inside Nanotechnology," AMPTIAC Quarterly, Volume 6, Number 1, Spring 2002, ITT Research Institute/AMPTIAC, 201 Mill Street, Rome, N.Y., 13440-6916

Sensors and Sensor Materials

Chapter 2

Receptor Protein-Based Bioconjugates of Highly Luminescent CdSe-ZnS Quantum Dots: Use in Biosensing Applications

J. M. Mauro[1], H. Mattoussi[2], I. L. Medintz[1], E. R. Goldman[1], P. T. Tran[1], and G. P. Anderson[1]

[1]Center for Bio/Molecular Science and Engineering and [2]Division of Optical Sciences, U.S. Naval Research Laboratory, 4554 Overlook Avenue, SW, Washington, DC 20375

We are developing new and useful ways to prepare bio-inorganic conjugates of highly luminescent semiconductor quantum nanocrystals (Quantum Dots; QDs) and proteins for use in biosensing applications. Conjugate assembly is driven by electrostatic interaction of negatively charged QD surfaces with positively charged proteins or protein subdomains. Conjugates retain the properties of both starting materials, i.e., biological activity of the proteins and optical characteristics of the QDs. We have used these hybrid bio-inorganic conjugates as tracking reagents in fluoroimmunoassays.

Colloidal semiconductor CdSe-ZnS core-shell quantum dots (QDs) are luminescent inorganic nanoparticles that have the potential to obviate some of the functional problems of fluorescent organic dyes in labeling applications. Problems with organic fluorophores include narrow excitation bands and broad emission spectra, which can make multiplex detection difficult due to spectral overlap, as well as low resistance to chemical and photodegradation (*1,2*). QDs have size-dependent tunable broad excitation and photoluminescence (PL) spectra with narrow emission bandwidths (full width at half maximum of ~ 30-45 nm) that span the visible spectrum (*1,2*); this allows simultaneous excitation of several particle sizes at a single wavelength. QDs also have exceptional photochemical stability, and their high quantum yield allows luminescence emission to be observed at concentrations comparable to organic dyes using conventional fluorescence methods. Finally, using polar bifunctional compounds such as 1-mercaptoacetic or dihydrolipoic acids to modify QD surfaces allows preparation of stable aqueous nanocrystal dispersions that can subsequently be used for bioconjugation (*1-5*).

In a departure from previous covalent chemistry techniques used for bioconjugate formation (*6-8*), we have developed a conjugation strategy based on the electrostatic interactions of negatively charged dihydrolipoic acid (DHLA)-capped CdSe-ZnS core-shell QDs with positively charged protein receptors which either occur naturally or are engineered to contain positively charged interaction domains. In this chapter, we describe the formation of water-compatible DHLA-capped quantum dots, methods of making electrostatically-stabilized QD-protein bioconjugates, and some of the properties of these materials. Subsequent preparation of QD-antibody conjugates and their employment in fluoroimmunoassays will also be described (*3-5*).

CdSe-ZnS Quantum Dot Preparation

In a typical preparation, a solution of dimethylcadmium ($CdMe_2$) and trioctylphosphine selenide (TOPSe), diluted in trioctylphosphine (TOP), is rapidly injected into a hot stirring solution of trioctylphosphine oxide (TOPO). (*1, and references therein*). The rapid introduction of these reagents and concomitant temperature drop result in discrete temporal nucleation of CdSe seeds. After reagent injection, the temperature of the solution is raised to 300-350°C in order to grow the particles (Figure 1). The high temperature growth promotes highly crystalline QD cores. Growth is monitored through UV/visible spectroscopy and when the desired size is reached (as monitored by the peak wavelength of the first absorption feature), the temperature is dropped below 100°C to arrest the growth.

Passivating the native CdSe QDs with an additional thin layer made of a wider band gap semiconductor (to make core-shell nanocrystals) improves the surface quality of the particles by providing better passivation of surface states resulting in dramatic enhancements of the fluorescence quantum yield (Figure 1).

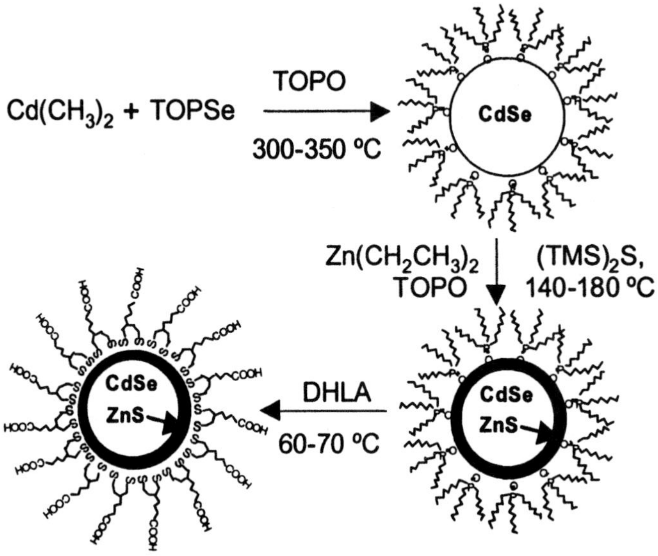

Figure 1. High temperature organometallic growth, ZnS overcoating, and DHLA capping of colloidal CdSe quantum dots.

Although the principle was previously known from semiconductor bandgap engineering (*1,2*), the optimal set of conditions for creating strongly fluorescent overcoated QDs was not realized until the published work of Hines and Sionnest (*9*), when they showed that overcoating CdSe QDs with ZnS improved quantum yields to 30% or greater. This was shortly followed by other studies describing additional characterization of CdSe QDs overcoated with ZnS (*10*) and CdS (*11*).

Cap Exchange for Aqueous Compatibility

TOP/TOPO capping groups can be subsequently exchanged with dihydrolipoic acid (DHLA) groups by suspending TOP/TOPO dots in dihydrolipoic acid and heating at ~ 60-70°C for several hours (Figure 1). Following deprotonation of the terminal lipoic acid –COOH groups with potassium-tert-butoxide (K-t-butoxide), the centrifugally sedimented cap exchanged nanocrystalline material can be dispersed in water (*1,3*). After removal of excess hydrolyzed K-t-butoxide by repeated dilution/concentration using an Ultra-free centrifugal filtration device (Millipore, MW cut-off of ~ 50,000 daltons), QD suspensions are obtained with the same emission characteristics of the initial nanocrystals and with photoluminescence quantum yields in the range of 10-20% (*1,3*).

Bioconjugate Formation and Purification

Two pioneering studies utilized CdSe nanocrystals bound to biological molecules (*6,7*). In one study, an avidin-biotin binding scheme was employed to attach CdSe-CdS core-shell nanocrystals to actin fibers. These dots were capped with an additional thin layer of silica in order to render them water compatible. The route to the final bioconjugate was complex, time-consuming and yielded a product with low quantum yield (*6*). The second study used CdSe-ZnS core-shell nanocrystals capped with mercaptoacetic acid groups. In this case, a conventional covalent cross-linking approach, based on 1-ethyl-3-(3-dimethylaminopropyl) carbodiimide hydrochloride (EDAC or EDC) condensation, was used to conjugate the nanocrystals to immunoglobulin G (IgG). Successful preparation of QD/IgG-nanocrystal bioconjugates was reported, albeit again with low PL quantum yields (7).

Our work in developing bioconjugated QD's has been focused in three principal areas. First, we developed an alternative conjugation strategy, based on self-assembly utilizing electrostatic interactions between negatively charged lipoic acid capped CdSe-ZnS quantum dot surfaces and engineered bifunctional recombinant proteins consisting of highly positively charged attachment domains genetically fused with desired biologically relevant domains, Figure 2A (*1,3*). Our initial prototype engineered protein consisted of a maltose binding protein-basic leucine zipper fusion (MBP-zb) which autoassembles into a dimer through an inserted cysteine residue. The highly positively charged attachment domains interact with the capped CdSe-ZnS QD surface while the MBP domain allows purification on amylose columns (*3-5*). This strategy of combining alkyl-COOH capped CdSe-ZnS nanocrystals and a two-domain recombinant protein cloned with a highly charged leucine zipper tail offers several advantages. (i) The alkyl-COOH terminated capping groups, which permit dispersion of the nanocrystals in water solutions at basic pH, also provide a surface charge distribution that can promote direct self-assembly with other molecules that have a net positive charge. (ii) The synthetic approach used to prepare the QDs can be easily applied to make a number of different sized core-shell nanocrystals, resulting in fluorescent probes with tunable emission over a wide range of wavelengths. This contrasts with the need for developing specific chemsitry

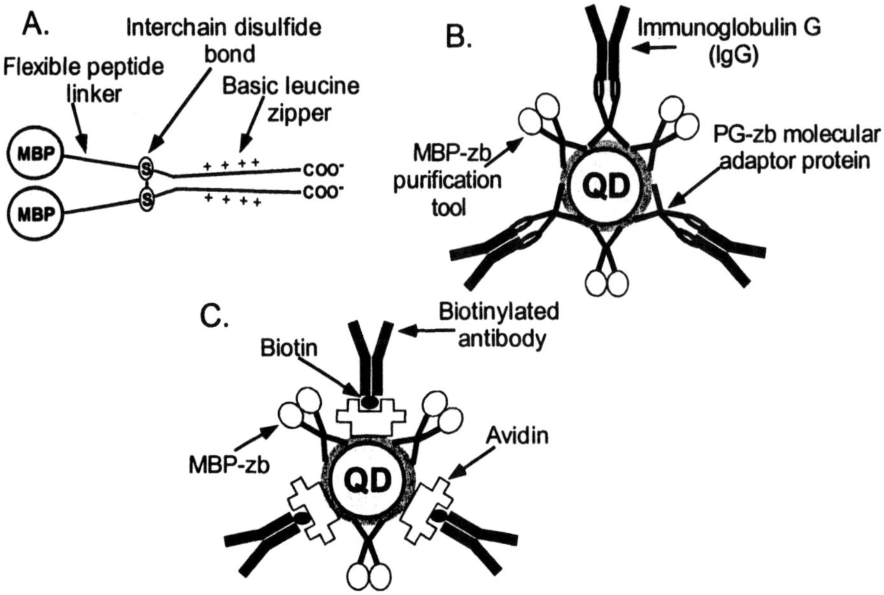

Figure 2. A. Cartoon of S-S linked MBP-zb homodimer consisting of the maltose binding protein fused with the dimer-forming positively charged tail (3). B. PG-zb (IgG-binding B2 domain of streptococcal protein G fused with the dimer-forming positively charged C-terminal tail) as molecular linker between QD's and the F_c region of IgG. MBP-zb is used for separating QD-IgG conjugate from excess IgG (4). C. Mixed surface QD-antibody conjugate with avidin bridging the QD and biotinylated antibody (5).

routes for individual organic fluorescent dyes. (iii) The fusion protein approach provides a general and consistent way to prepare a wide selection of biological macromolecules amenable to alterations in the interaction domain, such as charge, size, stability to pH and temperature. This allows control of the assembly of individual proteins into dimers and tetramers, a property that can be exploited in protein packing around the nanocrystals to form complex bioconjugates. Figure 3 shows the release of QD/MBP-zb bioconjugates from amylose beads after being allowed to self assemble and bind to the beads. Since unconjugated DHLA-QD's do not release from the amylose beads, we are able to routinely use amylose affinity columns in QD-bioconjugate purification.

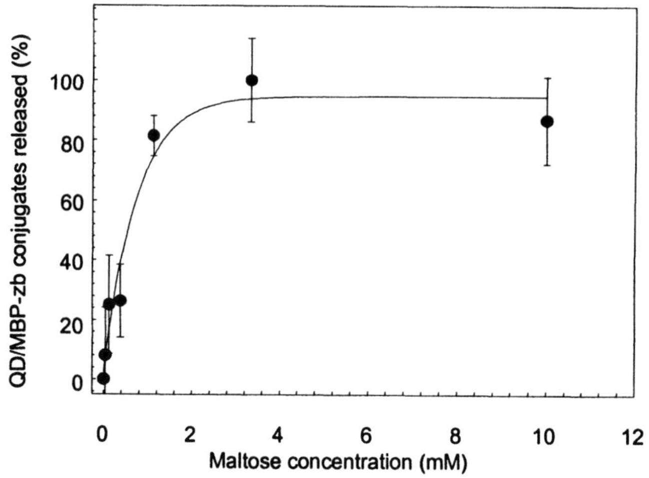

Figure 3. Release of QD/MBP-zb bioconjugates from amylose beads as a function of concentration of added maltose. Experiments were carried out in 5 mM sodium borate, pH 9 (3).

A second research area concerned engineering an adaptor protein to allow conjugation of the QD's to antibodies (Figure 2B) (*4*). This allows us to prepare bioconjugates of CdSe-ZnS QDs with antibodies using mixed composition conjugates containing both a molecular adaptor protein and a second protein used as a purification tool (Figure 2B). The engineered adaptor protein employs the immunoglobulin G (IgG)-binding β2 domain of streptococcal protein G (PG) modified by genetic fusion with the same positively charged leucine zipper interaction domain we previously developed and characterized using *E. coli* MBP. Using this novel PG based adaptor protein, QD-IgG bioconjugates can be formed readily, and the conjugates can be used in fluorescence-based assays to detect proteins and small molecules (*4*). We showed that both direct and sandwich fluoroimmunoassays using these antibody-conjugated QDs can be performed for detection of staphylococcal enterotoxin B (SEB), a causative agent of food poisoning. We also demonstrated the use of antibody-conjugated QDs in both plate-based and continuous-flow immunoassays for the detection of low levels of the explosive 2,4,6-trinitrotoluene (TNT) in aqueous samples (*4*). Details of some of these assays will be presented in a later section.

A third research effort in QD bioconjugation involves the direct use of avidin (a positively charged tetramer) as a bridge between QD's and biotinylated antibodies to form QD-antibody conjugates (Figure 2C) (*5*). Avidin, a glycoprotein found in avian egg white, is a homotetramer with a molecular weight of 68,000 Da, which interacts stoichiometrically with biotin, binding one biotin per subunit. Due to the specific nature and high affinity of the interaction (K_d = 10^{-15}M) between avidin and biotin, the system is a powerful tool for research and analysis. For example, the interaction between biotin and biotin binding proteins (such as avidin) has been widely exploited in various applications including immunoassays, selections, and purification schemes. Biotin-modified antibodies, proteins and DNAs are commercially available or can be easily prepared. Conjugating QDs with avidin potentially permits the attachment of these fluorescent inorganic particles to any biotinylated protein or DNA.

We prepared bioconjugates of CdSe-ZnS QDs with avidin that were then used to form QD-antibody conjugates by reaction with biotinylated antibodies (*5*). To facilitate purification of the conjugated QDs, a mixed surface conjugation strategy was employed. In this method both maltose binding protein genetically fused with a positively charged leucine zipper interaction domain (MBP-zb) and avidin were adsorbed to the QD surface. While avidin tightly binds the biotinylated antibody, the MBP-zb permits removal of any excess unbound antibody through affinity chromatography (Figure 2C) (*5*). These avidin conjugates were used for both direct and sandwich fluoroimmunoassays (see below). This work has shown the stepwise progress in functional bio-conjugation of QD's with both engineered and other proteins, demonstrating that QD-antibody conjugates can be readily formed and purified for use in fluorescence-based assays.

Bioconjugate Properties

Figure 4 (left) shows the absorption and PL spectra of unconjugated nanocrystals along with QD/MBP-zb conjugates dispersed in buffered solutions at pH 7.4, 8.1, 9.1 and 10. Fig. 4 (right) shows the PL intensity of the bioconjugates normalized to that of unconjugated QD suspensions for samples containing QD/MBP-zb bioconjugates over the same pH range. The bioconjugated QD material retains the spectroscopic properties of the starting nanocrystals with respect to light absorption and photoluminescent emission. Electrostatically conjugating these QDs with MBP-zb enhances the PL emission in comparison with that of unconjugated nanocrystals, with a somewhat more pronounced relative enhancement measured at higher pH (Figure 4 right panel). The PL quantum yield (QY), as compared to the emission from a Rhodamine 6G standard in methanol, increased from ~ 10% for unconjugated CdSe-ZnS QD's to ~ 20-30% for (CdSe-ZnS)/MBP-zb bioconjugates (*3*).

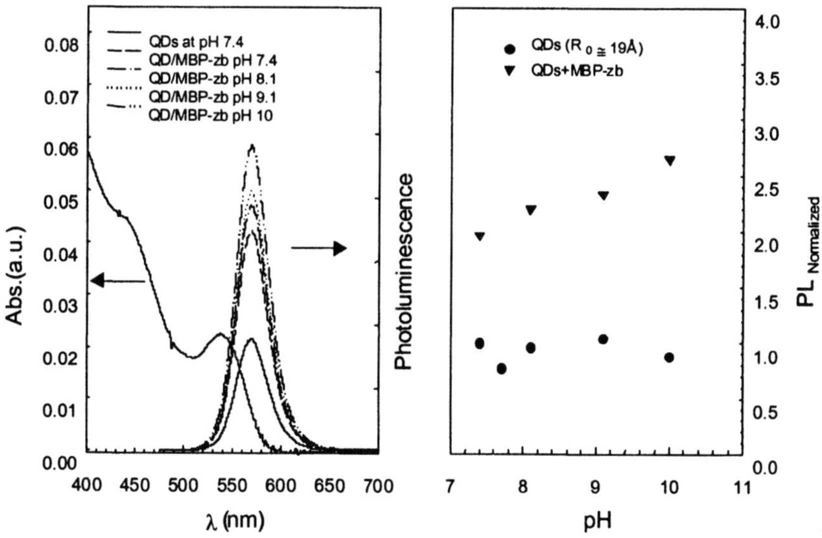

Figure 4. (Left panel) Absorption and PL spectra of DHLA-QDs and QDs conjugated with MBP-zb. Only one absorption spectrum is shown because all solutions have identical absorption spectra (same QD concentration in all samples). (Right panel) Variation of the PL total integrated intensity versus pH for DHLA-QDs and QD/MBP-zb conjugates. QD concentration was 150 nmol/L; concentration of MBP-zb protein was 900 nmol/L (ratio MBP-zb:QD 6:1). Samples were excited at 500 nm with a10 mm optical pathlength (3).

The enhancement of the PL emission of conjugated QD's in comparison to that of unconjugated nanocrystals is an interesting phenomenon. As described by Mattoussi et al., (*3*), upon bioconjugate formation between QD's and an increasing ratio of interacting MBP-zb protein, a corresponding increase in photoluminescent enhancement was observed. This enhancement demonstrates saturation behavior at higher MBP to QD ratios. This property has subsequently been corroborated in other reports (*8*).

Figure 5 shows the results of an experiment designed to investigate certain novel interactions that can occur between DHLA-QD's and proteins. In this experiment an engineered maltose binding protein bearing a pentahistidine (5•HIS) C-terminal extension or "tail" was compared with the identical MBP protein without the 5•HIS tail. QD's and the indicated amounts (increasing molar ratios of protein:QD) were incubated in 10 mM sodium borate buffer (pH 9.55) for 30 min. The resulting photoluminescence was measured and the data normalized against unconjugated QD's. The results demonstrate conclusively

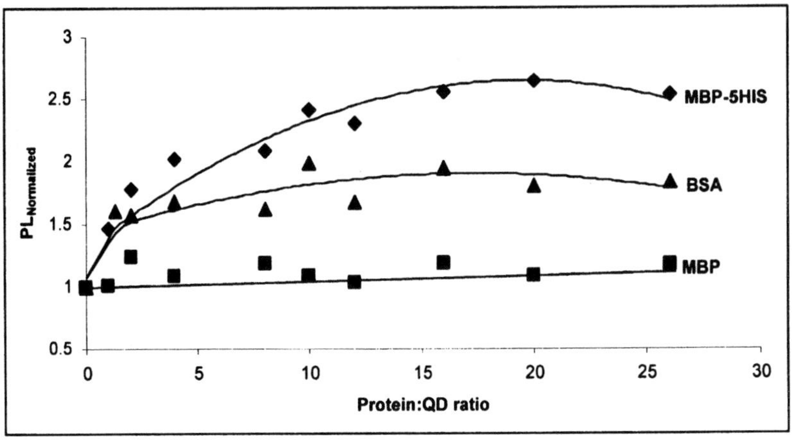

Figure 5. QD PL enhancement occurring upon bioconjugate formation with increasing ratios of 3 different proteins as described in the text.

that the MBP-5HIS interacts with the QD's through the 5•HIS tail, increasing PL ~2.5x while the non-HIS MBP has no effect on PL. The exact mechanism of how this protein interaction increases QD photoluminescence remains unexplained (*3*). BSA was used as a control protein. Although BSA is used commonly as a non-specific blocking agent, these results show that it too can interact with DHLA QD's, increasing their PL yield by as much as 50% at saturation. In previous work, Willard et al. (*8*) conjugated biotinylated BSA (bBSA) to QD's and then reacted them with fluorescently labeled streptavidin in a model protein-protein binding FRET assay. Their results were difficult to deconvolute because, as shown above, BSA itself can alter QD emission intensity, while bBSA can enhance the fluorescence output of labeled streptavidin even in the absence of QD's (Willard, D.M.; Carillo, L.L.; Jung, J.; Van Orden, A; personal communication). Taken together, these results suggest that a good deal of care must be taken in designing, conducting, and interpreting QD-protein experiments.

QD Bioconjugates in Fluoroimmunoassays

Both the PG based adaptor protein and avidin-bridged QDs have been used to generate QD-antibody conjugates reagents that were subsequently utilized in rabbit- anti-cholera toxin conjugate was measured over a range of concentrations of toxin from 4 ng/mL to 4 μg/mL (Figure 7A). The lowest con-

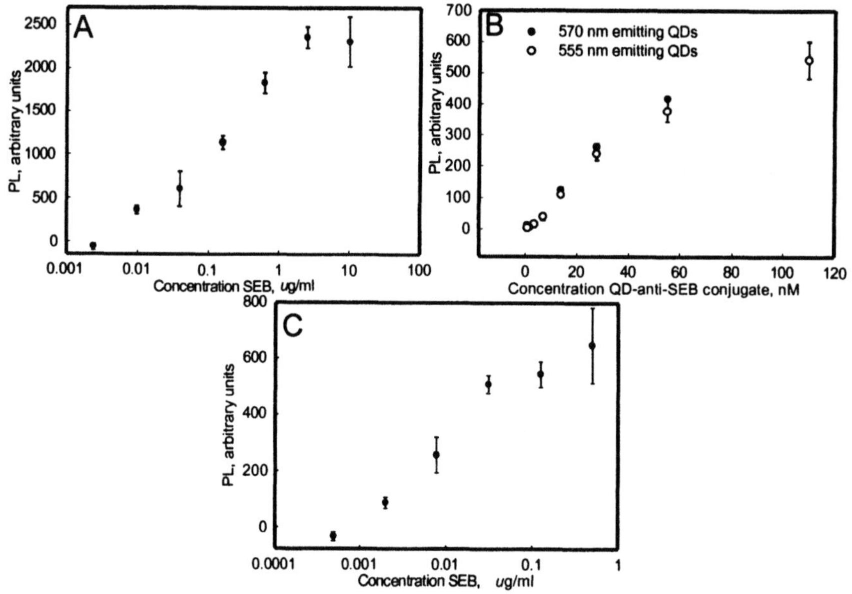

Figure 6. A. Direct detection of SEB by Mab 2b anti-SEB antibody- conjugated QDs at increasing concentrations of SEB (2.4 ng/ml to 10 μg/ml). B. PL intensity measured for binding of increasing concentrations of QD-PG-zb-IgG to wells coated with a saturating amount of SEB (10 μg/ml). A comparison of 555 nm emitting and 570 nm emitting QD conjugates is shown. C. Sandwich fluoroimmunoassay for detection of SEB where varying concentrations of SEB (0.49 ng/ml to 500 ng/ml) were captured by anti-SEB antibody immobilized in wells. Signals generated using QD-Mab 2b conjugate (4).

centration of cholera toxin that gave positive signal over background was approximately 60 ng/ml; signal increased until saturation was reached at about 1 μg/ml. QD-sheep- anti-SEB antibody conjugates (formed via avidin bridges) were similarly used in a sandwich assay (Figure 7B). The lowest concentration of SEB that gave useful signal over background was approximately 15 ng/ml, and the signal increased until saturation was reached at about 250 ng/ml. This is the first demonstration of the use of avidin conjugated DHLA-QDs in sandwich fluoroimmunoassays, and one of only a few of the use of QDs in sandwich fluoroimmunoassays. The limits of detection achieved in these experiments are fluoroimmunoassays. The PG based adaptor protein strategy was used to demonstrate binding of QD-antibodies to staphylococcal enterotoxin B (SEB) in a direct binding fluorometric assay (*4*). Direct binding assessment showed that QD-anti-SEB conjugate was able to bind to SEB protein toxin adsorbed to the wells of polystyrene microtiter plates (Figure 6A). Effects on fluorescent signals obtained by varying the amount of SEB immobilized in wells were investigated using a QD-Mab 2b (anti-SEB) conjugate. In these experiments, the fluorescent signal for bound conjugate was measured over a range of toxin concentrations from 2.4 ng/ml to 10 μg/ml. The lowest concentration of SEB that gave meaningful signal over background was approximately 0.010 μg/ml, and signal increased over the entire range of SEB applied to wells (*4*).

The effects in microtiter plate assays of increasing QD-Mab 2b conjugate concentration while keeping the amount of adsorbed SEB constant at a saturating amount was also investigated in the direct binding mode. Figure 6B shows the relative emission intensity measured as a function of increasing QD-Mab 2b conjugate concentration. The fluorescent signal increased relatively linearly through the concentration range up to about 30 nM QD-Mab 2b conjugate per well (3 pmol total QDs present), approaching saturation above about 120 nM (12 pmol total QDs present). This result was expected, since when all available SEB epitope sites are occupied with QD-Mab 2b the fluorescent signal should remain constant regardless of the addition of more QD-antibody conjugate. Experiments showed QDs with emission maxima at 555 nm and 570 nm behave essentially identically in the direct binding assay (Figure 6B). QD-anti-SEB antibody conjugates were also used in a sandwich assay format. A series of dilutions of SEB were applied to microtiter plate wells with adsorbed Mab2b anti-SEB capture antibody, followed by QD-polyclonal sheep anti-SEB conjugate as the signal producing reagent. Figure 6C shows an experiment where the fluorescent signal from bound sheep-anti SEB conjugate was measured over a range of concentrations of toxin from 0.49 ng/ml to 500 ng/ml. The lowest concentration of SEB that gave measurable signal over background was ~ 2 ng/ml, and the signal increased until saturation was reached at about 30 ng/ml.

For the use of avidin bridged QD-antibody conjugates, a sandwich assay format was also explored. QDs conjugated to biotinylated anti-toxin antibodies via avidin bridges were used in this format for the detection of cholera toxin and SEB. Anti-cholera toxin capture antibody was adsorbed to wells of a microtiter plate. A series of dilutions of cholera toxin were applied to wells containing adsorbed capture antibody, followed by QD- polyclonal rabbit anti-cholera toxin conjugate as the signal-producing reagent. Fluorescent signal from bound comparable to conventional fluoroimmunoassays using bright organic dye fluorophores.

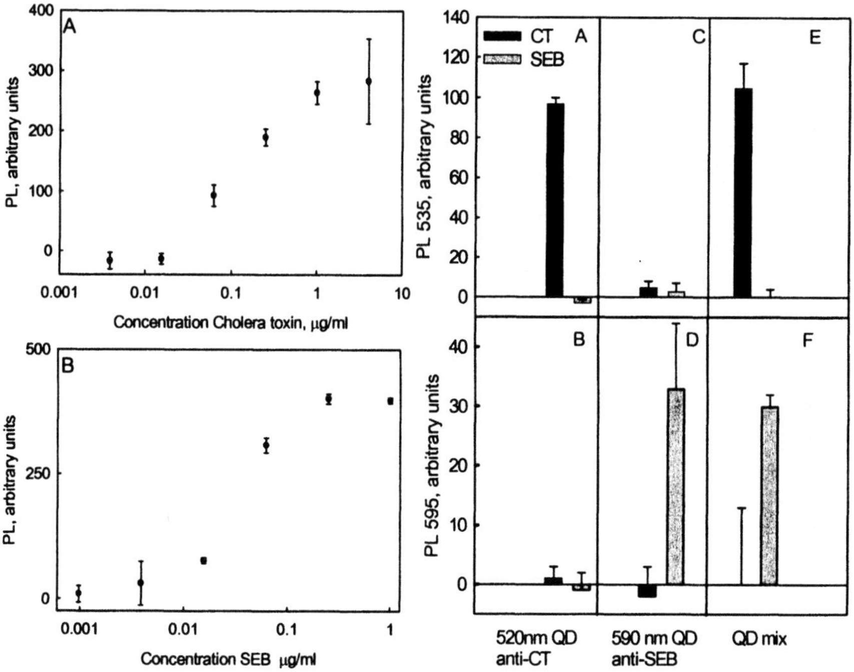

Figure 7. Sandwich assays for cholera toxin (CT) and SEB using QDs emitting at 570 nm. A. Cholera toxin (4ng/ml to 4 µg/ml) captured by immobilized anti-CT IgG was quantitated using QD-anti-CT conjugate. B. SEB (0.98 ng/ml to 1 µg/ml) captured by immobilized anti-SEB IgG and quantitated using QD-anti SEB conjugate. Right panels demonstrate dual analyte detection using anti-CT and anti-SEB coated QDs. QD-IgG conjugates were examined alone and mixed for binding to wells coated with saturating amounts of CT (5 µg/ml) or SEB (10 µg/ml). Black bars represent the signal from CT coated wells, gray bars from SEB coated wells. A. Anti-CT-QD conjugate incubated with CT and SEB coated wells (535 nm emission filter). B. Same wells as in Panel A, signal read using a 595 nm filter. C. Anti-SEB-QD conjugate incubated with CT and SEB coated wells. Signal read using the 535 nm filter. D. Same as Panel C, using a 595 nm filter. E. Mix of anti-CT-QD and anti-SEB-QD conjugates incubated with CT and SEB coated wells. Signal read using a 535 nm filter. F. Same as Panel E, using a 595 filter.

A major advantage of QDs over conventional fluorophore labels is that different color nanocrystals can be excited at a single wavelength. We examined the anti-SEB and anti-cholera toxin QD-antibody conjugates prepared via an avidin linker in dual-analyte tests aiming at eventual multianalyte detection. Anti-cholera toxin antibody was conjugated to 520 nm emitting QDs, and anti-SEB antibody was conjugated to 590 nm emitting QDs. These conjugates are spectrally separable using the appropriate plate reader. Figure 7 (right panel) shows signal generated when QD-anti-cholera toxin conjugates added to wells coated with saturating amounts of SEB or cholera toxin and examined with a 535 nm filter. Signal was observed only in wells coated with cholera toxin, indicating that there is no cross reactivity between the anti-cholera toxin antibody and SEB. When observed with a 595 nm filter these same wells show no signal Similarly, no signal was observed when QD-anti-SEB conjugate was added to wells coated with saturating amounts of SEB or cholera toxin and examined using a 535 nm filter (Figure 7-right panel C). This demonstrates that there was minimal cross talk between the 520 and 590 nm emitting QD conjugates. When the wells containing QD-anti-SEB conjugate were examined using a 595 nm filter (Panel D), signal was observed for only the SEB coated wells, demonstrating that this anti-SEB antibody has no cross reactivity towards cholera toxin. A mix of anti-cholera toxin conjugated QDs and anti-SEB conjugated QDs was incubated on wells coated with cholera toxin or SEB and imaged using the 535 or 595 filters. When assayed using the 535 nm filter, only signal from the cholera toxin coated wells was observed (Panel E). Using the 595 filter generates signal from only the SEB coated wells (Panel F). Our results suggest that improved instrumentation will allow QD-antibody conjugates to be used in multianalyte assays in both direct and sandwich formats with at least 3-5 different toxins and associated colored QD conjugates. The advantages of combining multiple assays in a single microtiter well include savings in reagents and consumables, decreased sampling errors, and higher throughput compared with single-assay systems.

Conclusions

We have developed and demonstrated a new and useful electrostatically driven method of preparing bioconjugated QD reagents for use in biosensing applications. After initially preparing a modified sugar binding protein (MBP-zb) protein for use in QD-bioconjugate purification, an engineered adaptor protein to attach antibody reagents to QD's was developed. The use of avidin as a bridging molecule for antibody attachement has also been demonstrated, which allowed us to utilize QD-antibody reagents in multiple fluoroimmunoassays.

We continue to focus on uses of QD bioconjugates as photostable substitutes for organic fluorophores in biosensing applications.

References

(1) Mattoussi, H.; Kuno, M.K.; Goldman E. R.; Mauro, J.M.; Anderson, G.P.; Colloidal semiconductor quantum dot conjugates. Chapter 17, 537-570. In Optical Biosensors: Present and Future. Ligler F.S., and Rowe Taitt, CA C.A. Eds. (2002) Elsevier, The Netherlands.

(2) Chan, W.C.W.; Maxwell, D.J.; Gao, X.H.; Bailey, R.E.; Han, M.Y.; Nie, S.M. Luminescent quantum dots for multiplexed biological detection and imaging. *Current Opinion in Biotechnology* **2002** *13*, .40-46.

(3) Mattoussi, H.; Mauro, J.M.; Goldman, E.R.; Anderson, G.P.; Sundar, V.C.; Mikolec, F.V.; Bawendi, M.G. Self-Assembly of CdSE-ZnS Quantum Dot Bioconjugates Using an Engineered Recombinant Protein. *J. Am. Chem. Soc.* **2000,** *122*, 12142-12450.

(4) Goldman, E.R.; Anderson, G.P.; Tran, P.T.; Mattoussi, H.; Charles P.T.; Mauro, J.M. Conjugation of luminescent quantum dots with antibodies using an engineered adaptor protein to provide new reagents for fluoroimmunoassays. *Anal. Chem.* **2002** *274*, 841-47.

(5) Goldman, E. R.; Balighian, E.D.; Mattoussi, H., Kuno, M.K.; Mauro, J.M.; Tran P.T.; Anderson; G.P. Avidin: A natural bridge for quantum dot-antibody conjugates. *J. Am. Chem. Soc.* **2002** *122*, 6378-6382.

(6) Bruchez, M.; Moronne, J. M.; Gin, P., Weiss S.; Alivisatos, A.P. Semiconductor nanocrystals as fluorescent biological labels. *Science* **1998** *281*, 2013-2018.

(7) Chan, W.C.W.; Nie, S. Quantum dot bioconjugates for ultrasensitive nonisotopic detection. *Science* **1998** *281*, 2016-2018.

(8) Willard, D.M.; Carillo, L.L.; Jung, J.; Van Orden, A. CdSe-ZnS Quntum Dots as Resonance Energy Transfer Donors in a Model Protein-Protein Binding Assay. *Nano Letters* **2001** *1*, 469-474.

(9) Hines, M.A.; Gnyot-Sionnest, P. Synthesis and characterization of strongly luminescing ZnS-Capped CdSe nanocrystals. J. Phys. Chem. B. **1996** *100,* 468-471.

(10) Dabbousi, B.O.; Rodriguez-Viejo, J.; Mikulec, F.V.; Heine, J.R.; Mattoussi, H.; Ober, R. Jensen, K.F., Bawendi, M.G. (CdSe)ZnS core-shell quantum dots: Synthesis and characterization of a size series of highly luminescent nanocrystallites. J. Phys. Chem. B. **1997** *101,* 9463-9475.

(11) Peng, X.G.; Schlamp, M.C.; Kadavanich, A.V.; Alivisatos, A.P. Epitaxial growth of highly luminescent CdSe/CdS core/shell nanocrystals with photostability and electronic accessibility. *J. Am. Chem. Soc.* **1997,** *119*, 7019-7029.

Note: The views, opinions, and/or findings described in this report are those of the authors and should not be construed as official Department of the Navy positions, policies, or decisions.

Chapter 3

Metal–Insulator–Metal Ensemble Gold Nanocluster Vapor Sensors

Arthur W. Snow[1], Hank Wohltjen[2], and N. Lynn Jarvis[2]

[1]Naval Research Laboratory, 4554 Overlook Avenue, SW, Washington, DC 20375
[2]Microsensor Systems, Inc., 62 Corporate Court, Bowling Green, KY 42103

A nanocluster Metal-Insulator-Metal Ensemble (MIME) chemical vapor sensor is a solid-state sensor composed of nanometer size gold particles encapsulated by a monomolecular layer of alkanethiol surfactant deposited as a thin film on an interdigital microelectrode. The principle by which this sensor operates is that vapors reversibly absorb into the organic monolayer which causes a large modulation in the electrical conductivity of the film. The tunneling current through the monolayer between gold particle contacts is extremely sensitive to very small amounts of monolayer swelling and dielectric alteration caused by absorption of vapor molecules. The nanometer scale of the particle domains and correspondingly large surface area translate into a very large vapor sensitivity range extending to sub-ppm concentrations. Selectivity of the sensor is regulated by incorporation of chemical functionalities in the structure of the alkanethiol surfactant or substitution of the entire alkane structure. The current focus of research is in mapping the selectivity and sensitivity of sensor elements made by incorporating these functionalities into the shell of the nanocluster. Targeted applications include detection of chemical warfare agents and explosives, and residual life indication of carbon filters and protective clothing.

Introduction

Detecting hazardous chemical vapors with highly miniaturized analytical devices is a present and future capability that is assuming an increasing importance in many DOD and civilian scenarios. These scenarios involve chemical weapons, concealed explosives, volatile organics in breathing air, residual life indication of gas mask filters and protective clothing, and electronic nose functions (identification of unknown substances by odor). Until 25 years ago detection of hazardous chemicals relied on large scale laboratory configured instrumentation such as mass and optical spectroscopies, gas chromatography, and to a smaller extent on colorimetric schemes utilizing wet chemistry. Since then, the impetus has been toward reduction in instrumentation size and field deployment of analytical instrumentation. Also during this time sensors as minimal-component, highly miniaturized, stand-alone devices emerged. These devices generally incorporated a chemically active surface interfaced to an electronic substrate. Examples include metal oxide semiconductor (MOS) devices, miniature electrochemical cells, surface acoustic wave (SAW) devices and chemiresistors. For vapor sensing these devices rely of a partitioning of an analyte vapor between the gas phase and that sorbed onto/into the chemically active surface. The chemically active surface then acts as a transducer. A property change caused by sorption of a vapor generates an electronic signal that may be processed into analytical information regarding the concentration of the analyte vapor. NRL has been active in microsensor research since 1981 with SAW and chemiresisor devices. Features which govern the sensitivity of these devices are the transduction mechanism, vapor partitioning, and the surface area to volume ratio of the chemically active adsorbent coating. The first feature is determined by the coupling between the electronic substrate and the chemically active surface or coating and is varied mostly by the substrate design. The second feature is determined by the chemical interaction between the analyte vapor and the adsorbent surface's chemical structure and is varied by the design and synthesis of this component. Over the past 20 years much research at NRL and elsewhere has focused on optimizing these features. The third feature (surface to volume ratio of the chemically responsive coating) has received very little attention as the thickness of the chemically active coatings were optimized to be thick enough for good sensitivity but thin enough for a fast response. A nanoscale materials approach changed this radically. A new type of chemical microsensor with distinct advantages derived from nanometer scale material domains emerged. The objective of this research is to understand the operating principles of this new sensor and exploit its advantages in practical applications.

MIME Sensor Concept

The new sensor is based on a metal nanocluster encapsulated by a single layer of organic molecules. Our original report *(1)* has been followed by other examples *(2-9)*. The materials concept is illustrated in Figure 1. A single cluster, depicted in the upper left corner, is composed of a gold core encapsulated by a shell of alkanethiol molecules. The gold core may range from 1 to 5 nm in diameter, and from it originates the electronic properties of this material. The alkanethiol molecules of the shell are bonded to the surface of the gold core by a gold-to-sulfur bond. This organic shell forms a very thin insulating barrier, and its thickness, which may vary from 0.4 to 1.0 nm, has an enormous effect on electron tunneling between adjacent clusters. The organic shell also imparts an organic character to the cluster which promotes solubility in organic solvents such that a cluster as much as 90% gold by weight will dissolve in toluene. This

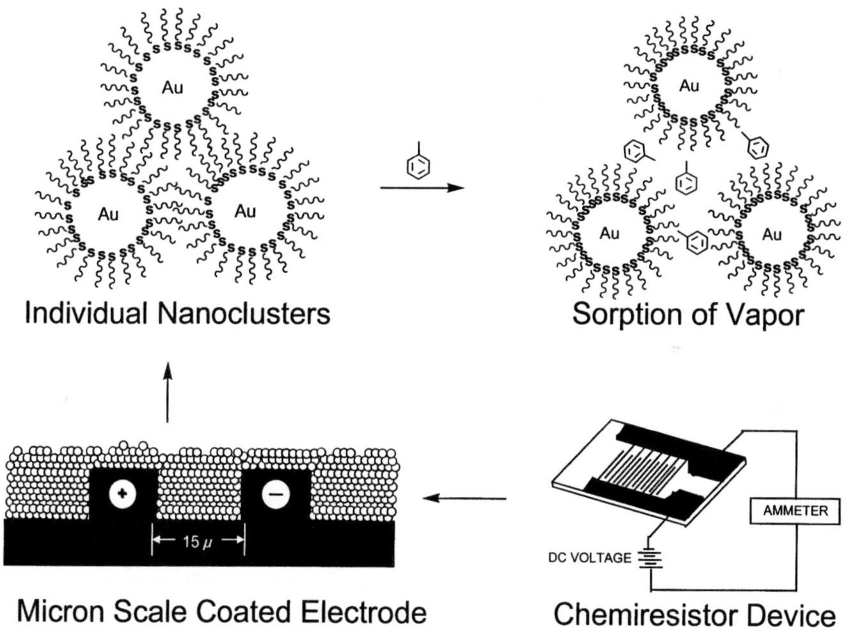

Figure 1. Schematic of the Metal-Insulator-Metal ensemble (MIME) sensor concept. A micron-scale interdigital electrode is coated with a film of alkanethiol stabilized gold nanoclusters and exposed to toluene vapor. The toluene adsorbs into the alkanethiol monolayer shell, and the consequent swelling causes an increase in the separation distance between gold cores and a reduction of electron tunneling between them. (Reproduced from NRL Review. U.S. Government work in the public domain.) *(See page 2 of color insert.)*

solubility makes the processing of these clusters into thin films very facile. The sensor is fabricated by depositing a film of these clusters onto a micron scale interdigital electrode substrate. When connected to a small bias (50 to 500 mV), a nanoamp current flows through the film. Exposure to vapors causes very large changes in the conductivity of the film. This results from a sorption of the vapor into the very thin shell, and the consequent swelling of the shell results in a small but very significant increase in the distance between cores of adjacent metal clusters. The tunneling current is extremely sensitive to the distance between cores. A final feature of significance in Figure 1 is the packing of the clusters in the film. Being spherically shaped clusters, any type of packing will have nanometer scale voids within its matrix. The size differential between a typical vapor molecule such as toluene and a nanocluster is approximately a factor of 10. As such, this network of voids in the cluster matrix makes for a rapid ingress and egress of vapors, much more so than the slow diffusion into polymer films used on typical microsensors. This provides a pathway for a much faster response and recovery for sensors based on a metal cluster ensemble. Combined with the cluster ensemble's fast kinetics for sorption and desorption is an extremely large surface to volume ratio which translates into a highly enhanced sensitivity for this MIME sensor. The MIME sensor derives its name from the Metal-Insulator-Metal Ensemble character of the cluster film. The critical features in its design are the dimensions of the core and shell and the chemical composition of the molecules composing the shell. These features are described in the following paragraphs.

Cluster Synthesis and Characterization

The dimensions of the core and shell of the cluster are determined by the conditions of its synthesis which are modified from those originally reported by Brust et al *(10)*. The alkanethiol-gold cluster synthesis is illustrated in Figure 2. The two critical reagents are the gold chloride and the alkanethiol. They are suspended in a common medium, and a reducing agent, typically $NaBH_4$, is added. The trivalent gold is reduced to neutral gold, and the gold atoms aggregate to form a particle nucleus. The gold particle grows by addition of gold atoms and smaller particles. As a competing process, the alkanethiol reacts with the neutral gold surface to form a sulfur to gold bond. The gold particle growth is terminated when its surface is encapsulated by complexation with the alkanethiol. The relative rates of gold particle growth and alkanethiol surface complexation are dependent on the concentrations of the respective gold chloride and alkanethiol reagents. Thus, the molar ratio of these reagents is a simple way to regulate the core size of the cluster. Typically, this ratio ranges from 1:3 to 8:1 and causes the corresponding core diameters to vary from 1 to 5 nm. The shell thickness of the cluster is determined by molecular chain length

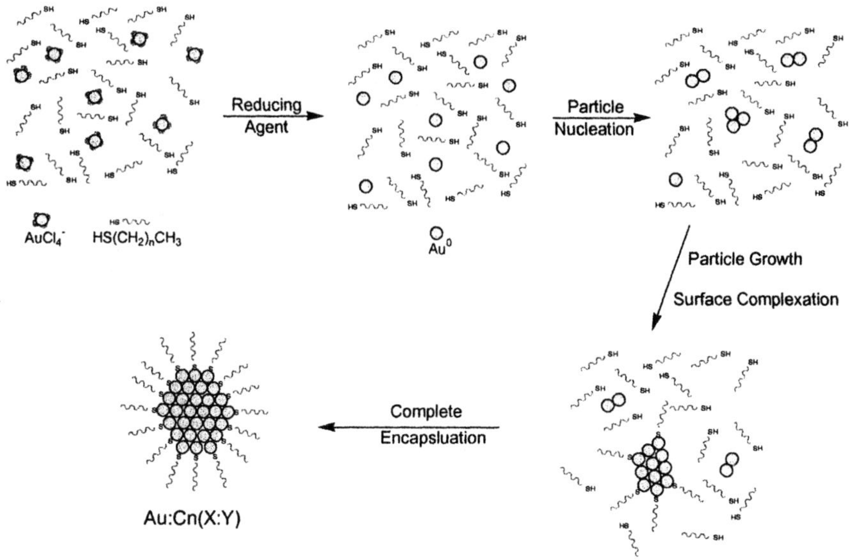

Figure 2. Schematic of the alkanethiol stabilized cluster synthesis. After reduction of the gold ions, the competitive processes of gold particle growth and alkanethiol surface complexation determine the size of the gold nanocluster. *(See page 3 of color insert.)*

of the alkanethiol selected as the reagent. The number of carbon atoms in the chain length typically ranges from 4 to 16 and generates a corresponding shell thickness ranging from 0.4 to 1.0 nm. A matrix of these clusters has been synthesized as is depicted in Table 1 where the varying core and shell relative sizes are represented pictorially by concentric circles. As a shorthand to designate individual clusters, the general abbreviation Au:C*n*(*X:Y*) is used where *X:Y* is the gold chloride:alkanethiol synthesis stoichiometric ratio that correlates with core size and *n* is the number of carbon atoms in the alkanethiol chain that correlates with shell thickness. The number beneath each cluster in Table 1 is the corresponding bulk DC electrical conductivity. An appreciation for the respective effects of the shell thickness and core size on the electrical conductivity can be obtained by examining the magnitude of conductivity variation down the column headed (1:1) and across the row headed C_{12}, For the former the variation is on the order of 10^7 which illustrates why such minute swelling effects on the shell have such dramatic effects on the conductivity.

MIME Sensor Response to Vapors

The response of a Au:C*8*(*1:1*) MIME sensor to five 60 sec exposure-purge cycles of toluene at high and low vapor concentrations is presented in Figure 3.

Table 1. Au:Cn(X:Y) Cluster Series Core and Shell Dimensions and Bulk Electrical Conductivity ((ohm cm)$^{-1}$)

Cn/(X:Y)	(1:3)	(1:1)	(3:1)	(5:1)	(8:1)	R_{Shell} (nm)
C_{16}		8×10^{-11}		1×10^{-8}	4×10^{-7}	1.0
C_{12}	2×10^{-9}	1×10^{-8}	5×10^{-8}	1×10^{-6}	3×10^{-6}	0.80
C_8	5×10^{-9}	1×10^{-6}	7×10^{-6}	2×10^{-5}		0.60
C_6	8×10^{-8}	6×10^{-6}	2×10^{-4}	6×10^{-4}		0.50
C_4	2×10^{-6}	2×10^{-4}				0.40
R_{Core}(nm)	0.65	0.88	1.2	1.8	2.4	

SOURCE: NRL Review. U.S. Government work in the public domain. See page 3 of the color insert.

The sensor response is a measured current change (right axis) which is electronically converted to a frequency (left axis) via precision current-to-voltage and voltage/frequency converters to allow data acquisition over a wide dynamic range using a computerized frequency counter. The toluene vapor causes a very large and rapid decrease in conductance of the sensor. Greater than 90% of the signal response occurs within 1 sec of its 30 sec exposure. The recovery is equally rapid and complete. The lower portion of Figure 3 indicates that detection limits well below 1 ppm are achievable.

The MIME sensor response to toluene displays a dependence on both the core and shell dimensions of its cluster component. When the matrix of clusters depicted in Table 1 is investigated, optimum sensitivities are found for both the core diameter and the shell thickness in the midrange of the matrix. Clearly two effects operate in each case. In the latter, a thicker shell requires more sorbed vapor to achieve an amount of swelling comparable swelling to that of a thinner shell, while a thinner shell has less of an organic character to solvate the

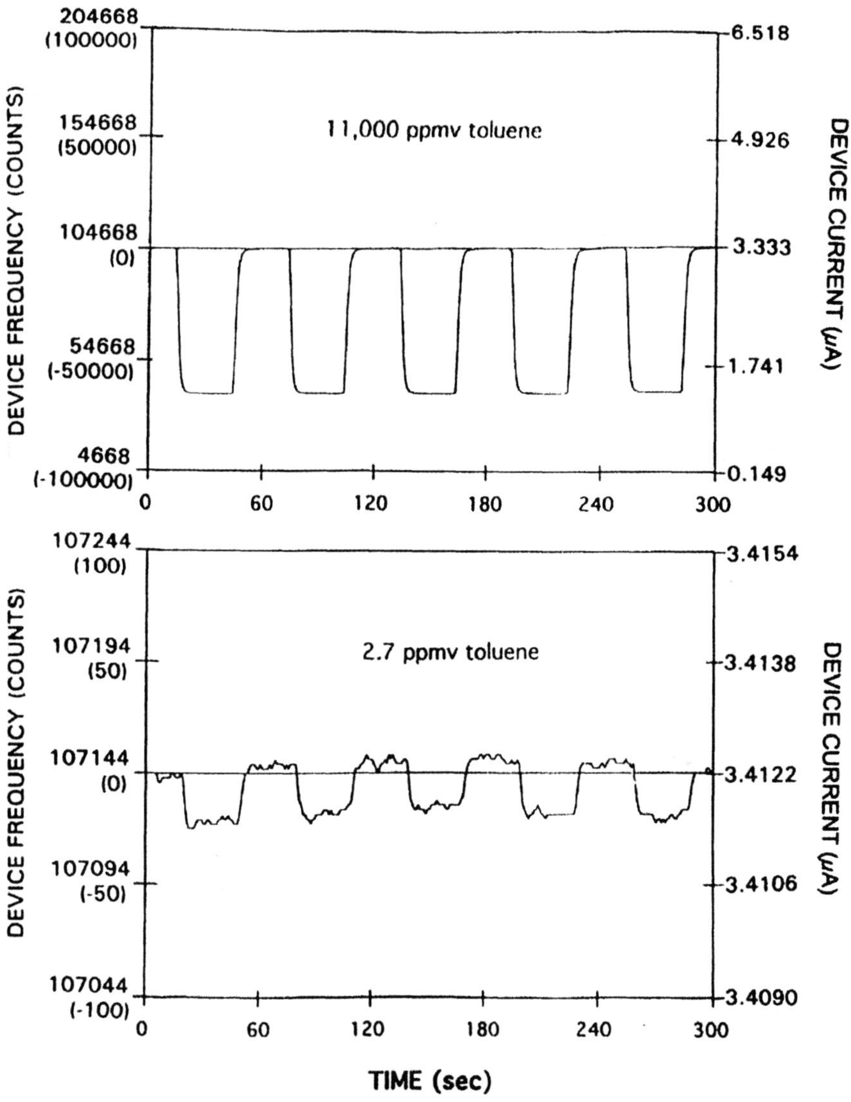

Figure 3. Response of Au:C*8*(*1:1*) MIME sensor to five 60 sec exposure-purge cycles of toluene at high and low vapor concentrations. The right axis indicates the current change in the MIME device, and the left axis is the sensor response when the current change is converted to a frequency. (Reproduced from NRL Review. U.S. Government work in the public domain.)

incoming toluene. In the former case, it is not clear why a variation from 1 to 5 nm would pass through an optimum in sensor sensitivity to toluene.

The dependence of the Au:C*8*(*1:1*) MIME sensor response to toluene vapor concentration is presented in Figure 4 along with responses to 1-propanol and water vapors. The inset depicts this response at the very low concentrations for toluene. Sensor response is expressed as the change in conductance normalized to the baseline (purge conditions) conductance, and vapor concentration is expressed as fractions of that corresponding to the vapor pressure of the pure liquid at 15°C. Since toluene, 1-propanol and water have nearly the same vapor pressures, the vapor concentrations of each are similarly comparable. The response to toluene vapor is very large and deviates slightly from linearity with the slope of the curve becoming greater at the low end of the concentration range. This sensor is remarkably insensitive to water vapor even at high concentrations. The lack of moisture sensitivity is of great importance for practical environmental sensing applications. Propanol vapor produces an initially unexpected result of a conductance increase at high concentrations. Propanol is a very polar organic vapor. Its sorption into the cluster shell increases the dielectric character between cores through which electrons tunnel.

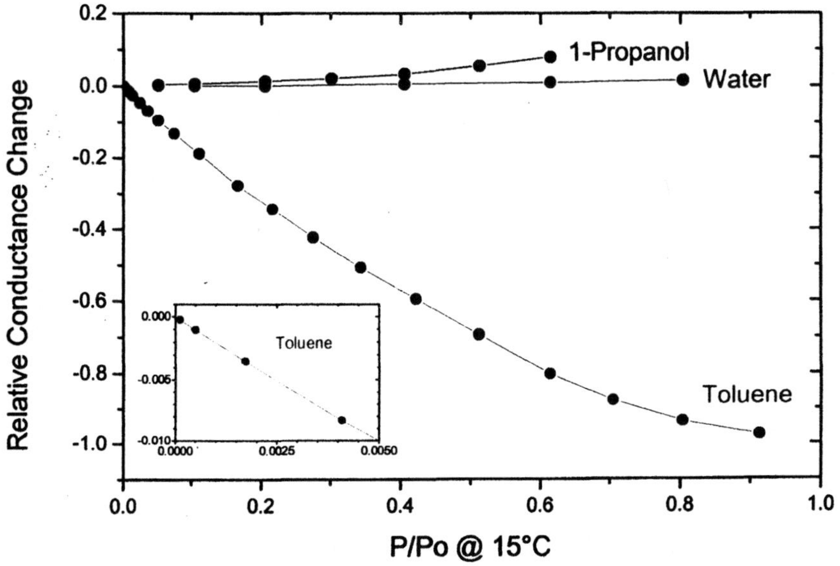

Figure 4. Vapor response isotherms of the Au:C*8*(*1:1*) MIME sensor to toluene, water and 1-propanol vapors based on 15°C vapor pressure. The inset displays the toluene response down to a 2.7 ppm concentration.

The observed conductance increase is believed to result from this increased dielectric character promoting charge transport between clusters.

While this Au:C8(*1:1*) MIME sensor can easily discriminate toluene vapor from water or alcohol vapors, it also responds to less polar organics to varying degrees. The selectivity of a MIME sensor for a particular coating-vapor combination is determined by the vapor's partition coefficient between the gas and coating phases. The partition coefficient is dependent on both physical properties of the vapor and its chemical interactions with the coating. Gas chromatography and early research with polymer coated SAW sensors *(11)* demonstrate that the partition coefficient , K (ratio of vapor absorbed in the coating to that remaining in the gas phase), is inversely dependent on the physical vapor pressure of the analyte vapor, p_2, and on the chemical activity coefficient for dissolution of the vapor in the coating, γ, as follows:

$$K = (RT)/(M_1 p_2 \gamma)$$

where R is the gas constant, T is the temperature and M_1 is the molecular weight of a coating molecule (M_1 becomes an undefined constant when the size differential between coating and vapor molecules becomes large). This indicates that sensors respond more strongly to analyte vapors with low vapor pressures than to those with higher vapor pressures. When the Au:C8(*1:1*) MIME sensor is challenged with a homologous series of alkane vapors (pentane through dodecane) wherein the chemical interaction is constant, the intensity of the sensor response displays a quantitative inverse dependence to vapor pressure as expected from the above equation. For this reason when we wish to investigate chemical interactions as a guide for enhancing sensor selectivity, care must be taken to either select vapors with very similar vapor pressures (as was done in Figure 4) or to invoke a vapor pressure compensating factor. In designing sensor coatings with vapor selectivities, the approach is to incorporate chemical structures that will have an interaction with a target vapor. Chemical interactions are quantified by an activity coefficient, γ, which appears in the above equation. The activity coefficient is a number between 1 and 0 having a value close to 1 for nearly "ideal systems" where no chemical interactions are occurring and progressing to smaller numbers as chemical interactions of increasing strength occur. Reversible chemical interactions range from weak van der Waals forces, to dipole attractions, to hydrogen bonding, to strong acid-base and charge-transfer interactions. Various models and schemes have been developed to correlate molecular structures with magnitudes of interactions. Qualitatively they are useful, but unfortunately attempts to quantify these models result in multi-parameter equations of 3 to 8 variables that sacrifice utility for precision. A reliable approach is to build a database with some systematic variations in structure. Some thiol ligand molecules with varied

Table 2. Functionalized Thiol Ligands used with MIME Gold Clusters

Unfunctionalized Thiol Ligand =  (Au:C_8(1:1) or -SC_8)

Terminal Functionalization

 (-SC_7OH)

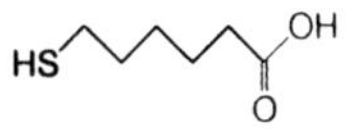 (-SC_5COOH)

 (-SC_5HFIP)

 (-SC_4HFIP)

 (-SC_6Cl)

Chain Structural Variation

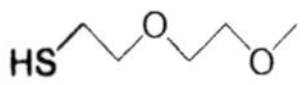 (-$SC_2OC_2OCH_3$)

Aromatic Functionalization

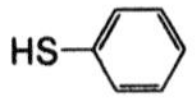 (-SC_6)

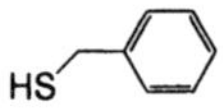 (-SC_1C_6)

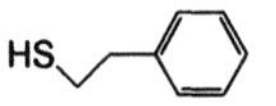 (-SC_2C_6)

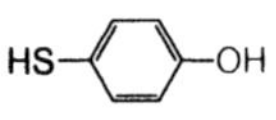 (-SC_6OH)

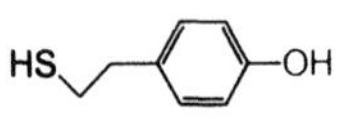 (-SC_2C_6OH)

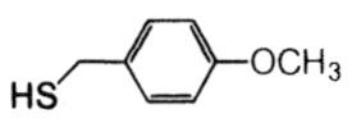 (-$SC_2C_6OCH_3$)

chemical functionality used for this purpose are depicted in Table 2. These include a variety of alkanethiols with different terminal functional groups, a case where an ethyleneoxide chain is substituted for the alkane chain and a variety of aromatically functionalized thiols. Preparation of the corresponding clusters has been accomplished, and a database exists. Some results in bar graph form are depicted in Figures 5 and 6 for sensor responses to a variety of vapors (All vapors except DMMP were chosen for a narrow range of vapor pressures.) at a concentration corresponding to one tenth of their vapor pressure. Figure 5 is a MIME vapor response pattern for an array of 7 different MIME coatings responding to the same vapor. Dimethyl methylphosphonate, DMMP, is a standard GB nerve agent stimulant and displays a significantly different pattern than toluene. It is these vapor response patterns that allow a sensor array to make identifications. Figure 6 is a MIME coating response pattern displaying the response profile of one coating to 7 vapors. It is clear that the fluoroalcohol functionalized cluster coating responds most strongly to basic and acidic vapors. The octanethiol cluster coating displays a particularly interesting pattern in that the polar vapors (nitromethane, 1-propanol, piperidine and acetic acid) are displaying a response where the relative resistance is decreasing (conductance

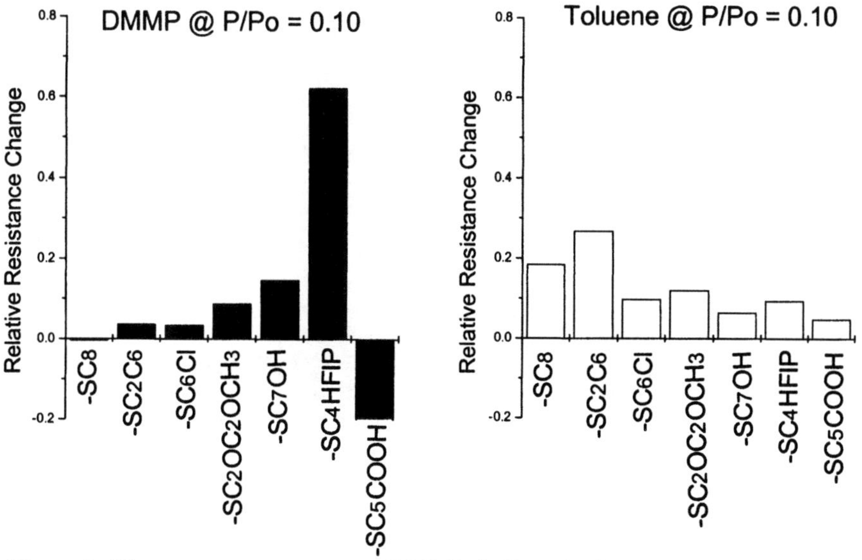

Figure 5. Vapor response pattern to DMMP (left) and toluene (right) for an array of 7 MIME sensors with cluster coatings composed of different alkanethiols.

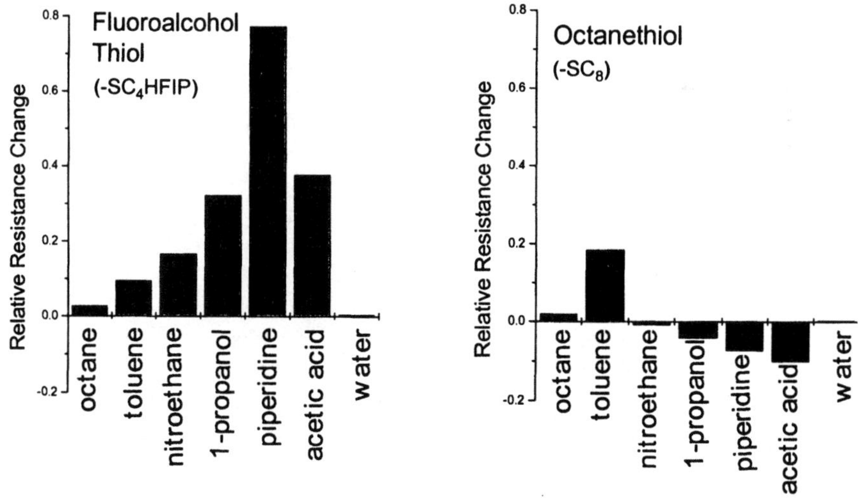

Figure 6. MIME sensor coating response profile to seven different vapors by a fluoroalcohol functionalized MIME sensor (left) and an octanethiol MIME sensor (right).

increase) as was observed for propanol example in Figure 4. The swelling mechanism described earlier does not accommodate a conductance increase. This clearly indicates a change in the transduction mechanism. The correlation of vapor polarity with a MIME sensor response in the direction of increasing conductance indicates that a change in the dielectric character of the medium between the metal cluster cores can influence tunneling current in a direction opposite to that resulting from the swelling mechanism. This dielectric effect is not as noticeable in a cluster with a more polar shell such as the fluoroalcohol functionalized cluster in Figure 6.

The charge transport through granular metal films has been studied since the 1960's and several models and mechanisms have been described. One that is particularly simple and parameterizes the cluster core and shell dimensions and the dielectric constant of the intercore medium is as follows *(12)*:

$$\sigma = \sigma_0 \exp[-E_a/(RT)] \qquad E_a = \left(\frac{1}{2}\right)\frac{e^2}{4\pi\varepsilon\varepsilon_0}\left[\frac{1}{r} - \frac{1}{r+s}\right]$$

where σ is the conductivity, E_a is the activation energy, R is the gas constant, T is temperature, e is an electron charge, r is the radius of the cluster core, s is the spacing between cluster core surfaces, ϵ is the dielectric constant of the medium between cluster core surfaces and ϵ_0 is the vacuum dielectric constant. An increase in the dielectric constant of the cluster shell has the effect of decreasing the activation energy for conductivity which in turn will increase the conductivity between clusters. This model is consistent with our observations of the sorption of very polar vapors resulting in conductivity increases.

This MIME sensor response to polar vapors with an increase in conductance is an added dimension in selectivity compared with other solid-state sensors such as SAW, MOS and electrochemical cell devices which respond in only one direction. Some vapors of particular interest are very polar and display exceptionally strong MIME responses in the positive direction. Such vapors include those from explosives based on nitrate chemistry. Trinitrotoluene, TNT, is an exceptionally polar compound with an extremely low vapor pressure (8 x 10^{-6} torr at 25°C with calculated corresponding concentration of 0.02 ppb by volume or 100 ng/liter). The MIME sensor response to this headspace concentration sampled by a short conduction line from a container of TNT is presented in Figure 7. The container is warmed from 25 to 35°C during the 60 sec exposure/180 sec purge cycles, and the response follows the increase in vapor pressure. The initial 5 sec response is very fast followed by a slightly slower but strongly continuous increase. After 30 sec exposure, a purge returns the signal to the baseline in less than 5 sec. Other nitrated organics (dinitrotoluene and dinitrobenzene) and explosives (RDX, urea nitrate, ammonium nitrate and black powders) cause similar responses. The response by

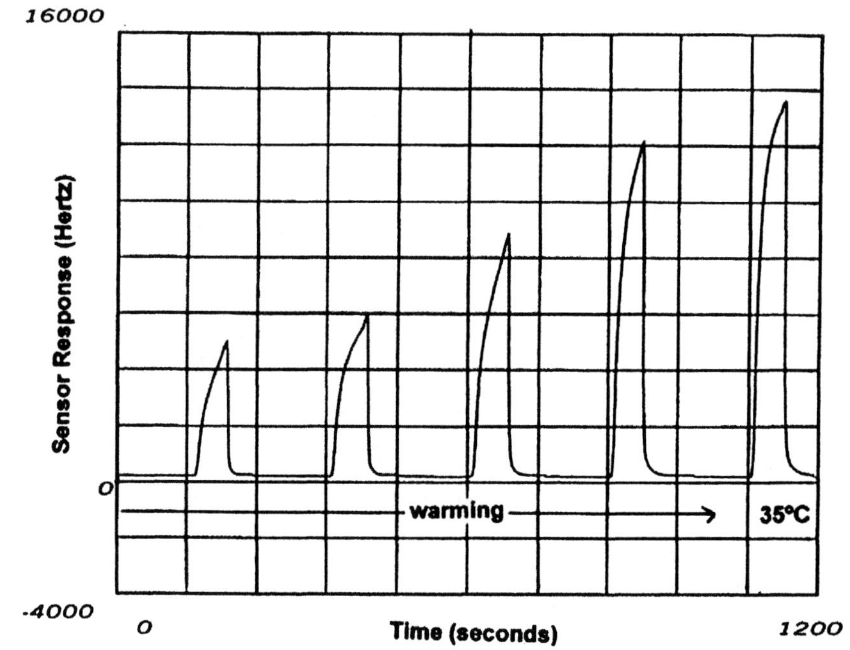

Figure 7. MIME sensor (SC_5HFIP cluster) response to TNT vapor illustrating the increase in signal intensity as the TNT source is warmed from 25 to 35°C during 5 purge-exposure cycles.

this class of compounds is unique by virtue of both its direction and strength. Of the compounds investigated to date, few types of vapors generate responses in the increasing conductance direction and none in this magnitude of strength. A promising application is in the area of explosives detection.

MIME Sensors Development and Future Work

This research was initiated in 1997 as a collaborative effort between NRL and Microsensor Systems, Inc. Over these years interfacing electronics, sensor housings, vapor sampling and transport mechanisms have been under development. An initial prototype is an electronic nose configuration known as VaporLab™ which is pictured in Figure 8. This is a hand-held, battery-powered rapid vapor identification system initially designed for SAW devices and reconfigured for a MIME sensor prototype. Future applications call for more highly miniaturized systems that may be packaged within the volume of a wrist watch or smaller. The MIME sensor fabrication integrates with planar silicon technology and future embodiments may see the sensor as a component on a chip that could be incorporated into breathing-air lines, inserted into filters in respirators or into protective clothing for residual life indication, attached to ground scanners for land mine detection, mounted on UAVs or deployed as an array of remote drop-off sensors for battle space information. Future basic research will investigate sensor fabrication by chemical self-assembly, reducing the sensor substrate size from the micron to the nanometer size, and investigating the frequency and capacitance of the sensor response.

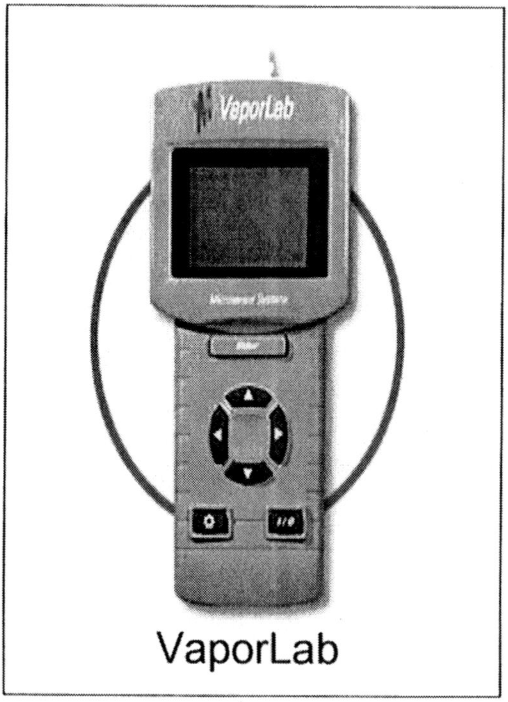

Figure 8. Photo of the VaporLab electronic nose prototype.

Acknowledgements

Support from NRL by way of the sabbatical program for advanced research (AWS) and from Microsensor Systems, Inc. for creation of special lab space and facilities access over the critical first year of this research are gratefully acknowledged. Sponsorship over subsequent years by ONR, Microsensor Systems, Inc. CRADA, DuPont Fabrics & Separations Systems, DTRA, Joint

Service Technology Panel for ChemBio Defense is also gratefully acknowledged. Appreciation for technical guidance/assistance with past and on-going work is expressed to Bruce Yost (DuPont), Mario Ancona (NRL), Ed Foos (NRL) and Richard Smardzewski (US Army ChemBio Center).

References

1. Wohltjen, H.; Snow, A. W. *Anal. Chem.*, **1998**, *70*, 2856-2859.
2. Shipway, A. N.; Lahav, M.; Blonder, R.; Willner, I. *Chem. Mater.*, **1999**, *11*, 13-15.
3. Kharitonov, A. B.; Shipway, A. N.; Willner, I. *Anal. Chem.*, **1999**, *71*, 5441-5443.
4. Lahav, M.; Gabai, R.; Shipway, A. N.; Willner, I. *Chem. Comm.*, **1999**, 1937-1938.
5. Lahav, M.; Shipway, A. N.; Willner, I. *J.Chem. Soc., Perkin Trans. 2,* **1999**, 1925-1931.
6. Evans, S. D.; Johnson, S. R.; Cheng, Y. L.; Shen, T. *J. Mater. Chem.*, **2000**, *10*, 183-188.
7. Lahav, M.; Shipway, A. N.; Willner, I.; Nielsen, M. B.; Stoddart, J. F. *J. Electroanal.Chem.*, **2000**, *482*, 217-221.
8. Labande, A.; Astruc, D. *Chem. Comm.*, **2000**, 1007-1008.
9. Han, L.; Daniel, D. R.; Maye, M. M.; Zhong, C.-J. *Anal. Chem.*, **2001**, *73*, 4441-4449.
10. Brust, M.; Walker, M.; Bethell, D.; Schiffrin, D. J.; Whyman, R.; *J. Chem. Soc., Chem. Commun.*, **1994**, 801.
11. Snow, A. W.; Wohltjen, H. *Anal. Chem.* **1984**, *56*, 1411-1416.
12. Abeles, B.; Sheng, P.; Coutts, M. D.; Arie, Y. *Adv. Phys.* **1975**, *24*, 407-461.

Chapter 4

Nanotechnology Challenges for Future Space Weather Forecasting Networks

Rainer A. Dressler[1,*], Gregory P. Ginet[1], Skip Williams[1], Brian Hunt[2], Shouleh Nikzad[2], Thomas M. Stephen[3], and Shashi P. Karna[4]

[1]Air Force Research Laboratory, Space Vehicles Directorate, 29 Randolph Road, Hanscom Air Force Base, MA 01731
[2]Jet Propulsion Laboratory, California Institute of Technology, Pasadena, CA 91109
[3]Department of Physics and Astronomy, University of Denver, Denver, CO 80208
[4]Army Research Laboratory, Aberdeen Proving Ground, MD 21005–5066

This chapter provides an overview of space weather and articulates the need for an affordable space weather forecasting network as communication and navigation become increasingly reliant on space-based systems. Space weather forecasting efforts face a number of technological challenges to provide global specification of the near-Earth and solar space environment. Nanotechnology is seen to play a pivotal role in technically enabling such an effort and in making it affordable. Concepts of future nano-structured energetic particle sensors are presented.

Introduction

We live in a world that increasingly relies on uninterrupted global communications and space-based navigation systems. Trans-ionospheric transmission depends on the health of communications satellites and the electrical properties of the ionosphere. Both can be seriously compromised by *space weather*, which refers to space environmental conditions as influenced by solar activity and, to a lesser degree, by cosmic radiation and interplanetary dust, such as cometary and asteroidal debris. The solar eruptions that result in the most prominent space weather effects experienced in Earth's surroundings have numerous detrimental consequences affecting both the civilian and military communities. Among the hazards associated with space weather are erroneous data transmissions, such as global positioning system (GPS) inaccuracies, total communication outages, power grid failures, satellite hardware degradation, and changes in the neutral density of the thermosphere, which alter spacecraft orbits. The commercial impact is real, as demonstrated by an exceptionally strong solar proton event on July 14, 2000 ("Bastille Day" event) that rendered a Japanese scientific satellite useless and disrupted instruments on many other satellites. It is, therefore, not a surprise that the space-weather threats to our current and future means of existence are increasingly recognized as a research and development priority by many industrialized nations. From a national defense point of view, the relevance of space weather and the associated space situational awareness is well understood and space weather is a core mission area of the Air Force Weather Agency and Air Force Space Command.

Forecasting space-weather events is daunting since specifying the space environment involves precise knowledge of the conditions of the surface of the sun, the solar radiative and particle outflow, the interplanetary medium, and the Earth's magnetosphere, thermosphere and ionosphere, all of which are strongly coupled. Given the task at hand and the pervasive impacts of space weather, a multi-agency National Space Weather Program (NSWP) was generated with involvement of the Department of Defense, the Department of Commerce (i.e., NOAA), the National Science Foundation (NSF), the National Aeronautics and Space Administration (NASA), and the Federal Aviation Administration (FAA).

A successful effort in forecasting the moments in time and global positions of space weather effects must involve a comprehensive effort combining space-monitoring, real-time data processing, and analysis based on physical space weather models. Currently, one of the main impediments to accurate forecasting is the serious lack of relevant data, due to the extended temporal and spatial ranges involved and the high cost of space instrumentation. Many proposals exist addressing these issues, involving both remote sensing approaches and in-situ sensors distributed on clusters of miniaturized satellites. At the current state of aerospace technology, the cost of a network of satellites that provides global coverage and continuous monitoring of vital parameters of both the Earth's space environment and solar activity would be prohibitive. The reduction of launch costs through miniaturization of both spacecraft and sensor technology must, therefore, be a high priority in attaining the objectives of the NSWP.

There is a general consensus that nanoscience and technology will play a fundamental role in enabling a future space weather network. The purpose of the present chapter is to identify some of the fundamental scientific questions that need to be addressed to pave the way for generation-after-next space-weather sensors. The identification of the underlying science undoubtedly requires breeding novel sensor concepts. This could be regarded as a premature effort, as nanoscience is still in its infancy. Our primary objective, however, is to portray space weather as an important and attractive technology driver for future government-sponsored nanoscience.

It is clear that this chapter cannot cover all aspects of space weather sensors that will benefit from nanotechnology and lead to improved forecasting systems. The importance of nanoelectronics in future sensors will only be marginally addressed. The reader is referred to the chapter of this volume by Karna et al. Although electromagnetic radiation is an important component of space weather, we prefer not to discuss optical nanosensors since efforts are already widespread in characterizing the optical response of nanomaterials.

In the following section, we provide a brief introduction to the space environment and space weather, and discuss current forecasting methodologies from a space-vehicle perspective. We then review the current state of micro-particle sensor technology followed by a delineation of scientific challenges posed in defining specific nano-structured sensors. Unlike research on optical properties and chemical reactivity or catalytic activity of nano-structures, essentially no efforts are known to us that investigate the response of nano-structured surfaces and materials to hyperthermal particles (i.e., ions, neutrals and electrons). We hope we can seduce a number of leaders of the field to study these interactions.

Overview of Space Weather

In this section, we provide a brief overview of the individual domains that govern space weather. For a more extensive treatise we recommend the Handbook of Geophysics and the Space Environment (*1*) distributed in CD-ROM form by the Air Force Research Laboratory Battlespace Environment Division at Hanscom AFB, MA.

Fig. 1 provides a schematic representation of the domains of space weather. While the preponderance of the solar energy affecting Earth is in electromagnetic form, space weather effects are induced by both the radiative and corpuscular components of solar output. The latter is referred to as the *solar wind*, consisting mostly of protons (~95%), alpha particles, and electrons that propagate at average velocities of 450 km s^{-1} away from the sun. Since the solar wind consists almost entirely of charged particles, the interaction between the fast moving plasma and the solar magnetic field causes the solar wind to pull the solar magnetic field away from the sun, producing the interplanetary magnetic field (IMF). The IMF plays an important role in particle transport through the interplanetary medium, and, more importantly, in the interaction of the solar wind with the terrestrial plasma.

Solar radiative and particle output, also summarized as *solar activity*, are highly variable on both short and long timescales. Sporadic large releases of energy, such as *solar flares* and *coronal mass ejections* (CMEs), can cause dramatic bursts of UV, EUV, and X-ray radiation and "gusts" of high density, more energetic solar wind. Solar flares are sudden releases of energy over the entire electromagnetic spectrum. When a solar flare is observed on the sun, it is already too late for satellite operators to conduct protective measures. Protection from solar flares must involve the observation of characteristic signatures on the solar surface, in the solar wind properties, and in coronal electromagnetic radiation that point towards the onset of a flare. Given the approximate solar rotational period of 26 days, long-time forecasting requires observations on the

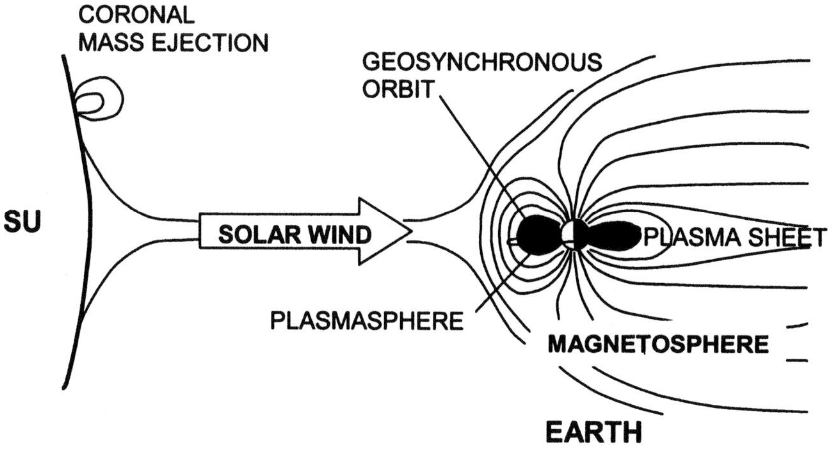

Figure 1. Schematic of the Sun-Earth Connection. The thin lines signify magnetic field lines.

backside of the sun. Missions involving clusters of satellites orbiting the sun are currently being discussed for this purpose.

Coronal mass ejections or *solar prominences* are large releases of solar mass. The total mass released can exceed 10^{13} kg, with ejection velocities up to 2,000 km/s. The observation of a coronal mass ejection normally precedes its impact on Earth by 2 to 3 days. Precise knowledge of the location, direction, speed of ejection, and solar and interplanetary magnetic field configurations are necessary to make forecasts of the moment of terrestrial impact. The frequency of energetic solar events such as flares and coronal mass ejections is strongly correlated with the number of sunspots observed on the photosphere. Sunspot numbers have been counted since the 16th century and exhibit a periodic trend with an average period of 11.4 years. Fig. 2 displays the recent *solar cycle* indicating that, as this chapter is being written, we have just passed a solar maximum.

Whereas knowledge of the moment of impact of a coronal mass ejection on Earth can be gleaned from solar and solar wind properties, as iterated above, prediction of the magnitude and precise location of space weather effects on Earth requires a comprehensive understanding of the response of Earth's plasma and neutral particle envelope to solar wind changes. If it weren't for the Earth's magnetic field, Earth's atmosphere would be directly exposed to the high energy particles of the solar wind, and life as we know it would not be possible. The interaction between the terrestrial and the solar wind magnetic fields and associated plasmas causes the solar wind to be diverted around our planet, thus sheltering us from a direct impact. The region of magnetized plasma surrounding Earth is referred to as the *magnetosphere*. The magnetic field lines of the magnetosphere are not shaped like those of a magnetic

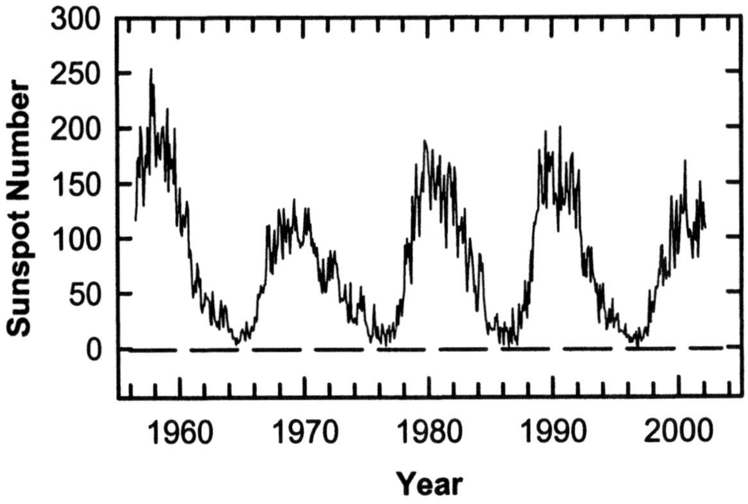

Figure 2. Sunspot numbers from 1955 to present depicting the recent solar cycle.

dipole, but exhibit a long tail, the *magnetotail,* at the night side of our planet. This is a direct consequence of the interaction between the solar wind and Earth's magnetic field. The diversion of the solar wind occurs at the *magnetopause* which, at quiescent solar conditions, is situated at a distance of approximately 10 Earth radii (1 R_E = 6.4 × 10^3 km) where plasma densities are approximately 10 cm^{-3}. Coronal mass ejections can produce solar wind gusts that momentarily push the magnetopause to radii well below 6 R_E, thus exposing geosynchronous satellites, orbiting at an altitude of 36,000 km above the equator, to the direct solar wind. Fig. 1 depicts the approximate location of a geosynchronous orbit (GEO).

Every so often, the solar wind and terrestrial magnetic fields couple in a way that allows energetic particles to penetrate the magnetosphere. Such *magnetic reconnection events* are still not understood well enough to predict the location and moment of their occurrence. These magnetic reconnection events are the source of *aurorae* and *geomagnetic storms* that can be so detrimental to satellites, communications, and terrestrial power grids. Aurorae are the low altitude signature of the Earth's plasma sheet, a layer of plasma extending from approximately 6 R_E on the nightside down through the magnetotail. This plasma poses charging hazards to geosynchronous orbiting satellites.

A region exists within the magnetosphere where energetic charged particles are trapped by the Earth's magnetic field. This region, normally within 4-7 R_E, has two overlapping components, the *plasmasphere* (Fig. 1) and the *radiation belts.* The radiation belts contain charged particles with energies ranging from 1 keV to hundreds of MeV. These trapped energetic particles are associated with the *extraterrestrial ring current* that counteracts the inherent magnetic field of Earth as observed on its surface. This current is intensified during a geomagnetic storm, resulting in important reductions of the geomagnetic field, similar to a barometric pressure drop during a tropospheric storm. The mechanism leading to a geomagnetic storm is currently hotly debated. A common sequence of events involves the diffusion of polar ionospheric ions (*ionospheric upwelling*) into the magnetotail region, where they are energized and injected into the radiation belts. The onset of a storm can be sensed by the change of ionic constituents in the ring current. During quiescent periods, the positive ion component of the ring current consists primarily of protons. During storms, O^+ densities can exceed proton densities. O^+ is the principal ion of the upper ionosphere.

Solar activity and geomagnetic storms alter electron density profiles of the ionosphere and can cause significant heating of Earth's upper neutral atmosphere, the *thermosphere.* Ionospheric perturbations, such as *ionospheric scintillation*, result in variations in electromagnetic signal propagation paths. These types of perturbations reduce the accuracy of single-frequency GPS receivers and geo-location systems, cause high frequency (HF) communications outages, and generate clutter for space surveillance radar.

Space Weather Sensing Approaches

Current efforts in specifying ionospheric scintillation involve mostly ground-based ionospheric sounding networks that provide forecasts of up to ~1/2 hour. Military operations, however, demand several hour forecasts. This requires a denser data set of the driving magnetospheric and solar wind parameters, as well as an improved understanding of the coupling between the ionosphere and magnetosphere. In the following we will focus primarily on the challenge for improved global specification of the magnetosphere and the solar wind.

Several concepts are being put forward to provide global specification of magnetospheric and solar wind properties. They can be separated in two approaches: *i)* a distributed set of micro, nano, or picosatellites with in-situ sensors (per unofficial definition, micro, nano, and picosatellites have mass ranges of 10-100, 1-10, and <1 kg, respectively); *ii)* remote sensing of the magnetosphere using a small number of satellites. Current visions of the former approach generally involve flotillas of spacecraft with limited sensing capabilities, i.e., each spacecraft provides measurements of only a few local plasma parameters. As miniaturization technology continues to mature, the sophistication of proposed satellite clusters is expected to increase.

The power of remotely sensing the space environment has recently been put on display by the NASA Imager for Magnetopause-to-Aurora Global Exploration (IMAGE) mission (*2*). Since this mission is trailblazing future missions, we will discuss it in greater detail. In a later section of this chapter, we will use it as an example where nanotechnology could have a major impact in future renditions of the methodology. The NASA IMAGE mission involves a single satellite with a suite of instruments that provide optical (EUV, FUV), radio-plasma, and energetic neutral [low (10 – 1000 eV), medium (1 – 30 keV), and high energy (20 keV – 500 keV/nucleon)] imaging, all of which combine to provide global 2-dimensional images of Earth's magnetosphere. The IMAGE spacecraft is in a highly elliptical polar orbit with a perigee of 1000 km and an apogee of 7 R_E. During the mission the orbit precesses, scanning the meridional orbital plane from dusk to dawn. The spacecraft rotates with a 2 minute period on an axis perpendicular to the orbital plane. This rotation sweeps the detector field-of-view (FOV) across the image plane.

A key novelty of this mission are the three energetic neutral atom (ENA) imagers. Energetic neutrals are formed in charge exchange collisions between energetic ions of the magnetosphere and neutrals of Earth's *exosphere*, the region of Earth's atmosphere where neutrals may escape to space with a negligible number of collisions. The neutrals of the magnetospheric *exosphere* are primarily hydrogen atoms, making up Earth's *hydrogen geocorona*. Neutral products resulting from charge-exchange collisions between energetic magnetospheric ions and the geocorona have a velocity distribution, in the center-of-mass reference frame, that is very close to that of the parent ions, since charge-exchange is primarily a long-range process. The neutral products are not affected by Earth's magnetic field and can thus leave the magnetosphere. Remote detection of these energetic neutrals can thus provide energy resolved images of the plasma densities of the inner magnetosphere. Both protons and oxygen ions have large charge exchange cross sections with hydrogen atoms. The ratios between protons and oxygen ions are an important signature of magnetospheric activity, and can, therefore, be effectively monitored with the neutral imaging technique.

The individual ENA detectors cover three energy ranges, high (HENA), medium (MENA) and low (LENA). The detectors target different magnetospheric areas. The HENA and MENA sensors are based on sweeping charged particles out of the line of sight with deflection potentials applied to collimators, followed by the transmission of ENAs through thin foils (e.g., Si-polyimide-C, carbon) that block a large fraction of UV photons. The ENA transmission through the foils generates secondary electrons that are detected and generate a start pulse for a time-of-flight (TOF) measurement. Subsequent position sensitive, time-resolved detection of the transmitted ENAs on a microchannel plate (MCP) or a solid-state detector utilizing pulse-shape and height analysis, provides the necessary energy, mass, and polar angle information of the detected ENA.

The correlation of start and stop signals provides the TOF information, as well as diminishes backgrounds due to EUV photons, which also produce secondary electrons at the foil. This approach is sufficient for the HENA detector, where ENA detection efficiencies exceed 0.2. For the reduced efficiencies of lower energy ENAs, the MENA detector combines the foil with a nanoscale grating (*3*) that substantially increases the rejection of the main source of EUV, H Ly-α emissions at 121.6 nm, while not greatly affecting the transmission of ENAs.

The low energy neutral atom sensor, LENA, must use a different approach because the ENAs of interest have insufficient energy to pass through foils. In the LENA sensors, the ENAs, following collimation and charged particle rejection, are converted to negative ions through grazing incidence scattering on a polished tungsten surface. The negative ions are then electrostatically accelerated to energies exceeding 10 keV by the ion extraction lens. The ion extraction lens focuses the ions formed at the conversion surface, dispersed in energy, at the entrance to an electrostatic energy analyzer. The ions pass through a broom magnet at this point, eliminating electrons produced on the conversion surface. At the exit slit of the analyzer, the negative ions are dispersed in energy and pass through a C foil, similar to the HENA and MENA analyzers, This triggers a TOF measurement, followed by position sensitive detection that yields polar angle and energy information.

Following its launch in March, 2000, IMAGE has successfully produced ENA images of the magnetosphere with a temporal resolution of 4 minutes, thereby providing first views on a global scale of how the magnetosphere changes during geomagnetic storms. Although the existence of the equatorial ring current has been postulated for decades, IMAGE provided first actual images of it (*4*). The ongoing success of IMAGE, presently scheduled to operate for 3 years, will undoubtedly call for new missions. A follow-on experiment, the Two Wide-angle Imaging Neutral-atom Spectrometers (TWINS) mission, will involve two MENA detectors on widely spaced, identical spacecraft providing the first 3-dimensional images of Earth's magnetosphere. The ultimate magnetospheric imager would involve several ENA imaging satellites providing continuous, 3-D images of the entire magnetosphere from dedicated orbits. Obviously, the cost of such a venture would be prohibitive at the current state of technology. The HENA, MENA, and LENA sensors, including electronics, weigh 19, 14, and 23 kg, respectively, and the entire spacecraft has a mass of 494 kg. In the following, we make an attempt at identifying the challenges of reducing the mass of space weather sensors, such as the ENA detectors described above, thereby suggesting pertinent lines of research in nanoscience.

Nanosensors for Energetic Particles

Current Particle Sensing Techniques

Trajectory Sensors

We shall refer to particle sensors that characterize particles by analyzing their trajectories *trajectory sensors*. Trajectory sensors can be subdivided into dispersion and time-of-flight sensors. In dispersion sensors, charged particles are subjected to electric and magnetic fields that separate the particle trajectories with respect to energy or velocity, providing information on the energy and mass of the particle. The dispersion device is coupled with a detector, and particle spectra are obtained by scanning the dispersion field, and/or using a position sensitive charged-particle detector such as a microchannel plate detector. Energy analyzers are usually realized through homogeneous electric fields generated by deflection electrodes in geometries such as parallel plate condensers, cylindrical and hemispherical analyzers. The latter geometries are, however, not optimally suited for imaging purposes. Pin-hole approaches coupled with retarding potential analyzers (RPAs) are the more common approach. In RPAs, a blocking field applied to two proximate, parallel grids prevents the passage of particles with energies below a set potential. The derivative of the transmitted current as a function of blocking potential then provides the energy distribution of the observed source particles. The mass of ions can be determined by dispersing the incoming particles with a homogenous magnetic field, through the use of inhomogeneous radiofrequency fields, such as those of a quadrupole, or through the evaluation of particle flight times within an instrument. Rf quadrupoles lose resolution with ion energy and are thus not well suited for ions with energies exceeding 1 keV. Time-of-flight mass analysis requires knowledge of the energy of the particle. The timing can be triggered by an electrostatic gate at the entrance of the flight path. This is only practical in high-signal situations. As described in the previous section, energetic particles with energies exceeding 1 keV can traverse thin foils without significant loss of momentum while simultaneously ejecting from the foil electrons that provide a start pulse to the timing logic.

Important difficulties arise when reducing the size of trajectory analyzers. A miniaturized electrostatic energy analyzer is limited with respect to the maximum energy it can disperse while maintaining safe voltages across the reduced electrode distance. The magnetic field of a magnetic sector mass filter scales as $R^{-0.5}$, while rf frequencies of quadrupole devices become impractically high at small dimensions. Time-of-flight devices, as practiced today, dramatically lose resolution as the flight path is reduced. Finally, trajectory sensors usually involve a charged-particle detector such as a channel electron multiplier or a microchannel plate which introduces further high-voltage problems. Consequently, it can be generally concluded that trajectory sensors have limited miniaturization capacity.

Solid-State Sensors

Solid-state detectors are compact, robust, and relatively inexpensive with high position resolution and low power consumption. Because they produce a signal proportional to the incident particle energy, solid-state detectors also provide the energy of individual particles in addition to incidence rate and position information. These features make these detectors very attractive for space science missions. Their main drawback is their loss of sensitivity at low particle energies. In the case of protons current state-of-the-art solid-state detectors lose sensitivity at energies below ~10 keV.

Solid-state sensors are widely used in nuclear physics, high-energy particle physics, and in space physics for energy-resolved detection of particles with energies above 10-50 keV. These detectors are based on the principal of reverse-biasing a pn or pin junction of a diode, thereby creating a charge depletion region at the junction. High-resistivity silicon wafers are often used for fabrication of these particle detectors. The wafer thickness is usually several hundred micrometers to several millimeters depending on the wafer resistivity that varies in the range of 1-10 kΩ -cm (doping concentration ~10^{10}-10^{12} dopants per cubic centimeter). These low doping concentrations allow full depletion of the wafer throughout its entire thickness with approximately 100 volts applied to highly doped electrodes that are added to each side of the wafer by diffusion, implantation, or deposition. When an energetic particle passes into or through the silicon, it will lose energy, in part by generating a cloud of electron-hole pairs along its path. Particles with energies greater than 10 keV generate on average one electron-hole pair for every 3.65 eV of energy lost. The signal charge that can be measured as a pulse is, therefore, related to the energy lost by the particle in the silicon and, therefore, is a function of the particle's mass and initial energy. The integrated signal is thus proportional to the particle energy, while the pulse shape can provide mass information.

The lack of sensitivity of solid-state detectors to lower energy particles is due to the "dead layer" of the detector. In the high resistivity detectors, this dead layer is due to the thickness of the electrode which can be on the order of 1 μm. Charge carriers generated by particles that do not have sufficient energy to penetrate significantly beyond this silicon layer are trapped there.

Although more commonly applied to photon detection, another class of solid state detectors or imagers, charge-coupled device (CCD) detectors, can also be used to detect energetic charged particles. To allow detection of low-penetrating ionizing radiation such as UV photons or low-energy particles, CCDs are back illuminated thereby avoiding the material thickness associated with the VLSI circuitry of the CCD pixels. In backside illumination, the silicon wafer is etched off all the way to the 10-20 μm thin epilayer silicon in which the CCD structure has been processed. Back-illuminated CCDs are also affected by a dead layer situated at the interface between the sensitive silicon and its native (~2 nm) oxide layer where charge accumulates, thereby causing formation of a potential well that traps carriers generated within 500-1000 nm of this interface. Thus, particles to be detected must have energies that allow them to penetrate deeper than 500-1000 nm.

Reducing the low energy particle detection threshold of silicon detectors requires reducing the thickness of the dead layer. This can be done either by reducing the depth of the surface potential well, or by reducing the thickness of the surface electrode, as appropriate for the device in question. Attempts to modify the surface potential using high-work-function metals or applied voltages have resulted in various degrees of success (*5,6*). Shallow ion implantation has been applied to CCDs, strip detectors and diode arrays leading to a commercially quoted ~10 keV detection threshold for protons (*7,8*). Other groups have also achieved some success detecting low energy particles using detectors capped with ultrathin oxides (*9-12*).

The low-energy detection limitation of solid state detectors can be addressed by "delta-doping," a technique that was invented and developed at JPL (*13,14*). In this technique, silicon epitaxy is used to grow a very highly doped layer in a single atomic layer, less than a nanometer from the surface. This highly doped layer acts as the surface electrode, and replaces the micrometer-thick conventional electrode, making the device sensitive to any particle that can pass more than a nanometer into the detector. The delta-doping technique was originally developed to extend the response of visible charge-coupled devices (CCDs) into the UV, and it has proved to provide superior performance and reliability than alternative technologies such as surface chemical treatments or low-energy shallow ion implantation.

Extension of CCD applications to detection of lower energy particles has been successfully demonstrated using delta-doping technology at NASA's Jet Propulsion

Laboratory (MBE) (*14-17*). The resultant delta-doped back-illuminated CCD can detect particles with a low-energy threshold improvement of at least an order of magnitude for protons to below 1 keV, and for electrons down to 100 eV. Extension of delta doping to high resisitivity detectors has also been demonstrated more recently. Further work in this area is underway. With delta-doped detectors, we have demonstrated the reduction of the deadlayer in silicon solid-state detectors to ~ 1 nm. The low-energy detection threshold, high efficiency, and the stability of performance of these detectors make them ideal candidates for future spacecraft skin sensors.

Nano Space Weather Sensor Challenges

General Concepts

The example of the nano-scale Hα rejection grating used by the MENA sensor of IMAGE demonstrates that nanotechnology is already applied in space. The problems described for current particle sensing technologies call for additional nanotechnological solutions for improved sensor performance, as well as for sensor miniaturization. Performance focuses on sensitivity, mass and energy resolution, while miniaturization addresses reducing the mass, volume and energy requirements of the complete sensor package (i.e., sensor head, particle optics, and electronics). Given current sensor designs, however, miniaturization counteracts sensitivity and resolution. If the signal levels of the particle to be detected is inherently low, as is the case for energetic neutrals of the magnetosphere, a reduction in aperture by scaling existing sensing technology could incapacitate a miniaturized sensor, even if the detection efficiency of the sensing surface or material approaches 100%. Consequently, miniaturization of the electronics and optical manifold is a primary challenge for low-signal sensing. The ultimate solution is the development of molecular electronic amplifier circuits situated in the direct vicinity of sensing surfaces. This capability would allow for a combination between a CCD detector and a solid-state detector, where each CCD pixel has its own processor for pulse-shape analysis, thereby obviating the need for gating circuitry. It is needless to say that this molecular electronic capability would also revolutionize optical imaging technology.

One solution to overcoming aperture reductions of a miniaturized sensor is the incorporation of networks of nanosensors in the "skin" material of the spacecraft and distributing them over the entire spacecraft surface. Thus, one can envision miniaturized pressure sensors spread over the surface of a low earth orbit (LEO) spacecraft that continuously registers the impact of thermospheric neutrals, thereby providing real-time drag and neutral density information. Nanoelectronic devices, early versions of which are rapidly emerging (*18*), may provide the key sensing elements to register such hyperthermal impacts. In parallel to the development of nanoelectronic sensing components, however, must be a concerted effort in understanding the fundamental physics behind the hyperthermal interaction between space environmental particles and nanostructured surfaces. While most efforts involving novel nanostructures examine their optical properties, there are essentially no research efforts that systematically investigate the response of nanostructured surfaces to particle beams, apart from electron diffraction studies. A primary issue will be whether such hyperthermal collisions induce irreversible changes in the surface, in which case the device is expendable, or whether the device can be "reset" electrically or through some chemical self-repair process. It is worth noting that current charged-particle detectors gradually lose sensitivity due to internal erosion, so performance degradation in itself does not present a new problem. We close this chapter with two space-weather sensing concepts that illustrate the potential of novel nanomaterials in contributing to greatly improved sensors and identify the required research directions.

Grazing Incidence Scattering on Nanostructured Surfaces

As mentioned above, the LENA sensor uses a grazing incidence W surface to convert low-energy neutrals (<1 keV) to negative ions that are subsequently mass and energy analyzed. This remote sensing capability can observe the onset of a geomagnetic storm through the observation of O^+ upwelling from the polar regions of the ionosphere. The current W surface has a very low conversion efficiency, and the dispersive ion optics are of considerable size. Improved conversion surfaces are already being explored, and novel nanomaterials are being considered. Fig. 3 is a conceptual display of a hypothetical energetic neutral detector using grazing incidence scattering from a nanostructured surface. The surface in the figure is representative of an array of quantum dots, but other structures such as self-assembled monolayers, or carbon nanotube assemblies, could just as well be considered. Also, the nanostructured film need not be deposited on a planar surface, but could also coat microcavities or tubes that define the solid angle of incidence. The incident energetic neutral O atom or H

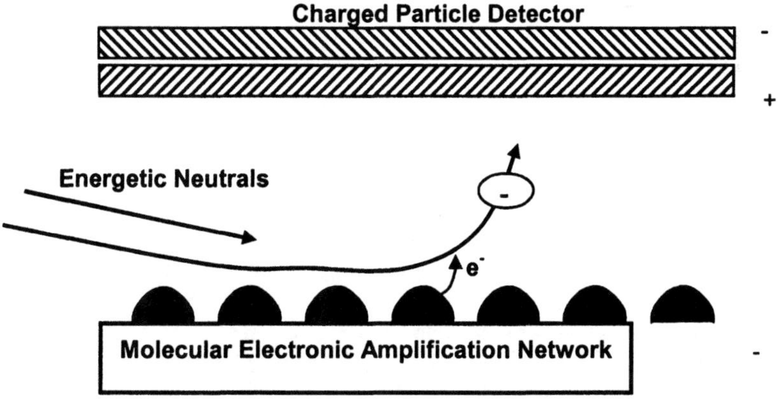

Figure 3. Schematic representation of a low-energy energetic neutral detector involving a nanostructured gracing-incidence negative-ion conversion surface.

atom, the most abundant of ENAs, will convert to a negative ion with a certain probability, after which it is accelerated to a charged particle detector. The schematic sensor omits the energy selective ion optics, which are part of the LENA detector, for simplicity.

Negative ion formation on metal surfaces can be described by a curve-crossing model that is analogous to the harpoon mechanism first observed for gas-phase alkali-halogen collisions, where very large ion pair formation cross sections are observed (*19,20*). The incident atom encounters curve crossings between the weakly attractive atom-surface potential and the strongly attractive potential between the negative ion and its surface image charge given by $-0.25R^{-1}$. The crossing coincides with the distance at which the negative ion–surface potential energy is equal to the metallic Fermi level.

Recently, it has been discovered that insulator surfaces, such as LiF crystals, can exhibit substantially higher negative ion conversion efficiencies than metal surfaces for glancing scattering of atoms, including atomic oxygen (*21*). This is somewhat surprising because the band gap of such materials is ~10 eV. The high efficiency is attributed to the fact that the exit channel interaction now consists of a "full" Coulomb interaction, -R^{-1}, between the negative ion and an electron hole from the donor atom that remains localized during the collision. This rapidly reduces the energy gap between entrance and exit channels. Nevertheless, in most cases there is still a considerable energy gap at the closest point of approach, and conversion efficiencies at low projectile energies, where nonadiabatic transition probabilities become small, are still quite low. Also, insulators have the disadvantage of exhibiting charging effects, however, it has been shown that such effects can be suppressed on ionic lattices such as LiF by mild surface heating (*21*).

Nanomaterials and nanostructured surfaces provide new opportunities to control the electronic band structure of materials. For example, nanoparticles, or quantum dots, such as metal atom clusters, have the unique property that the energy required to remove an electron from the particle depends on the size of the particle. They have thus been referred to as designer atoms (*22*). 2-dimensional arrays of quantum dots can be constructed with controlled spacing between the dots through the choice of passivating ligands (for example variation of ligand chain length) that prevent dot coalescence (*22-25*). Since the electronic coupling between dots is directly a function of dot distance, quantum dot arrays can be continuously tuned between insulator and metallic properties. Thus, it can be envisioned that a surface can be tailored to have optimal band gap and charge-immobilization characteristics resulting in high negative ion conversion efficiencies at specific projectile energy ranges.

Ligand effects, however, are anticipated to play an important role in current passivated quantum dot assemblies. A first glimpse at the possible effect of typical organic ligands was recently obtained in the laboratory of D. Jacobs at Notre Dame, where a decanethiol self-assembled monolayer (SAM) was deposited on a flat gold surface and irradiated with a hyperthermal O^+ beam (*26*). Efficient negative ion formation was already observed at incident energies of ~5 eV, including and O^-, OH^- and H^-. Assuming a sequential mechanism, the observation implies that O atoms efficiently produce negative charges, thus making SAMs interesting candidates for future sensor conversion surfaces. The hyperthermal gas-phase interaction between O^+

and O atoms and small alkanes is currently being investigated in the laboratories of R. A. Dressler (AFRL) and T. Minton (Montana State University), respectively, to further elucidate the mechanism. Obviously, the use of hydrocarbons in space raises questions as to the durability with respect to space EUV and atomic oxygen exposure. Fluorination of hydrocarbons may be an effective countermeasure to hydrocarbon breakdown. The group of Jacobs is currently investigating scattering on perfluorinated decanethiol SAMs.

While novel nanomaterials could make an impact on conversion surfaces in the not too distant future, miniaturization of the ion optics and associated electronics is far more challenging. Fig. 3 alludes to the possibility that the sensor surface is in direct contact with a molecular electronic amplifier network. Such a network could register single electron transfer events and localize its occurrence. The electron transfer signal would initiate a time-of-flight measurement cycle which is terminated when the negative ion is detected by a position sensitive detector at a defined distance from the surface, as in Fig. 3. Position and time-of-flight could provide the information to determine the identity and energy of the energetic neutrals, provided the product negative ions and their respective recoil velocity distributions have been well characterized for the particular nano-structured surface as a function of neutral atom identity and incident energy.

Nanosensors for Low-Earth Orbit Density Measurements

It was mentioned earlier that high-precision nano pressure sensors dispersed in the surface material of a spacecraft could continuously measure the instantaneous drag forces on a satellite. This would, for example, enable real-time propulsion adjustments for high-precision orbit trajectories (drag-free orbits), a crucial element of satellite formation flying. The ever more frequent discoveries of novel nanoelectronic devices (*18*) and first demonstrations thereof as chemical sensors (*27*) suggest that such devices will also play a role in future hyperthermal particle sensing.

The intriguing electrical and mechanical properties of carbon nanotubes makes them prime candidates for atomic and molecular sensing (*28*). Carbon nanotube electrical resistance has been demonstrated to be very sensitive to oxygen (*29*) and other gases as well as metal vapors (*30,31*). Meanwhile, bimorph-carbon nanotube structures are being explored as potential nano-mechanical atmospheric drag sensors. This concept (*32*) is shown in Fig. 4a and a sample Scanning Electron Microscope (SEM) image of a bimorph structure (without electrodes) produced in chemical vapor deposition growth experiments is shown in Figure 4b. Individual, vertically oriented nanotubes can be produced on e-beam-defined catalyst dots with integrated electrodes. The adjacent tubes are attracted and held together by van der Waals forces to produce the bimorph structure shown in Fig. 4a. An applied bias shrinks one tube and lengthens the other, causing a lateral deflection of the tube ends. Conversely, a force acting to deflect the tube end will produce a voltage signal. Thus, signals from hyperthermal space environmental atoms and molecules colliding with arrays of such nanotube structures could be processed to provide the thermospheric density and composition, the incident velocity, and the drag force experienced by the satellite. Obviously, an extensive research and development effort is necessary to determine the optimal carbon nanotube dimensions for characterizable impact responses, the effects of rarer energetic events on the device response, general stability of sensing materials, EUV effects, etc.

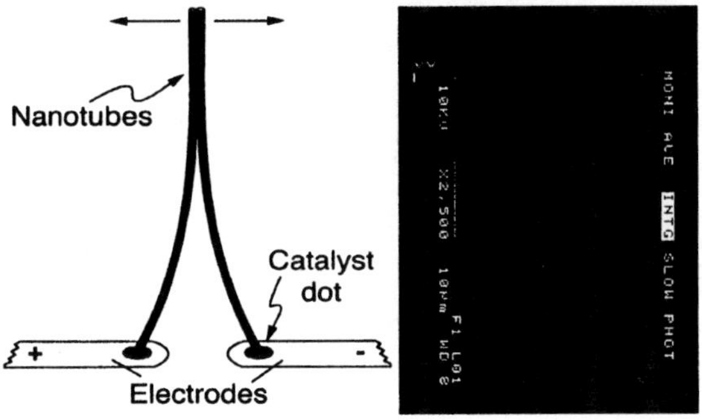

Figure 4. (a) Schematic representation of a nanotube bimorph-geometry force sensor. A force acting to deflect the nanotube end will produce a voltage signal. (b) SEM of a prototype carbon nanotube bimorph structure grown in chemical vapor deposition experiments.

Conclusions

The future of space technology and its protection from space environmental hazards will depend heavily on successful miniaturization of existing and novel capabilities. Nanomaterials and nanoelectronics will play a fundamental role in the miniaturization research and development efforts. This chapter serves to illustrate how novel nanomaterials, nanoparticles and devices are poised to revolutionize the way future microspacecraft monitor the space environment and mitigate its adverse effects. The development of novel nanosensors of space particles, in particular sub keV particles, will depend on a concerted research effort to explore the interaction between energetic particles and nanostructures.

Acknowledgments

The authors wish to thank Prof. Dennis Jacobs for communicating his first results on ion-SAM scattering prior to publication. RAD and SPK were supported by AFOSR under task 2303EP02 (M. R. Berman, program manager).

References

1. Jursa, A. S. *Handbook of Geophysics and the Space Environment*; National Technical Information Servic, 1985.
2. Burch, J. L. *Space Sci. Rev.* **2000,** *91*, 1-506.
3. Beek, J. T. M. V.; Fleming, R. C.; S.Hindle, P.; Prentiss, J. D.; Schattenburg, M. L.; Ritzau, S. *J. Vac. Sci. Technol.* **1998,** *B16*, 3911-3916.
4. Burch, J. L.; Mende, S. B.; Mitchell, D. G.; Moore, T. E.; Pollock, C. J.; Reinisch, B. W.; Sandel, B. R.; Fuselier, S. A.; Galagher, D. L.; Green, J. L.; Perez, J. D.; Reiff, P. H. *Science* **2001,** *291*, 619-624.
5. Janesick, J.; Elliott, T.; Fraschetti, G.; Collins, S.; Blouke, M.; Corrie, B. *Proc. SPIE* **1989,** *153*, 1071.
6. Tassin, C.; Thenoz, Y.; Chabbal, J. *Proc. SPIE* **1988,** *142*, 1447.
7. Hohansen, G. A.; Stadsen, J.; Søraas, F.; Hansen, T. E. *Nucl. Instrum. Meth. B* **1991,** *62*, 162.
8. Ray, J. A.; Barnett, C. F. *IEEE Transaction on Nucl. Science* **1969,** *16*, 82.
9. Hartmann, R.; Struder, L.; Kemmer, J.; Lechner, P.; Fries, O.; Lorenz, E.; Mirzoyan, R. *Nucl. Instrum. & Methods in Phys. Res. A* **1997,** *387*, 250.
10. Hartmann, R.; Hauff, D.; Lechner, P.; Richter, R.; Struder, L.; Kemmer, J.; Krisch, S.; Scholze, F.; Ulm, G. *Nucl. Instrum & Methods in Phys. Res.A* **1996,** *377*, 191.
11. Funsten, H. O.; Suszcynsky, D. M.; Ristzu, S. M.; Korde, R. *IEEE Trans. on Nucl. Sci.* **1997,** *44*, 2561.
12. Maisch, T.; Gunzler, R.; Weiser, M.; Kalbitzer, S.; Welser, W.; Kemmer, J. *Nucl. Instrum.& Methods in Phys. Res. A* **1990,** *288*, 19.
13. Hoenk, M. E.; Grunthaner, P. J.; Grunthaner, F. J.; Fattahi, M.; Tseng, H.-F.; Terhune, R. W. *Appl. Phys. Lett.* **1992,** *61*, 1084.
14. Nikzad, S.; Hoenk, M. E.; Grunthaner, P. J.; Grunthaner, F. J.; Terhune, R. W.; Winzenread, R.; Fattahi, M.; Hseng, H.-F. *Proceedings of the SPIE* **1994,** *335*, 221.
15. Smith, A. L.; Yu, Q.; Elliott, S. T.; Tombrello, T. A.; Nikzad, S. *Proc. of the MRS* **1996,** *448*, 177-186.
16. Nikzad, S.; Yu, Q.; Smith, A. L.; Jones, T. J.; Tombrello, T. A.; Elliott, S. T. *Appl. Phys. Lett.* **1998,** *73*, 3417-3420.
17. Nikzad, S.; Croley, D.; Elliott, S. T.; Cunningham, T. J.; Proniewicz, W. K.; Murphy, G. B.; Jones, T. J. *J. Appl. Phys.* **1999,** *75*, 2686-2688.
18. Tseng, G. Y.; Ellenbogen, J. C. *Science* **2001,** *294*, 1293-1294.
19. Levine, R. D.; Bernstein, R. B. *Molecular Reaction Dynamics and Chemical Reactivity*; Oxford University Press: New York, 1987.
20. Gadzuk, J. W. *Comments At. Mol. Phys.* **1985,** *16*, 219-240.
21. Winter, H. *Progress in Surf. Sci.* **2000,** *63*, 177-247.
22. Remacle, F.; Levine, R. D. *Proc. Nat. Acad. Sci.* **2000,** *97*, 553-558.
23. Collier, C. P.; Saykally, R. J.; Shiang, J. J.; Henrichs, S. E.; Heath, J. R. *Science* **1997,** *277*, 1978-1981.
24. Remacle, F.; Collier, C. P.; Heath, J. R.; Levine, R. D. *Chem. Phys. Lett.* **1998,** *291*, 453-458.

25. Remacle, F.; Levine, R. D. *ChemPhysChem.* **2001**, *2*, 20-36.
26. Jacobs, D., private communication.
27. Cui, Y.; Wei, Q.; Park, H.; Lieber, C. M. *Science* **2001**, *293*, 1289-1292.
28. Baugham, R. H.; Zakhidov, A. A.; Heer, W. A. d. *Science* **2002**, *297*, 787-792.
29. Collins, P. G.; Bradley, K.; Ishigami, M.; Zettl, A. *Science* **2000**, *287*, 1801-1804.
30. Kong, J.; Franklin, N. R.; Zhou, C.; Chapline, M. G.; Peng, S.; Cho, K.; Dai, H. *Science* **2002**, *287*, 622-625.
31. Zhou, C.; Kong, J.; Yenilmez, E.; Dai, H. *Science* **2000**, *290*, 1552-1555.
32. Hunt, B.; Noca, F.; Hoenk, M. In;, 2002.

Chapter 5

Molecularly Imprinted Polymers for the Detection of Chemical Agents in Water

Amanda L. Jenkins[1,3], Ray Yin[1], Janet L. Jensen[2], and H. Dupont Durst[2]

[1]U.S. Army Research Laboratory, Building 4600, Aberdeen Proving Ground, MD 21005
[2]Edgewood Chemical and Biological Command, Blackhawk Road, Aberdeen Proving Ground, MD 21010
[3]Current address: Jasco Inc., 8649 Commerce Drive, Easton, MD 21601

The use of chemical weapons as agents of war has been banned. Nevertheless, several countries, including the United States and the former Soviet Union, are known to have manufactured and stockpiled such weapons. Two nerve gases of particular concern are the organophosphorus species Sarin and Soman. These are the agents used on troops in the Iran/Iraq War, and currently leaking from stockpiles of aging weapons in the United States. Many pesticides and insecticides are so chemically similar to nerve agents that they trigger the same decrease in the enzyme plasma cholinesterase as the agents themselves. Growing concerns over possible contamination of water supplies by nerve agents, insecticides and pesticides has prompted the desire for small portable devices that can quickly detect trace amounts of these substances in water. Novel polymeric materials based on molecular imprinting techniques have been constructed for detection of phosphonate-containing species in water.*(1,2)* Detection is based on sensitized luminescence that occurs when the analyte selectively binds a luminescent europium reporter molecule incorporated in the polymer. This interaction is so specific it can discriminate two almost identical species. A miniature fiber-optic spectrometer monitors the changes in luminescence that result when the analyte is reversibly bound to the co-polymer. Several of these fibers, each for a specific analyte will be mulitplexed together. The limit of detection for these sensors are in the low parts per trillion (ppt) in solution with linear ranges from low parts per trillion to parts per million. The sensors exhibit the same recognition characteristics over several months of use with a response time of less than 5 minutes. Selectivity of the sensors against other pesticides and chemically similar compounds has been demonstrated.

Introduction

The search for highly selective, rugged, low-cost, portable sensors with specificity for toxic chemical, biological, and environmental analytes has attracted enormous interest after the recent terrorist attacks. Molecular recognition occurs in a variety of living systems, and has been used in the detection of many chemical and biological species. However, these natural recognition systems such as antibody/antigen binding reactions although very selective are of limited use due to poor stability, assay concentration range, and cost. Therefore, the use of artificial or biomimetic recognition systems have been proposed and among them, some of the most promising alternatives are the molecularly imprinted polymers (MIPs). These polymers are highly stable and exhibit high selectivity approaching their natural antibody counter-parts. Molecularly imprinted polymers, originally developed for separation purposes are now beginning to be used for enhancing the specificity of a variety of chemical sensors. Quartz crystal microbalances have been coated with MIP films for the detection of analytes such as terpenes*(3)*, dansylphenylalanine*(4)*, and 2-methylisoborneol*(5)*. MIPs have also been used to detect materials with biological significance. These include metalloporphyrins*(6)*, the enhancement of radioligand binding assays for morphine and cholesterol*(7)*, as well as the detection of glucose using MIPs by capacitance.*(8)*

A growing area for MIP materials has been in the area of pesticide detection and remediation. MIP based solid-phase extraction media have been made to remove triazine herbicides like terbumeton *(9)* and combined with LC-MS liquid chromatography/ diode array detection LC/DAD and LC-MS to detect triazines in natural water *(10,11)*. A sensor system for the herbicide 2,4-dichlorophenoxy-acetic acid has been developed using electrochemical detection with screen-printed electrodes. *(12)* Bulk acoustic wave sensors coated with MIP films have been used for detection of the herbicide 2,4-dichlorophenoxyacetic acid.*(13)*

In the United States alone, the use of pesticides has grown from about 500 million pounds in 1960's to over 1.1 billion pounds in 1990's. These pesticides are used for controlling weeds, insects, and other organisms in a wide variety of agricultural and non-agricultural settings. The use of pesticides has helped to make the United States the world's largest producer of food, but has also been accompanied by concerns about their potential adverse effects on the environment and human health. The greatest potential for unintended adverse effects of pesticides, in many respects, is through contamination of the Earth's hydrologic systems, which supply water for both humans and natural ecosystems. *(14)*

Pesticide contamination of water systems through unintentional means such as agricultural applications or by intentional methods of terrorism is an issue since ground water is used as drinking water by about 50 percent of the nation's population. Concerns over pesticides in ground water are especially acute in some areas where over 95 percent of the population relies upon ground water for drinking and other uses. The widespread detection of pesticides in ground water coupled with recent findings linking pesticides with different types of cancer indicates that acceptable levels of these materials in water are likely to decrease. As a result a wider range of pesticides and their transformation products will need to be detected. One of the most commonly used and most dangerous class of pesticides are the organophosphates. Organophosphate pesticides whose chemical structure is very similar to their more deadly counterparts the nerve agents, affect the nervous system by irreversibly binding the active site of acetylcholine esterase, (AChE), an essential enzyme needed for proper nervous system function.

Current methods for the detection of these pesticides include gas chromatography - atomic emission detection *(15,16)* (GC-AED), gas chromatography - mass spectrometry *(17)* (GC-MS), and high performance liquid chromatography with either mass spectrometric or diode array detection (LC-MS or LC-DAD). *(18-20)* These techniques generally have high sensitivity, but are complex and costly, require skilled technicians and large laboratory based instrumentation. Additionally, most of these techniques require sophisticated and time consuming sample preparation including extraction procedures. Biochemical determination of organophosphate pesticides based on the ratio of this inhibition using AChE *(21)*, cholinesterase (ChE), *(22)* or phosphotriesterase (PTE) *(23)* is growing in popularity but suffers from many of the same limitations as antibody technologies in that they have irreversible binding processes and have limited environmental stability.

Recent advances in the field of molecular imprinting have provided polymers that are extremely sensitive and specific to organophosphate pesticides and the hydrolysis products of the nerve agents. *(1,2)* The MIPs that have been developed in this work have been integrated into a fiber optic luminescence system for the determination of these toxic chemicals in water. The MIPs consist of an artificial biomimetic recognition site with a europium metal ion center that allows signal transduction of the binding events. The work presented here suggests that lanthanide based MIPs and their corresponding sensors can be prepared for almost any organophosphorus containing compounds. Thus far, over 20 species have been successfully imprinted and incorporated into sensors for water detection. All of the sensors developed in this work have demonstrated superior sensitivity and selectivity and have detection limits below 10 ppt with long linear dynamic ranges. Positive analysis occurs almost immediately and quantitative results are obtained within 15 minutes. The resulting sensors are reusable and can be used repeatedly by simply rinsing them in water.

Experimental:

Reagents. Unless otherwise indicated, materials were obtained from commercial suppliers and used without further purification. Analytical reagent grade chemicals were used along with deionized water to prepare solutions. Europium (III) oxide, styrene, vinyl benzoate, AIBN, methyl phosphonic acid and pinacolylmethyl phosphonate were obtained from Aldrich (Aldrich, Milwaukee, WI 53233). Neat analytical standards of the pesticides and insecticides were obtained from Radian (Radian International, Austin, TX 78720). Parathion, methyl-parathion, thionazin, and dibutyl chlorendate were obtained as neat standards from Supelco (Supelco Chromatography Products, Bellefonte, PA 16823).

Apparatus. Luminescence was excited using a model 60X-argon ion laser (MWK Industries, Corona, CA). A 488nm holographic filter (Kaiser Optical Systems, Ann Arbor, Michigan) turned to pass the 465.8nm line was used to exclude all other laser lines. Spectra were collected using an f/4, 0.5 m DKSP240 monochromator with a direct fiber coupler (CVI Laser Corp., Albuquerque, NM 87123) equipped with a Model ST-6 CCD (Santa Barbara Instruments Group, Santa Barbara, CA) using Kestrel Spec Software (K&M Co., Torrance, CA, USA). Spectra were plotted and calculations performed using Igor Pro Software (WaveMetrics Inc., Lake Oswego, OR 97035).

Complex Preparation. Europium complexes were synthesized, using a stoichiometric ratio of one mole europium, to one mole of pesticide/insecticide and 3 moles of the vinyl benzoate-ligating molecule. Eu $(NO_3)_3$ was prepared by dissolving the oxide in water with just enough nitric acid to produce a clear solution. The calculated amount of each pesticide/insecticide was diluted/dissolved in a 50/50 water-methanol mixture to which the vinyl benzoate monomer was subsequently added. The resulting solution was added to the $Eu(NO_3)_3$ and the pH adjusted to approximately 9.5 with 1.0M NaOH. The resulting solutions were slowly stirred on low heat for approximately 4 hours, then covered with a watch glass and left to crystallize. The spectra of the crystals were taken, examined and interpreted to determine the symmetry changes associated with analyte inclusion. A control complex "blank" containing only europium and vinyl benzoate was also prepared as described above.

MIP Preparation. The key step in making a molecularly imprinted polymer is to form a complex that will survive the polymerization process leaving behind a suitable set of binding sites when the templating species is removed. To form such a complex, ligands must be chosen that exhibit sufficiently high affinities to resist dissociation. The polymerization process must provide sufficient rigidity to effect structural "memory" but be sufficiently flexible to allow removal of the template ion. In the case of sensors, the ligating monomers must provide a means of signal transduction. The success of the end product hinges on the selection of the ligating monomer. In this application the lanthanide ion europium was chosen not only for its luminescence properties, but also for its affinity for the phosphonate functional group present in the materials imprinted.

Polymers were prepared by dissolving 3 mole percent complex compound in 95 mole percent styrene with approximately 0.1 mole percent azobisisobutyronitrile (AIBN) added as an initiator, and 2 mole percent divinyl benzene (DVB) added as a crosslinking agent. (Traditional imprinting uses higher levels of crosslinking up to 90%, however we have found that for sensing applications, much lower amounts of crosslinking allow better accessibility to the site and faster diffusion of the target molecules. The structure of the site is stabilized by the formation of an organometallic complex between the vinyl benzoate and the europium ions. The latter may provide directional bonding in the cavity. The resulting solutions were placed in glass vials, purged with nitrogen, and sealed using parafilm and screw on tops. The polymers were sonicated until they became viscous (usually 1-2 hours) in a Cole Parmer Ultrasonic cleaner model 08849-00 that maintained a constant temperature of 60°C. (Sonication appears to help maintain homogeneity.)

The fiber optic sensors consisted of an SMA terminated 400 micron optical fiber (Thor Labs, Newton, NJ, 07860). The sensing portions of the fibers were prepared by removing the cladding, heating the stripped end in an air/acetylene flame and manually pulling the fibers into tapers to permit more efficient coupling of the evanescent field to the polymer. The polymers used in the construction of optical sensors were prepared *in situ* on the distal end of an optical fiber whose surface was prepared by directly binding the polymerizable agent on the surface. The final thickness of the coating was 200 um. Methanol was used to swell the polymer pores and 1.0 M nitric acid was used to facilitate the removal of the imprinting molecules and allow greater access to the imprinted recognition sites containing the coordinated europium ions. This step is necessary since, unlike most commercial ionic resins that often have a large amount of functionalization (ionogenic sites) which readily swell in polar solvents, the templated resins have a relatively low degree of functionalization and are primarily nonionic matrices. An unimprinted polymer was also constructed using 3 mole percent of the blank complex with 2 mole percent DVB in styrene.

Analysis. Measurements for the calibration data, response time and interference testing within each class were performed using the same fiber to demonstrate the reversibility of the sensors. Analytical figures of merit were obtained with serial dilutions of the standards in 0.01M NaOH maintaining the pH at 10.5. Luminescence was excited using 1mW of the 465.8 nm line of the argon laser and the active end of the sensor placed in one of the sample dilutions. The performance of the fiber optic sensors was evaluated using the CVI monochromator set with 200 micron slits, and an exposure time of 25 seconds. Spectra were collected at each concentration with various equilibration times. The sensors were rinsed with deionized water between each sample. To demonstrate the reversibility of the sensor standards were analyzed in order of both increasing and decreasing concentration. Calibration curves based on an exposure time of 15 minutes were obtained and linear regressions performed. The response time of the sensors was evaluated using a method similar to the one described above. The sensor was placed in a cuvette with each solution and spectra collected at a variety of exposure times. Response was evaluated through a comparison of peak area at each time.

Pesticide and insecticide standards were tested as possible interferences against each other. Standard 100 ppm solutions were prepared by the dissolution and/or dilution of the samples in deionized water when possible. The pesticides/insecticides with limited solubility in water were prepared using 50:50 water/methanol or 50:50 water/acetone. The pH of each of the solutions was adjusted to 9.5 using 1M sodium hydroxide. Spectra for each analyte against the sensors were taken at regular intervals for 30 minutes, and then compared with the response from the imprinted analyte. The sensors were rinsed with deionized water between each analysis. The response of the blank sensor to each of the pesticides was also evaluated. The sensors were also evaluated using a variety of water sources including tap water, seawater (Ocean City, MD), pond water (Baltimore, MD), and deionized water.

Results and Discussion

Sensor Characteristics: The preparation scheme for all the imprinted polymers were identical with the exception of the target imprinted and in a few cases the solvent used in the initial dilution of the pesticides. The molecules that were imprinted along with their corresponding detection limits are shown in Table I. The entire series of sensors function in the pH range between 6.5-13 with faster response times at the higher pHs. An initial positive response from the sensors occurs within 2 minutes with the 80% response time varying from 11-18 minutes. The response times are affected by the affinity of the coordinated lanthanide for the specific analyte as well as the overall size of the analyte. In all cases, the binding of the target analyte to the MIP polymer was reversible and the analyte removed simply by rinsing with deionized water at a pH below 6.5. Each of the sensors evaluated had a detection limit below 19 parts per trillion (ppt) with the sensors for pinacolyl methyl phosphonate and chloropyrifos having detection limits below 1ppt. The upper detection limit of the sensors is 10 ppm, which is determined by the number of active sites available within a given thickness of the polymer layer.

Each of the sensors was evaluated against the entire pesticide panel to determine cross reactivity. All of the MIP sensors were 50-1000 times more selective for the imprinted molecule even when interrogated against members within the same chemical class. The most remarkable example of this was the ability to distinguish and discriminate methyl and ethyl chloropyrifos, which differ by a single methyl verses ethyl group on the side chain. Even in the case of these two very similar analytes, it takes 10 times as much of the chloropyrifos ethyl to elicit the same response intensity as the imprinted chloropyrifos (Figure 1). The spectra can be resolved using the shift in the band at 611 nm and the chloropyrifos ethyl peaks at 622.5 nm. In this case it was also noted that the polymers that were imprinted for the smaller pesticides within a class were more selective than those imprinted for the larger pesticides.

A mixture of pesticides (glyphosate, dimethoate, diazinon, and parathion) all at the 1ppm level was evaluated using the chloropyrifos sensor to determine mixtures of chemically similar species would affect the detection capabilities. As shown in Figure 2, the spectrum from the mixture does have a higher background but the individual signature peaks are still clearly visible at 611.5 nm, 614.5nm, and 616.7 nm. In this case, as a result of the higher background, the limit of detection would be raised from 1ppt to 23ppt. Samples were also prepared in several different water matrices including seawater, tap water, and pond water to determine levels of interference from other ions in the solution (such as sodium in seawater). The sensors provided the same detection limits and selectivity regardless of the water system. Characteristics of the individual sensors are detailed in the following sections.

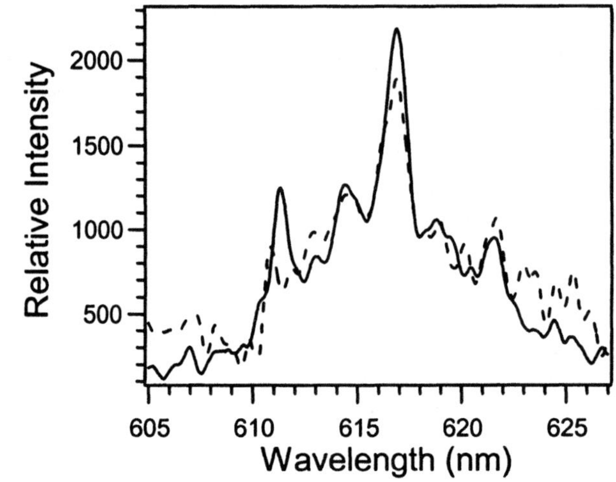

Figure 1. Response of chloropyrifos sensor to 1 ppm chloropyrifos (solid line) and 10 ppm chloropyrifos ethyl (dotted line).

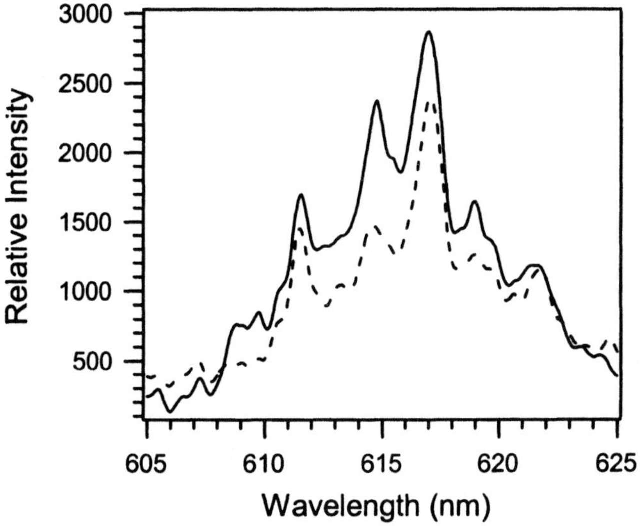

Figure 2. Response of the chloropyrifos sensor to 1 ppm chloropyrifos (dotted line) and a mixture of 1 ppm chloropyrifos with 10 ppm each of glyphosate, dimethoate, diazinon, and parathion (solid line).

Table I. Materials Imprinted and their Detection Limits

Name	Detection Limit	Response Time	Interferents Screened
1. Chloropyrifos	1ppt	11 minutes	2,3,5,7-15, 17, and 18
2. Chloropyrifos-ethyl	4ppt	12 minutes	1,3,5,7-15, 17, and 18
3. Coumaphos	18ppt	20 minutes	1,2,4,7-14, 16, 17, and 18
4. Diazinon	3ppt	12 minutes	1-3, 7-15, 17 and 18
5. Dibutyl Chlorendate	9ppt	15 minutes	1-4, and 7-18
6. Di-isopropyl fluorophosphates (DFP)	10ppt	15 minutes	1-5, and 7-18
7. Dimethoate	8ppt	13 minutes	1-4, 7-13, 17, and 18
8. Disulfoton	1ppt	10 minutes	1,2,4,7, 9-14, 16, 17, and 18
9. Ethion	8ppt	13 minutes	1-5,7, 8, and 10- 18
10. Glyphosate	9ppt	14 minutes	1-5,7-9, and 11- 18
11. Malathion	5ppt	12 minutes	1-5,7-10, and 12- 18
12. Methyl parathion	8ppt	13 minutes	1-5,7-11, and 13- 18
13. Parathion	11ppt	14 minutes	1-5,7-12, and 14- 18
14. Phorate	6ppt	13 minutes	1-4, 7-13, 15, 17, and 18
15. Phosphamidon	7ppt	13 minutes	1-4, 7-14, 17, and 18
16. Pinacolyl methyl phosphonate (PMP)	0.5ppt	11 minutes	1-5, 7, 9-13 and 17-18
17. Pirimiphos-ethyl	12ppt	15 minutes	1-7, 9-13 and 18
18. Thionazin	2ppt	12 minutes	1-5, 7, 9-13, 16 and 17

Pyrimidine Organothiophosphate Pesticide Sensors: Diazinon, is the most common insecticide found in surface waters. It has a moderately acute toxicity, is readily absorbed through the skin, and synergistic with other chemicals. In 1999, the EPA reported that diazinon was one of the leading causes of acute pesticide poisoning in the US. Its neurotoxic properties have caused its sale to be phased out by 2004. Diazinon as with many of the organophosphorus pesticides has a strong odor. The diazinon-imprinted polymer has a slight yellowish color, and is characterized by peaks at 610, 615.3, and 623nm. The sensor has a limit of detection of 3ppt and a response time of 12 minutes. Pirimiphos ethyl is another less commonly used member of this class of pesticides. The imprinted polymer for this analyte is beige in color, and characterized by peaks at 611.3, 615, 617.2, and 620.8 nm. Figure 3 shows the spectral response of this sensor to 1 ppm of the imprinted pirimiphos ethyl and to 10 ppm of several of the other pesticides. None of the pesticides evaluated were determined to cause interference. This sensor has a limit of detection of 12ppt and a response time of 17 minutes.

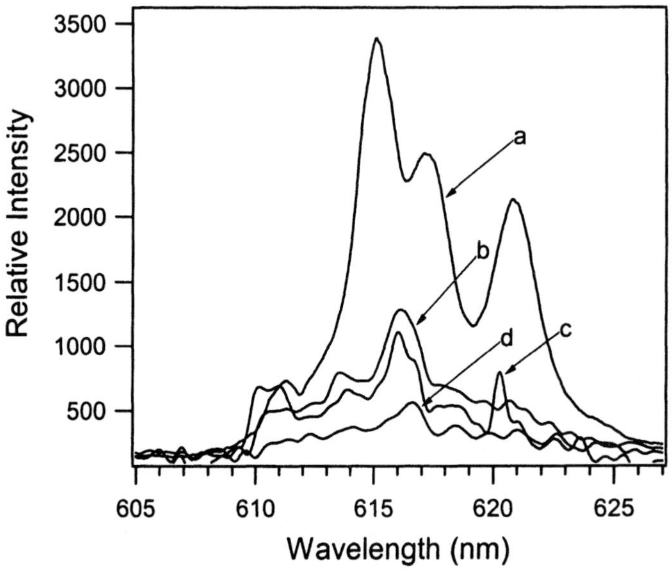

Figure 3. Response of the pirimiphos-ethyl sensor to a) 1 ppm pirimiphos ethyl and to 10 ppm of the interferences b) malathion c) methamidophos and d) dimethoate.

Pyridine Organothiophosphate Pesticide Sensors: Chloropyrifos is the 3rd most commonly used pesticide in the world and is another leading contributor to acute pesticide poisoning in the US that has been linked to thousands of poisoning incidents. It is a broad-spectrum chlorinated organophosphate pesticide with over 25 million pounds applied to agricultural lands annually. The sale of chloropyrifos for residential use will be banned after December 2001, and may be completely banned within a few years because of its toxicity. The imprinted polymer is white in appearance and characterized by peaks at 612, 613.5, and 616 nm as shown in Figure 1. The sensor is sensitive to 4 ppt and has a response time of 12 minutes. The chloropyrifos-methyl polymer is also white and characterized by peaks at 611.5, 614, 617, and a small peak at 621.5 nm. The sensitivity for this material is 1 ppt with a response time of 11 minutes. A series of other pesticides were tested with the sensor and none of them interfered with the performance. The most likely candidate for false positives chloropyrifos-ethyl does react with the sensor but to a much lower level. The two can be discriminated using the unique spectral signatures from each analyte.

Phenyl Organothiophosphonate Pesticide Sensors: Parathion and methyl parathion are insecticides that come in two forms, white crystals or a brownish liquid. They smell like rotten eggs and are similar to nerve gas. They have been made in the United States since 1952. Both parathion and methyl parathion are used to kill insects on farm crops, especially cotton. They are restricted-use pesticides, which means that only specially trained people are allowed to mix, load, and spray them. The imprinted polymers for both of these chemicals are bright yellow and have a strong smell even after imprinting. The parathion sensor has a detection limit of 11ppt and is characterized by peaks at 612, 617.6, and 625.5 nm with a response time of 14 minutes. Methyl parathion is characterized by peaks at 611, 617.1, 624.5 nm, has detection limit of 8ppt, and a response time of 13 minutes. Figure 4 shows the response of the sensor to 1 ppm methyl parathion compared with 10 ppm of three other similar pesticides ethion, phosphonazin, and chloropyrifos. None of the other pesticides tested against the sensor interfered with its performance.

Thionazin another member of this family is a persistent pesticide that can be found in soil up to one year after its application. Thionazin was also imprinted to give a white polymer characterized by a series of peaks at 611.5, 614, 616.5, 618.8, and 621.5 nm. The LOD of this sensor was 2 ppt with a response time of 12 minutes.

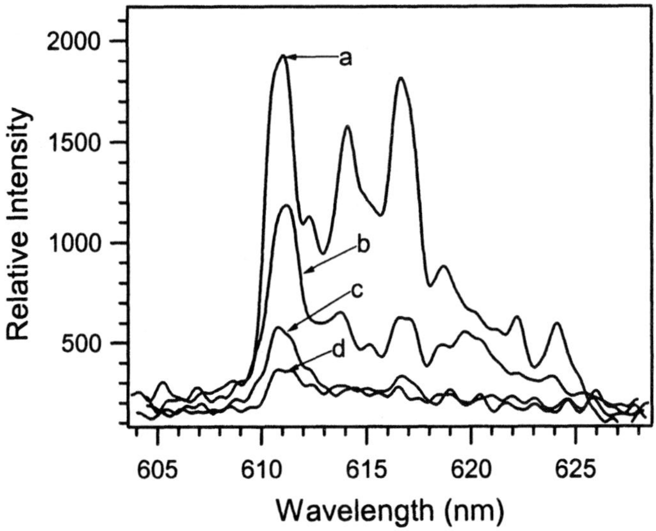

Figure 4. Response of the methyl parathion sensor to a) 1 ppm methyl parathion and to the possible interferents b) 10 ppm parathion, c) 10 ppm glyphosate and d) 10 ppm diazinon.

Aliphatic Organothiophosphates: Ethion is a pesticide used in agriculture, mainly to control insects on citrus trees, but also on cotton, fruit and nut trees, and some vegetables. It may also be used on lawns and turf grasses, but it is not used in the home for pest control. Pure ethion is a clear-to-yellowish liquid with an unpleasant sulfur type of smell. Most of the ethion used in pest control is diluted with other liquids and used as a spray. It is also sometimes used as a liquid adsorbed on dust or granules. Ethion is sold under many trade names including Bladan®, Rodicide®, and Nialate®. The ethion present at hazardous waste sites will most likely be in a liquid solution or adsorbed on solid granules. Ethion has been found in at least 9 of 1,577 National Priorities List sites identified by the Environmental Protection Agency (EPA). The ethion MIP developed for this work is whitish in color and characterized by a series peaks at 611.2, a very small peak at 612.4, 614, 616.8, and 624.3 nm. The spectra of 1ppm ethion along with the response to10 ppm of the pesticides methamidophos, phosphonazin, and coumaphos are shown in Figure 5. This sensor can detect 8ppt with a response time of 13 minutes.

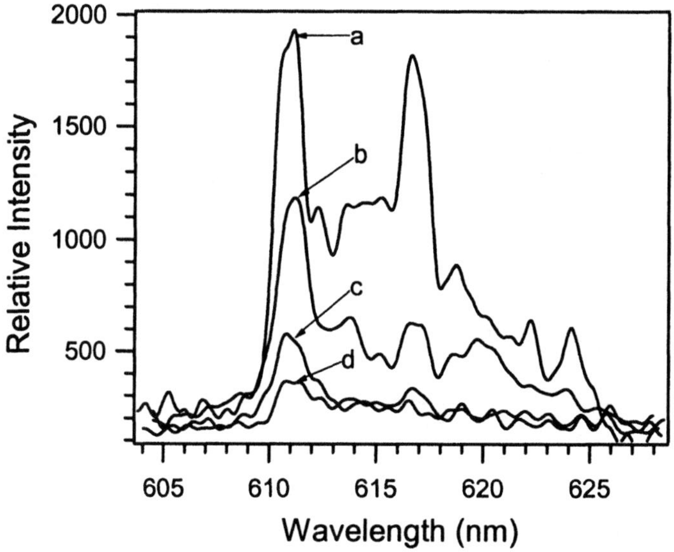

Figure 5. Response of the ethion sensor to a) 1 ppm ethion and to 10 ppm b) methamidophos, c) phosphonazin, and d) coumaphos.

Disulfoton is a substance used as a pesticide to control a variety of harmful pests that attack many field and vegetable crops. The common trade names for disulfoton are Disyston®, Disystox®, Frumin AL®, and Solvirex®. Pure disulfoton is a colorless oil with no identifiable odor and taste. The technical product is dark yellowish, and has an aromatic odor. It does not easily dissolve in water or evaporate to air. Disulfoton is used to protect small grains, sorghum, corn, and other field crops including some vegetables, fruit, and nut crops; and ornamental and potted plants against certain insects. Although it is used primarily in agriculture, small quantities are used on home and garden plants. Small quantities are used for other purposes such as mosquito control in swamps. The use of disulfoton has decreased in recent years. The disulfoton MIP is a white polymer characterized by a small peak at 611.5 and a large broader peak at 616.7 nm. This is the most sensitive material prepared thus far with a limit of detection at 1ppt and a response time of 10 minutes. The third species imprinted in this category was phorate. This polymer had a mustard yellow appearance. The spectrum for this polymer consists of peaks at 611.2, 614.5, 616.8, and 620.2 and the sensor has an LOD of 6ppt with a response time of 13 minutes.

Organophosphate Pesticides: Glyphosate and dichlorvos were selected for imprinting within this class. Dichlorvos is an insecticide used to control parasites in pets and is primarily found in flea collars. Dichlorvos is sold under many trade names including Vapona®, Atgard®, Nuvan®, and Task®. Dichlorvos may also be called DDVP, which is an abbreviation for its full chemical name. Pure dichlorvos is a dense colorless liquid that evaporates easily into the air and dissolves slightly in water. Dichlorvos has a sweetish smell and readily reacts with water. Dichlorvos is manufactured by reacting chloral with trimethyl phosphite. Other uses of dichlorvos include insect control in food storage areas, greenhouses, and barns, and for parasite control in livestock. Dichlorvos is generally not used on outdoor crops, but it is sometimes used for insect control in workplaces and the home. The dichlorvos-imprinted polymer is characterized by a light orangish color and a spectrum consisting of peaks at 611.3 and 616.2nm. Figure 6 demonstrates the spectral response to 1ppm of the imprinted dichlorvos as compared with the sensor's response to 10 ppm chloropyrifos, diazinon, and ethion. None of the other pesticides evaluated interfered with the detection of the imprinted analyte. The sensor has a limit of detection of 7ppt and a response time of 14 minutes. The glyphosate polymer is a white material with a spectrum consisting of peaks at 611, 613.5, 617, and 620 nm. The glyphosate sensor has a limit of detection of 9ppt and a response time of 14 minutes.

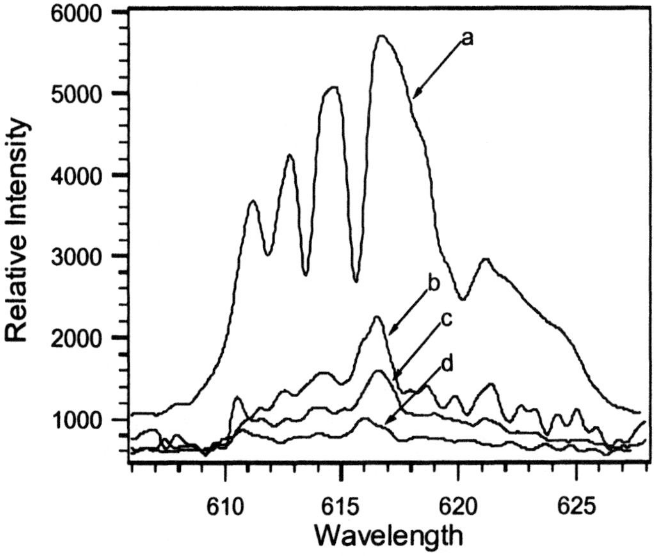

Figure 6. Response of the dichlorvos sensor to a) 1 ppm dichlorvos and to 10 ppm of the possible interferences b) chloropyrifos, c) diazinon and d) ethion.

Other Organophosphate Pesticides: Coumaphos is a heterocyclic organothiophosphate insecticide/acaricide registered to control arthropod pests, primarily ticks, scabie mites, lice, flies, fleece worms and screwworms on livestock (i.e., dairy cattle, beef cattle, swine, and horses) and swine bedding. The USDA uses coumaphos in its Tick Eradication and Import Programs. There are no registered uses of coumaphos on agricultural crops or in/around residences. The coumaphos imprinted polymer is white with spectral features at 610.3, 614.6, and 618.1 nm. The complex was very difficult to make, and after several attempts using different solvent systems and pHs, the complex was finally made by mixing the coumaphos and the vinyl benzoate in acetone/water, then adding this to the europium solution and adjusting the pH to just above 8. This complex when added to the styrene polymerized very rapidly in less than 30 minutes. The resulting sensor had a limit of detection at 18ppt with a response time of just under 20 minutes.

Dimethoate is an aliphatic amide organothiophosphate insecticide primarily used on a variety of field and orchard agricultural crops, and ornamentals. The currently labeled uses consist of, non-crop land adjacent to vineyards, farm premises and equipment, sewage treatment, food processing plants and equipment, outdoor residential and commercial/institutional/industrial premises and outdoor refuse. Dimethoate is a general use chemical applied using ground and aerial equipment, with an estimated 2.6 million pounds used annually. The dimethoate imprinted polymer is orange colored with a spectrum consisting of a peak at 611nm, a broad peak from 612 to 614.5 nm, and a sharp peak at 616.6 nm. The sensor can selectively detect 8ppt with a response time of 13 minutes.

Nerve Agent Simulants: Two nerve agent simulants di-isopropyl fluorophosphate (DFP) and pinacolylmethyl phosphonate (PMP) were also imprinted using the vinyl benzoate monomers. PMP, the hydrolysis product of the nerve agent soman has been imprinted using divinylmethyl benzoate with a detection limit of 7 ppt. The current PMP sensor has a detection limit of 0.5ppt with a response time of 12 minutes. The characteristic peaks are at 610, 612.5, and 615 nm. The DFP sensor has a detection limit of 10 ppt and a response time of 16minutes. The spectral features of this sensor are found at 613.7, 618, and 625.9 nm. Each of the pesticides used in the previous analyses was evaluated against these sensors. Although some pesticides do enter the imprint sites better than others as demonstrated by an increase in overall luminescence, none of the pesticides had a spectral signature identical to or a binding as strong the imprinted analyte.

Conclusions

The search for highly selective, rugged, low-cost, portable sensors with specificity for toxic chemical, biological, and environmental analytes such as pesticides has attracted enormous interest after the recent terrorist attacks. Molecular recognition occurs in a variety of living systems, and has been used in the detection of many chemical and biological species. However, these natural recognition systems such as antibody/antigen binding reactions although very selective are of limited use due to poor stability, assay concentration range, and cost. Therefore, the use of artificial or biomimetic recognition systems have been proposed and among them, some of the most promising alternatives are the molecularly imprinted polymers (MIPs). These polymers are highly stable and exhibit high selectivity approaching their natural antibody counter-parts. Molecularly imprinted polymers, originally developed for separation purposes are now beginning to be used for enhancing the specificity of a variety of chemical sensors.

The combination of molecularly imprinted polymers and europium signal transduction has proven applicable as a generic scheme to develop materials for the detection of hydrolyzed and non-hydrolyzed organophosphate containing compounds such as pesticides and nerve agents. These polymers can be coated onto optical fibers and used as sensors for the detection of these species in aqueous environments. Similar functional polymers can also be used for enhancing the sensitivity and selectivity of other detection devices such as surface acoustic wave sensors.

The imprinted polymer based sensors have been shown to work without interference in tap, sea/ocean, pond, river, and deionized waters. A series of these imprinted sensors can be bundled together to create an array that can detect multiple analytes in water simultaneously. The existing technology uses an argon ion laser for excitation, but new materials are under development that will allow excitation with blue diode lasers or 470 nm light emitting diodes (LEDs). Our previous work has already demonstrated the applicability of using a small fiber optic spectrometer to decrease the size of the device.[1,2] The superior stability, sensitivity, selectivity, and reversibility of molecularly imprinted polymers provide a real time aqueous sensing solution for organophosphate containing chemicals like pesticides and nerve agents.

Acknowledgements

The authors would like to express their sincerest appreciation to Dr. Michael Ellzy, and Michael Lochner of the Edgewood Chemical and Biological Command Center, and Mr. Ryen Hydutsky and Mr. Robert Schafer at the US

Army Research Laboratory on Aberdeen Proving Ground, Maryland for financial support and technical assistance .

References

1. Jenkins, A. L.; Yin, R.; Jensen, J.; "Molecularly Imprinted Polymers for the Detection of Pesticides and Insecticides in Water" Analyst, **2001**, 126, p. 798-802.
2. Jenkins, A.L.; Uy, O. M.; Murray, G. "Polymer Based Lanthanide Luminescent Sensor for the Detection of the Hydrolysis Product of the Nerve Agent Soman in Water, " Analytical Chemistry, **1999**, 71(2), p. 373-378.
3. Percival, C.J.; Stanley, S.; Galle, M.; Braithewaite, A.; Newton, M.; McHale, G.; Hayes, W; Anal. Chem. **2001**, 73, 4225-4228.
4. Cao, L.; Zhou, X.; Li, S.; Analyst, **2001**, 126(2), 184-8.
5. Ji, H.; McNiven, S.; Lee, K.; Saito, T.; Ikebukuro, K.; Karube, I.; Biosens Bioelectron, **2000**, (7-8), 403-409.
6. Takeuchi, T.; Mukawa, T.; J.; Higashi, M; Shimizu, K.; Anal. Chem, **2001**, 73, 3869-3874.
7. Ansell, R.; Kriz, D.; Mosbach, K.; Curr. Opin. Biotechnol., **1996**, 7(1), 89-94.
8. Cheng, Z.; Wang, E., Yang, X.; Biosens Bioelectron., **2001**, 16(3), 179-185.
9. Sergeyev, T.; Matuschewski, H.; Piletsky, S.; Bendig, J.; Schedler, U.; Ulbricht, M.; J Chromatogr A, **2001**, 907(1-2), 89-99.
10. Koeber, R.; Fleisher, C.; Lanza, F. Boos, K.; Sellergren, B.; Barcelo, D.; Anal. Chem, **2001**, 73(11), 2437-2444.
11. Ferrer, I.; Lanza, F.; Tolokan, A.; Horvath, V.; Sellergren, B.; Horvai, G.; Barcelo, D.; Anal. Chem. **2000**, 72(16) 3934-41.
12. Kroger, S.; turner, A.; Mosbach, K.; Haupt, K.; Anal Chem, **1999**, 71(17), 3698-3702.
13. Liang, C.; Peng, H.; Nie, L.; Yao, S.; Fresenius J Anal Chem, **2000**, 367(6), 551-555
14. U.S Geological Survey Fact Sheet FS-244-95
15. "Pesticides in Ground Water Database: 1988 Interim Report," Office of Pesticide Programs, Environmental Protection Agency, Washington, DC **1988**.
16. Bernal, J.; del Nozal, M.; Martin, M. T.; Jimenez, J.J.; J. Chromtogr. A, **1996**, 754, 245-256.

17. Becker, G.; Colmsjo, A.; Ostman, C; Anal. Chim Acta, **1997**, 340, 181-189.
18. Maruyama, M.; Fresenius J. Anal. Chem., **1992**, 343, 890-892.
19. Martinez, R.C.; Gonzalo, E. R.; Amigo Moran, M.J.; Hernandez Mendez, J.; J. Chromatogr., **1992**, 607, 37-45.
20. Martinez, R.C.; Gonzalo, E. R.; Garcia, F.G.; Mendez, H.J.; J. Chromtogr., **1993**, 644, 49-58.
21. Giang, P.; Hall, S. Anal Chem, **1951**, 23, 1830-1832.
22. Bhattacharya, S.; Alsen, C.; Kruse, H.; Valentin, P.; Env. Sci. Tecnol., **1981**, 15, 360-365.
23. Sode, K., Togo, H.; Yamazaki, T.; Tsugawa, W.; Ohuchi, S.; Narita, M.; Yagiuda, K.; 1996, 64(12), 1234-1238.

Nanocomposites andPolymers

Chapter 6

Nanocomposites for Extreme Environments

Derek M. Lincoln[1,3], Hao Fong[2], and Richard A. Vaia[1]

[1]Air Force Research Laboratory, Materials and Manufacturing Directorate, 2941 P Street, Wright-Patterson Air Force Base, OH 45433
[2]Universal Technology Corporation, Air Force Research Laboratory, Wright-Patterson Air Force Base, OH 45433
[3]Current address: Department of Chemistry, U.S. Air Force Academy, Colorado Springs, CO 80840

Nanoscale dispersion of only a few percentage of layered silicate (montmorillonite) in poly(caprolactam) (Nylon 6) results in the formation of a uniform self-passivating inorganic layer upon exposure to extreme environments. This ceramic-like silicate layer provides an overcoat to the nanocomposite, and can significantly retard the penetration of extreme environments into virgin material. This response originates from the large areal density of inorganic afforded by the uniform dispersion (exfoliation) and preferential orientation with respect to the sample surface of large aspect ratio aluminosilicate sheets, and thus should be a general property of polymer layered silicate nanocomposites, not only restricted to nylon 6 nanocomposites. Nanocomposite concepts, therefore, should enhance the survivability of polymeric materials in aggressive environments, such as thermo-oxidative, reactive-oxidative, and UV-electron radiation environments encountered in space launch, Low Earth Orbit (LEO) and geostationary orbit (GEO).

Introduction

The U.S. Air Force's ever increasing employment of space based platforms, NASA's strong commitment to space exploration, and rapidly increasing commercial satellite production has resulted in an urgent need for new lightweight, space-resistant multifunctional materials with extended lifetimes. Unfortunately, aggressive environments encountered during launch and in orbit (1) have continually challenged the integrity of high-performance materials, especially organic materials. Reported space environment damage to man-made bodies in orbit are staggering with numerous studies, such as those from STS missions and NASA Long Duration Exposure Facility (LDEF), reporting that both radiation (atomic oxygen, vacuum ultraviolet, proton, electron and particle) and thermal cycling contribute to material degradation (2,3) drastically reducing the lifetime of orbiting bodies.

Polymers are attractive and desirable for use in launch and space applications, in particular because they are lightweight, easily processable at relatively low temperatures, adaptable to fiber, film and monolith forms, and can potentially satisfy multifunctional requirements. However, thermo-oxidative degradation during launch and reactive-oxidative and electromagnetic radiation (EM) degradation on orbit (100 km to 36,000 km) is a prominent concern. Because of the wealth of flexibility polymers offer, there remains a large motivation to overcome stability issues.

In launch systems ablative materials provide critical protection for the propulsion systems, payload, and ground systems from the severe effect of extremely high combustion temperatures. In particular, the internal insulation in solid rocket motors, while containing the combustion of the solid rocket fuel and protecting the integrity of the rocket motor casing, experiences temperatures that can exceed 2000°C and pressures of 1000 psi in a harsh thermo-oxidizing atmosphere. This insulation also experiences various mechanical extremes as it is violently compressed from fuel combustion and sheared as gas and particulate velocities climb upwards of Mach 10+. Current state-of-the-art internal insulation materials are made of EPDM (ethylene-propylene-diene monomers) rubber with 30 wt% micron scale silica filler or 15 wt% chopped Kevlar® fiber. These materials are often very difficult to process, which makes them expensive and unreliable. Since the material cannot be manufactured as a single part, strips of the insulation must be made and laid up separately inside the rocket casing; this practice predisposes it to failure from burn-through at the seams.

For space systems, a different aggressive environment is encountered depending on orbit. For example, survivability in low earth orbit must address very high atomic oxygen fluxes (~10^{15} atoms/cm^2·s for orbital speed of 8 km/s (1,2,4)), where as survivability in geosynchronous earth orbits (GEO) must deal with a complex interplay of simultaneous UV-vis radiation, electrons and charged particles. Many current strategies are based on deposition of inorganic coatings on the polymer surface or enrichment of inorganic precursors at the polymer surface in response to exposure to the aggressive space environment.

The vast majority of efforts have focused on polymer durability to atomic oxygen, which is encountered in low earth orbit, and have shown varying degrees of success.

Nanocomposites are a novel class of materials that offer the attractive advantages of polymers, as they are easy to process and lightweight, but are modified to provide marked enhancements in physical properties and potentially to perform well in launch and space environments. To this point, nanocomposites have found use in sporting goods, such as athletic shoes, and under-the-hood components in automobile engines but can potentially be used in a much broader range of applications including aerospace applications.

Over the last decade, the utility of layered silicate nanoparticles as additives to enhance polymer performance has been established (5-17). These nanoscale fillers result in physical behavior that is dramatically different from that observed for conventional microscale counterparts. For instance, increased modulus (7,8), decreased permeability (9-11), reduced coefficient of thermal expansion (CTE) (12,13) and impact strength retention (7,14) are observed with only a few volume percent addition of exfoliated layered silicate; thus maintaining polymeric processability, cost and clarity.

The chapter will discuss recent work demonstrating the utility of nanocomposites as self-passivating materials for use in aggressive environments. This response originates from the large areal density of inorganic afforded by the uniform dispersion (exfoliation) and preferential orientation with respect to the sample surface of large aspect ratio aluminosilicate sheets, and thus should be a general property of polymer layered silicate nanocomposites. The concept of self-passivation response of nanocomposites to aggressive thermo-oxidative (ablation), reactive oxidative (oxygen plasma) and UV radiative environments is examined using nylon 6/layered silicate nanocomposite as a model system.

Experimental

Materials

5 wt% (NCH5) and 2 wt% (NCH2) layered silicate/nylon 6 in-situ polymerized nanocomposite materials and pure nylon 6 polymer (Nyl6) were acquired from Ube Industries, Ltd. (Japan) in pellet form. The in-situ polymerization has been detailed elsewhere (15). In addition, melt processed nylon 6/layered silicate (montmorillonite) nanocomposites (NLS) with different percent layered silicate (1, 2.5, 5 and 7wt%) were received from Southern Clay

Products, Inc. in the form of extruded films and pellets. Solution cast films from 1,1,1,3,3,3-Hexa-fluoro-2-propanol (used as received from Aldrich) solution, and thermo-compressed films (compressed at 230°C, 2000 pound per square inch, using 10 layers of as-received films) were prepared. Morphological characterizations indicate that the montmorillonite in these nylon 6 systems is extensively exfoliated (16,17).

Exposure Tests

Ablation tests were carried out at Edwards AFB, CA in a mock solid rocket motor firing rig. 7" by 3" triangular samples were compression molded at 246°C and 200 kPa pressure for 5 minutes. Tests were then performed in a 4" diameter char motor as shown in Figure 1 operating at a Mach number range of 0.25 to 0.012 and a mass flux range from 13.8 to 0.69 kPa/s and a chamber pressure of 4.4 MPa and 3000°C. The ablation rate is determined by the change in thickness of the samples after an 8-second exposure in the combustion chamber. Additional details can be found in reference 23.

Oxygen plasma was generated using GSC-200 plasma generator (March Instruments, Inc. Concord, CA.) Oxygen plasma concentration was approximately 10^{18} ions (or radical) per liter. Oxygen plasma erosion rates are determined by measuring sample mass loss after treatment for a predetermined length of time.

The Space Combined Effects Primary Test and Research facility (SCEPTRE) at Wright-Patterson AFB, OH was employed to conduct a simulated space exposure study with combined UV radiation and low-energy electrons. Five samples (7.5, 5.0, 2.5, 1.0 and 0 wt% layered silicate) were mounted on the exposure platen in the SCEPTRE sample holders in positions 1-5, respectively, with the surface normal to incident radiation and equidistant from the center of the exposed surface as shown in Figure 2. During this exposure study the specimens were simultaneously irradiated with UV radiation equivalent to 2.2 times that of the sun in a geostationary orbit and low-energy electrons (1keV at 6.4 x 10^{15} e^-/cm^2 and 10 keV at 3.2 x 10^{15} e^-/cm^2) with a total fluence of 9.6 x 10^{15} e^-/cm^2 over 282 hours at a vacuum of 10^{-7} torr.

Characterization

X-ray Photoelectron Spectroscopy (XPS) spectra were recorded using a Surface Science Instrument SSX-100. This system produces monochromatic Al Kα X-ray with energy of 1486.6eV. This instrument was operated using a 300 μm X-ray spot for all experiments. In the experiment, Ar^+ ion etching was used to remove material to examine the relative composition within the film. The

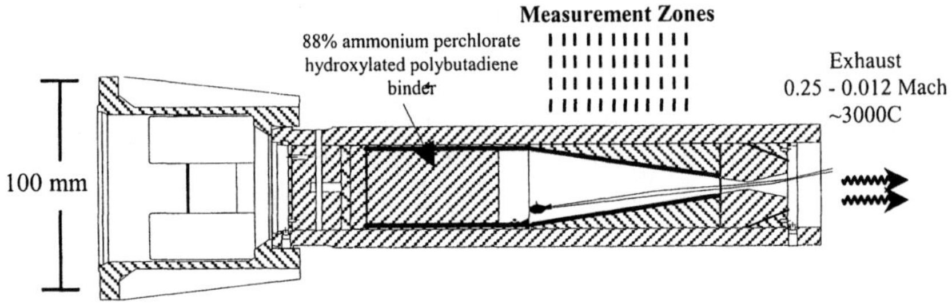

Figure 1. Mock solid rocket motor firing rig.

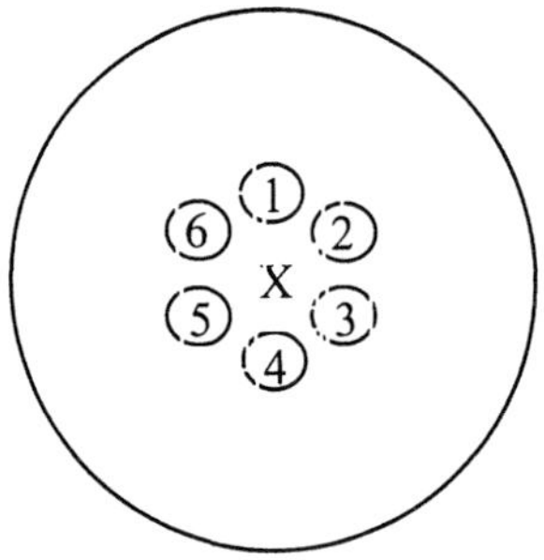

Figure 2. SCEPTRE sample arrangement (1: 7.5% NLS; 2: 5% NLS; 6; 3: 2.5% NLS; 4: 1% NLS; 5: Nylon 6; 6: control PMMA). "X" indicates the center of the exposed surface where a control specimen of Kapton was placed.

etching rate of the Ar^+ ion was 0.07nm per minute calibrated by pure silicon, but note that calibration of the etch time to an etch depth is difficult for these heterogeneous systems, and thus changes in chemical composition is discussed relative to the surface.

Attenuated Total Reflection infrared (ATR) spectra were obtained on a Bruker Equinox 55 FT-IR spectrometer using a SpectraTech Model 300 ATR. By varying the types of crystals and the angles of incidence, information from different depths within the sample can be obtained. Transmission FT-IR spectroscopy was also performed on the samples.

The Transmission Electron Microscope (TEM) employed in the work was a Philips CM-200 TEM with a LaB6 filament operating at 200 kV. For TEM, nanocomposites before and after the plasma treatment were embedded in epoxy, and 50-70 nm thick sections were microtomed at room temperature using a Reichert-Jung Ultracut Microtome and mounted on 200 mesh copper grids.

Results and Discussion

In addition to the numerous physical property enhancements reported for polymer nanocomposites, nanoscale dispersion of the inorganic particles has the potential to provide unique chemical properties, such as self-passivation and self-healing. Figure 3 summarizes the conceptual origin of the self-passivating/self-healing response of nanocomposite materials.

Neat polymers are vulnerable to aggressive environments throughout the thickness, rapidly degrading. Coatings provide outstanding protection, however they are susceptible to failure, especially in applications where coating inspection and repair is not possible, such as on-orbit. Ideally a self-passivating / self-healing response is desirable. Self-passivation refers to a self-limiting response arising from the generation of a protective coating on the surface of the polymer in response to exposure to a degrading environment. Since the precursors to this protective coat are integral to the polymeric material, failure of the initial coating will expose underlying material that can respond to the aggressive environment, again forming the protective coating, and effectively self-healing.

Two general material strategies for self-passivation are 1) blending with inorganic polymers and micron-scale fillers (18,19), and 2) copolymerization with phosphazenes (20), siloxanes or silsesquioxanes (21-23). Issues associated with these approaches include decreased mechanical and optical properties associated with blending, formation of volatile degradation products and increased cost associated with copolymerization of inorganic monomers. Nonetheless, direct chemical modification of the polymers holds great promise; however, in many situations a lower cost, additive approach, such as a filler, is very desirable.

Traditional filled polymer systems with correctly selected micron-scale inorganic particles have been demonstrated to offer increased protection. However, relatively large loadings (>30 %) are necessary to provide uniform areal density of particles with respect to the surface – at lower loadings, such as necessary to maintain processibility and optical clarity, the system remains unprotected and susceptible to effects from the aggressive environment because of the low areal fraction of precursor filler. The nanoscopic size and large aspect ratio of nanoscale plates most efficiently utilizes the volume of filler added, enabling large areal fractions at low (<10%) loading and thus efficient self-generation of a protective coating. With regard to self-generation of the protective coat, at least three roles can be envisioned for the nanoscale plate. The nanoparticle may 1) act as a precursor, chemically reacting with the environment to form the coat, 2) be chemically inert, increasing surface composition with polymer removal or 3) selectively absorb detrimental radiation or chemical species, minimizing exposure of the underlying inorganic material.

The following three examples of thermo-oxidative (ablation), reactive oxidative (oxygen plasma) and UV radiative environments will demonstrate the

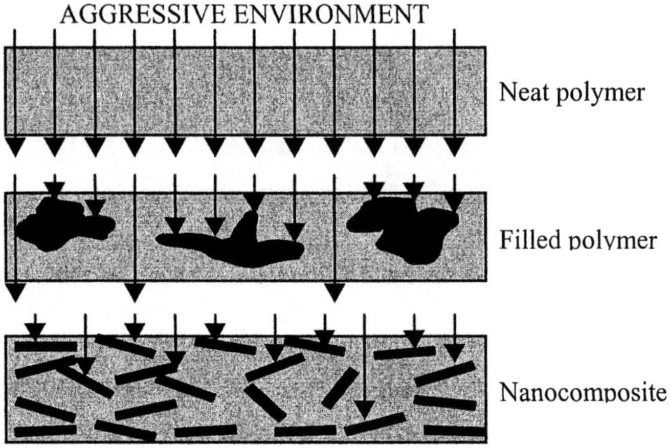

Figure 3. Comparison of areal fractions for a neat polymer, traditional filled polymer and nanocomposite. The large areal fraction of the nanocomposite shields the underlying material from aggressive environments.

potential of layered silicate nanoparticle to function as the critical element to a self-passivating polymer system and highlight issues.

Thermo-oxidative Ablation

Results of ablation tests show that the nanocomposites perform comparably to current state-of-the-art insulation materials with only 2 wt % addition of nanoscale filler (17). Table 1 shows the erosion rates for two state-of-the-art internal solid rocket motor insulation materials versus neat Nylon 6 and nanocomposites derived from Nylon 6 at two mass fluxes. As expected, neat Nylon 6 does not survive this extreme environment, completely being removed within the timescale of the test. In stark contrast, the erosion rates for the nanocomposites with as little as 2 wt% filler content are comparable to the rates for the current materials. The reduction in required filler content will offer a reduction in weight, an increase in reliability because the material can be manufactured as one single part, and ultimately, a drastic reduction in cost.

As Figure 4 shows, an inorganic char layer forms on the surface of the ablation test specimen in less than 8 seconds in the combustions chamber. The virgin material lying less than 200 um below the char surface remains unaffected by the harsh thermo-oxidizing environment of the combustion gases.

Table 1. Ablation rates for current state-of-the-art insulation materials versus nanocomposite materials (17).

	Erosion Rate low mass flux (0.1 lb/in^{2}s), mils/s	Erosion Rate high mass flux (0.4 lb/in^{2}s), mils/s
EPDM/15% Kevlar®	6	10
EPDM/30% Silica	6	11
Nylon 6	nvmr[a]	nvmr[a]
Nylon 6/2% LS (PI mat[b])	4	9
Nylon 6/5% LS (PI mat[b])	3	10

[a]nvmr – no virgin material remaining
[b]The PI mat was used to enhance material retention during the test

During the ablation experiment, temperature within the char layer exceeds 1000°C and approach 2000-2500°C at the surface. At these temperatures, any carbonaceous residue from the polymer will contain graphite. Additionally, mica-type layered silicates, such as montmorillonite, irreversibly transform into other aluminosilicate phases. Between 600 and 1000°C, montmorillonite dehydroxylates and has been observed to initially transform into spinel, cristobolite, mullite and/or pyroxenes (enstatite) (24). At temperatures greater than 1300°C, mullite, cristobolite and cordierite form and subsequently melt at temperatures in excess of 1500°C (mullite 1850°C, pure cristobolite 1728°C and cordierite ~1550°C) (25). The presence of an inorganic that transforms into a high viscosity melt on the surface of the char will improve ablation resistance by flowing to self-heal surface flaws. This is known to occur in silica-filled ablatives (26).

Another possible advantage of the nanocomposite versus the traditional filled system is retention of the char layer formed on the surface. Because of the high degree of dispersion a more uniform char layer is formed over the entire exposed surface (Figure 4). This prevents mechanical forces from eroding away the surface as the combustion proceeds. In traditional filled systems the dispersion is not as uniform and islands of char will be formed on the surface as opposed to the uniform layer allowing mechanical forces to gain a foothold and flake away the char layer exposing more virgin material beneath.

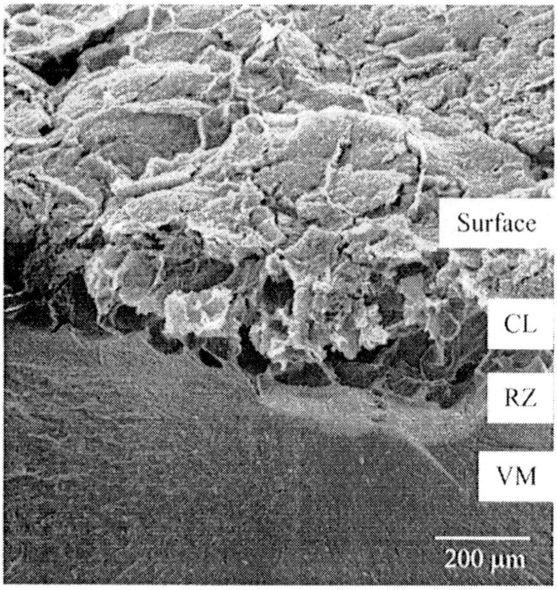

Figure 4. Scanning electron micrograph of nylon 6/5wt% layered silicate nanocomposite after the ablation test. CL: char layer; RZ: porous reaction zone; VM: virgin material

Oxygen Plasma Treatment

Oxygen plasma contains many species including free radicals, cations, anions, electrons, ozone molecules and electrons making it an extremely strong oxidant and a harsh environment for organic materials. Generally, unfilled organic materials such as polymers containing only C, H, O, N, and S react with approximately the same efficiency, or volume loss per atom of 1 to 4 x 10^{-24} cm^3/atom (27). Direct exposure of polymers to oxygen plasma leads to rapid degradation and removal of organic material. This is commonly exploited to clean surfaces of organic materials and as a reactive ion etch in microfabrication.

The optical micrographs in Figure 5 show the effect of oxygen plasma exposure to pure nylon 6 and a nylon 6/7.5wt% layered silicate nanocomposite. Both are melt-processed samples recast from the 1,1,1,3,3,3-hexa-fluoro-2-propanol solution. The nylon 6 sample experiences almost complete deterioration after 8 hours (480 minutes) of continuous exposure. In contrast, deterioration of the nanocomposite is minimal, with no significant decrease in thickness. Buckling of the nanocomposite sample after exposure arises from differences in thermal expansivity of the self-generating ceramic surface and the bulk polymer nanocomposite.

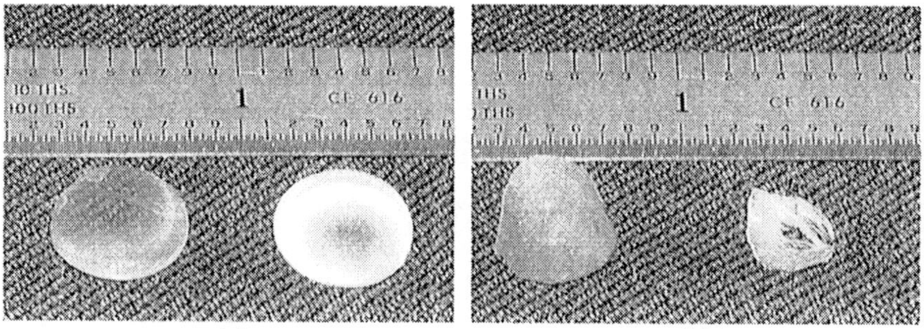

(a) (b)

Figure 5. Photographs of nylon 6 / layered silicate nanocomposite (left in both pictures) and neat nylon 6 (right in both pictures) before (a) and after (b) plasma treatment for eight hours.

In addition to the recast samples, as-received, melt extruded films of the nylon 6 and nylon 6/layered silicate nanocomposite (7.5wt%) were exposed to the oxygen plasma. The nylon 6 nanocomposite develops of a colorful, interferometric surface reflection upon exposure (not shown). The optical interference effects suggest highly uniform alteration of the surface composition on the order of 200-500 nm.

Erosion rates under oxygen plasma exposure of the nylon 6 and nylon 6 nanocomposites are shown in Figure 6. At long times the equilibrium erosion rates decrease with increased weight percent of montmorillonite (60 nm/min, 15 nm/min and 10 nm/min for nylon 6, nylon 6/5.0wt%, and nylon 6/7.5wt%, respectively). However, initial erosion rates for all three samples are a comparable >150 nm/min. This suggests that the oxygen plasma initially removes organic material from the near surface irrespective of the presence of inorganic. As the top organic layer is removed, inorganic concentration increases and the effective erosion rate decreases, indicating that the underlying material is being protected by the inorganic material. It comes as no surprise that this transition is dependent on the concentration of layered silicate in the nanocomposite, decreasing with increased concentration of montmorillonite (120 min for nylon 6/5.0wt% montmorillonite nanocomposite and 60 min for nylon 6/7.5wt% montmorillonite nanocomposite). The general effect of this self-generating passivation layer is to decrease erosion rate, and increase life time, by greater than an order of magnitude (~15 times). Despite having no inorganic precursor material the nylon 6 also makes this transition stemming form the formation of carbonaceous char giving a modest protection to the underlying material.

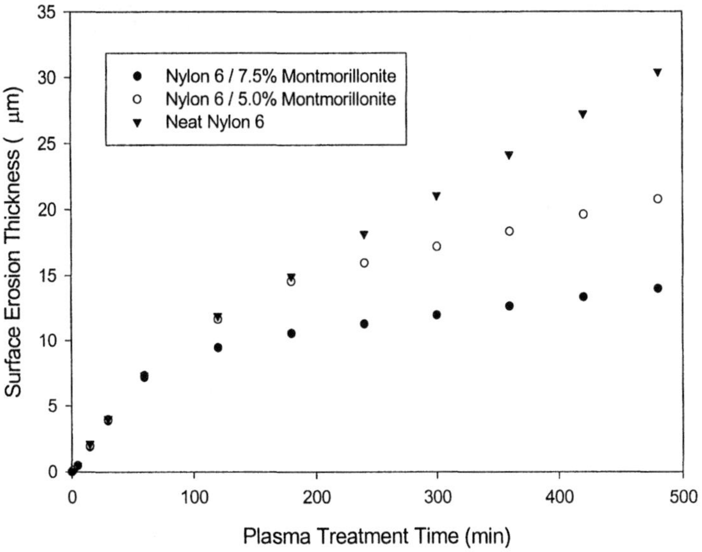

Figure 6.Oxygen plasma surface erosion rate of neat nylon 6, nylon 6/5.0% layered silicate nanocomposite and nylon 6/7.5% layered silicate nanocomposite.

Argon ion etching coupled with XPS analysis can give information on the composition of the exposed films to help examine the effects through the thickness of the sample. Note that for these heterogeneous systems calibration etch time to etch depth ratio is difficult, and thus changes in chemical composition are discussed relative to the surface (28).

Figure 7 shows the relative chemical compositions during continuous Ar^+ etching the film. The atomic percent composition in Figure 7 is based on the four most abundant elements (Si, Al, C, and O) – hydrogen, nitrogen and other elements have been neglected. Detailed depth profiling reveals three distinct regions: (i) surface region (i.e. within 500 seconds etching time), containing high concentration of Si, Al, Mg and O and very little carbon; (ii) transition-region (i.e. 500 to 3000 seconds etching time), containing increasing amounts of carbon and decreasing amounts Si, Al, Mg and O with increased depth; and (iii) bulk-region (i.e. longer than 3000 seconds etching time), largely unaffected by the oxygen plasma, and corresponding to the original composition of the nanocomposite. The slight unbalanced increase in percentage of carbon and oxygen prior to Ar^+ etching is due to surface adsorption of atmospheric carbon dioxide and other carbon containing compounds. In the transition-region, the formation of a carbonaceous tar due to etching may enhance the carbon

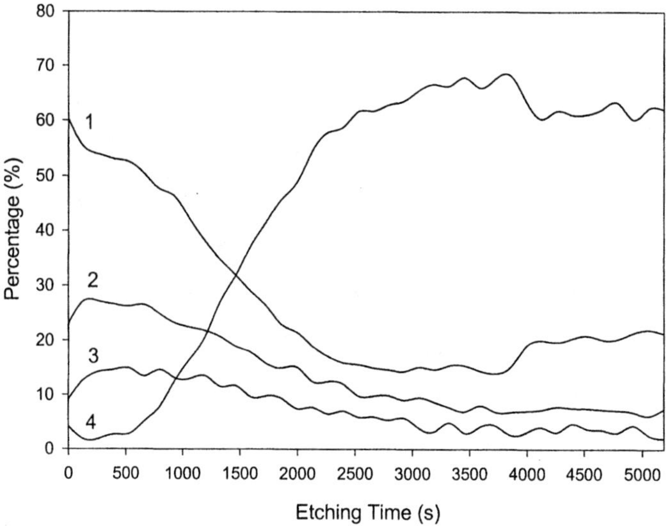

Figure 7. Elemental depth profile after one hour plasma treatment of nylon 6/7.5% layered silicate nanocomposites. (1. oxygen, 2. silicon, 3. aluminum, 4. carbon).

percentage relative to the unexposed and bulk (etching time > 3750 sec) nanocomposite.

The ratio of silicon, aluminum, and oxygen (Si:Al:O) at the very surface of the exposed nylon 6 nanocomposite is 22.96%:9.14%:60.2%, which is 2.5:1:6.5. This ratio corresponds well to that generally anticipated for low-magnesium montmorillonite (2.3:1:6.8, $(Al_{3.5}Mg_{0.5})Si_8O_{20}(OH)_4$ (5)). This further indicates a thin layer of aluminosilicate covers the surface of the oxygen plasma treated nanocomposite film.

Figure 8 shows bright field transmission electron micrographs of the cross-section of nylon 6/7.5wt% montmorillonite nanocomposite before and after 1 hour of oxygen plasma treatment. The unexposed sample displays a morphology consistent with previous studies the dark lines being individual silicate layers oriented perpendicular to the sample surface (17). The exposed sample shows the same morphology below the exposed surface. At the surface of the exposed sample a "coating" forms as the average distance between individual silicate layers decreases forming a region varying from 0.1 to 0.5 μm thick.

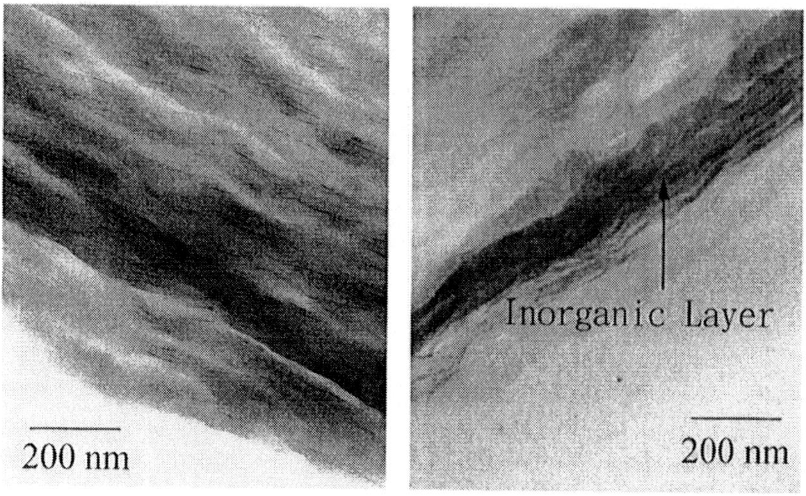

Figure 8. TEM images of nylon 6/7.5% clay nanocomposites before (left) and after (right) one hour oxygen plasma treatment.

UV-Low Energy Electron Irradiation

In contrast to the outstanding potential demonstrated by layered silicate nanocomposites in the thermal and chemical oxidative environments, the behavior and potential benefits toward UV-vis and low-energy electron exposure are not as drastic.

As with the previous examples, exposed nanocomposite samples showed enrichment of the silicate near the exposed surface. Figure 9 shows the FT-IR spectrum for a thin 7.5 wt% nanocomposite before and after the simulated space exposure. As seen in the post exposure FT-IR spectra (spectra 2), the relative intensity of the peak centered around 1000 cm^{-1}, indicative of silicate absorption, is much larger after than before UV exposure, indicating enhancement of silicate species in the exposed sample with regard to selective removal of organic material. In corroboration, sample mass decreased with exposure.

Supporting the FTIR results, x-ray absorption near edge structure (XANES) spectra (29) of an unexposed and exposed nanocomposite indicated surface enhancement of silicate. Figure 10 displays the spectra as recorded in total electron yield (TEY) mode – an extreme surface sensitive mode that captures information from the top 100 nm of the material. The peak at about 1846 eV is indicative of SiO_x species. The relative increase in the peak in the exposed

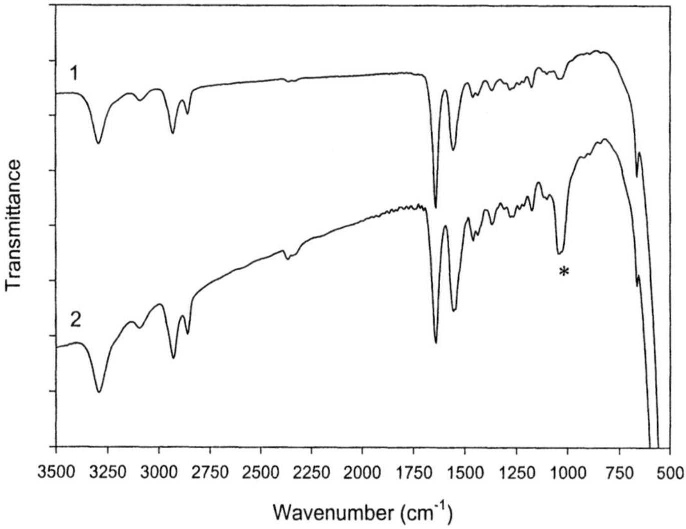

*Figure 9. FT-IR characterization of NLS-7.5wt% samples (1) before and (2) after space exposure. *indicates Si-O-Si vibration around 1000 cm^{-1}*

sample indicates an increase in the SiO_x species at the material surface, presumably arising from selective erosion of a thin surface layer of polymer, which leaves an enriched inorganic layer.

XPS enables an estimate of the composition and thickness of the in-situ generated surface layer. Table 1 lists the mole percents of carbon, oxygen, nitrogen and silicon as a function of argon etching time for the NLS/7.5 wt% sample after simulated space exposure. The composition of silicate near the exposed surface (0-30 seconds etching time) is about 3 times that of the bulk regions (>2 minutes etching time), but the thickness of the silicate enriched layer is very thin as it quickly decreases after 30 seconds etching time. This indicates that the silicate enhanced surface is thinner (<< 100nm) and the total enhancement through the thickness is much less (~16 times) than that observed in the plasma exposure (based on relative Ar+ etching time required to reach bulk material). This is most likely due to the shallow penetration depth of the UV-electron radiation compared to that of the oxygen plasma. The unbalanced high percentage of oxygen at the surface prior to etching is due to surface absorption of atmospheric CO_2.

Although enrichment of silicate at the surface occurred, coloration of the sample, and thus increased solar-absorbtivity, occurred. Optical micrographs in Figure 11 show a 2.5wt% nylon nanocomposite sample before (a) and after exposure (b) in the SCEPTRE facility. After the space simulated exposure, the

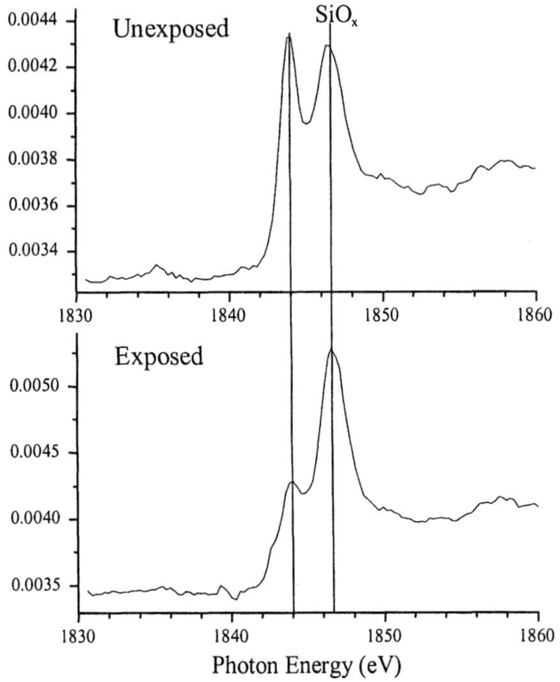

Figure 10. XANES Si-K edge spectra for unexposed and UV/electron exposed Nylon 6/2 wt% layered silicate.

Table 1. Composition profile (mole percentage) of NLS-7.5wt% film

Ar^+ etching time	C	O	N	Si
0 second	79.75	12.87	4.82	2.56
30 second	88	4.8	4.0	3.2
2 minutes	92	2.5	4.2	1.8
5 minutes	94	1.6	3.7	1.0
15 minutes	94	1.1	4.1	0.7
30 minutes	94	1.1	3.9	0.6
60 minutes	95	0.9	3.6	0.4
120 minutes	95	0.5	3.9	0.4

color of the nylon 6/montmorillonite film turned substantially darker, initially browning and then developing a dark purplish hue, indicating severe degradation within the film. The other nanocomposites (1, 5 and 7.5 wt% montmorillonite) appeared comparable to the 2.5 wt% sample before and after exposure, with a very slight increase in the final extent of coloration with silicate concentration. Pure Nylon 6 behaved very similarly, although the rate of coloration was not as fast. For the nanocomposites, the major changes occurred within the first 96 hrs, where as complete coloration of the pure Nylon 6 required upwards of 160 hrs.

The coloration that occurred however, was restricted to the near-surface region. Figure 12 shows an on-edge view of the pure polymer sample with delamination or peeling of the exposed surface. Similar observations were made for the nanocomposite samples.

The ultra-high vacuum environment of the simulated environment necessitates energy dissipation from the sample through black-body radiation (Plank's Law). During the test, the average surface temperature of each sample, despite being equidistant from the center of the exposure platen, was different, generally decreasing with layered silicate content. The 7.5 wt% specimen experienced the lowest average temperature of 86°C while the 1.0 wt% specimen experienced the highest at 153°C. The average temperature of the pure polymer specimen was 131°C. The initial emissivity between 2-30 μm of the all the samples was comparable (0.88±0.02). Thus the increased surface temperature arises from a combination of increased energy absorption associated with coloration and changes in surface emissivity associated with compositional changes at the sample surface. The sample heating will be synergistic with the UV and electron absorption and associated chemistries, greatly complicating detailed investigation of decomposition pathways and rates from the current experiment. This failure mechanism should be similar to that observed for coatings (30), where the combination of heat with UV and low-energy electrons leads to an increased decomposition rate, which increases color and absorption, resulting in a cascade that continually accelerates degradation.

Since the initial absorption (i.e. color) of the nanocomposite increases with montmorillonite addition, future investigations should focus on examining synthetic aluminosilicates that do not contain isomorphic substitution from transition metals such as Fe. Alternative types of inorganic fillers include laponite, fluorohectorite or synthetic silicic acids (31). Note that the UV exposure is not only an on-orbit concern, but will drastically affect life-cycle and long-term environment stability and performance of nanocomposites used in external terrestrial environments.

Conclusions

In summary, polymer layered silicate nanocomposites are able to self-generate a compositionally-graded silicate passivation layer upon exposure to

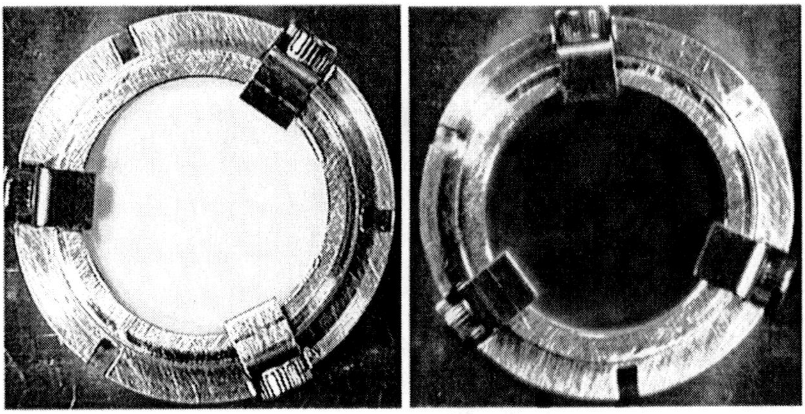

(a) (b)

Figure 11. Nylon 6/montmorillonite (2.5 wt%) (a) before and (b) after SCEPTRE exposure.

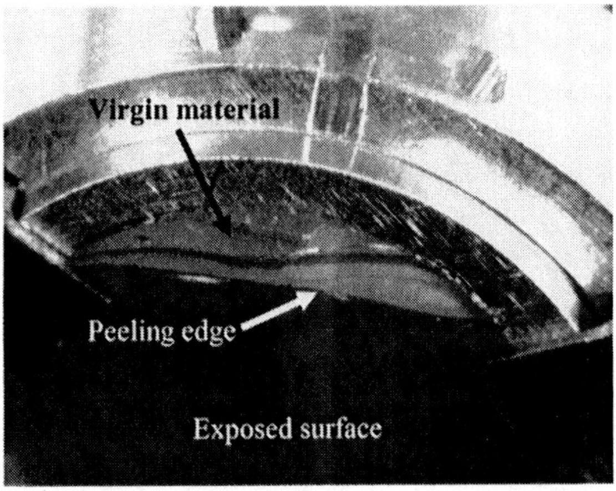

Figure 12. Edge view of pure nylon 6 polymer after SCEPTRE exposure.

extreme thermo-oxidizing, reactive oxidative, and UV environments. Results indicate that depending on the aggressive environment the thickness of the layer may vary from less than a hundred nanometers to greater than 100 microns. The chemical composition of the uppermost region is comparable to the base montmorillonite. The formation of the layer is due to the preferential oxidation/degradation of the polymer and the corresponding enrichment of the nanoscale layered silicate on the surface. Furthermore, the compositionally-graded-surface will be more durable to thermal cycling and mechanical flexure, which leads to failure in conventional rigid coatings on flexible substrates.

The process to form the self-passivation layer is dependent on the environment. Within thermal-oxidative environments such as rocket motor exhaust, the nanoparticle serves as a precursor, chemically reacting and/or thermally decomposing to actively form the enriched inorganic coating. In contrast, exposure to oxygen plasma, UV or electrons preferentially removes the organic polymer, increasing the montmorillonite composition at the surface to passively form the protective inorganic coating. In these later cases the nanoparticles may also selectively absorb detrimental radiation or chemical species, minimizing exposure of the underlying inorganic material.

Complete understanding of the behavior of the nanocomposites in the complex environments discussed will necessitate investigation of potential synergism between the components of the environment (chemical, radiative and thermal) and the nanoparticles. In other words, it should not be expected a priori that all nanofillers will provide the responses observed for the montmorillonite systems. Degradation mechanisms may use the nanoparticles as an amplification center, decreasing system durability. Furthermore, controlled morphology of the nanocomposites is critical. Poorly aligned systems will be less likely to form uniform inorganic passivation layers leaving the material open to mechanical degradation such as in the ablative environment and further attack from the chemically reactive environment. In addition, the mechanical performance and durability of the in-situ generated layer will depend on its morphology and thus the initial dispersion and orientation of the nanoparticles, especially in the case of rods and plates.

As a final comment, the self-passivation response is not only advantageous for enhancing durability in service, but also provides unique opportunities for post-processing modification of nanocomposites. For example, pretreatment of the nanocomposite in oxygen plasma may enhance the materials resistance to thermal-oxidative or UV-electron environments. Alternatively, self-generation of the inorganic surface on commercial nanocomposites films may enhanced barrier properties before introduction to service.

Acknowledgements

This work was supported in part by funding from AFOSR. The Synchrotron Radiation center is supported by the National Science foundation under award #DMR-95-31009. The authors would like to thank Drs. Shawn Phillips, Jeff Sanders, Peter J. John, Clifford A. Cerbus and Hong G. Jeon for UV/electron exposure studies, XPS, TEM work and helpful discussion. This work was supported by Entrepreneurial Research of U.S. Air Force Office of Scientific Research. The authors are grateful to Southern Clay Products, for supplying the nylon 6/ layered silicate nanocomposites.

References & Notes

1. Champion, K. S. W.; Cole, A. E. and Kantor, A. J. in Handbook of Geophysics and the Space Environment, Jursa, A.S. ed, Air Force Geophysics Lab., U.S. Air Force, National Technical Information Service, Springfield, VA, 1985, 14-1.
2. Koontz, S. L.; Leger, L. J.; Visentine, J. T.; Hunton, D. E.; Cross, J. B. and Hakes, C. L., J. Spacecraft and Rockets, 1995, 32, 483.
3. deGoh, K. K. and Banks, B. A., J. Spacecraft and Rockets, 1994, 31, 656. Tribble, A.S. The Space Environment, Implications for Spacecraft Design, Princeton University Press, Princeton, NJ, 1995.
4. Mark, J. E., Lee, C. Y. and Bianconi, P. A., Hybrid Organic-Inorganic Composites, ACS Symposium Series 585, Washington D. C., 1995.
5. Vaia, R. A., in Polymer Clay Nanocomposite, Pinnavaia, T. J. and Beall, G. W. eds., , John Wiley & Sons, New York, 2000.
6. Collister, J., in Polymer Nanocomposites, Vaia, R. A., & Krishnamoorti, R. eds., ACS, Washington D. C., 2001.
7. Kojima, Y.; Usuki, A.; Kawasumi, M.; Okada, O.; Fukushima, Y.; Kurachi, T.; Kamigaito, O., *J. Mater. Res.*, 1993, **8**, 1185.
8. Ke, Y.; Long, C.; Qi, Z., *J. Appl. Polym. Sci.*, 1999, **71**, 1139.
9. Messersmith, P.B.; Giannelis, E.P., *J. Polym. Sci. Part A: Polym. Chem.*, 1995, **33**, 1047.
10. Akelah, A.; Moet, A., *J. Mater. Sci.*, 1996, **31**, 3589.
11. Kojima, Y.; Usuiki, A.; Kawasumi, M.; Okada, A.; Kurachi, T.; Kamigaito, O., *J. Appl. Polym. Sci.*, 1993, **49**, 1259.
12. Yano, K.; Usuki, A.; Okada, A.; Kurachi, T.; Kamigaito, O. J. Polym. Sci.: Part A: Polym. Chem. 1993, 31, 2493.
13. Yano, K.; Usuki, A.; Okada, A. J. Polym. Sci.: Part A: Polym. Chem. 1997, 35, 2289.
14. Liu, L.; Qi, Z.; Zhu, X., *J. Appl. Polym. Sci.,* 1999, **71**, 1133.
15. Usuki, A.; Kojima, Y,; Kawasumi, M.; Okada, A. Kukushima, Y.; Kurauchi, T.; Kamigaito, O., *J. Mater. Res.*, 1993, **8**, 1179.
16. J.W. Cho, D.R. Paul Polymer 2001, 42, 1083.
17. Vaia, R.A.; Price, G.; Ruth, P.N. Nguyen, H.T.; Lichtenhan, J., *Appl. Clay Sci.*, 1999, **15**, 67.
18. Thorne, J.A.; Whipple, C.L. in The Effects of the Space Environment on Materials, Society of Aerospace Material and Process Engineers, North Hollywood, CA, 1967, 243.
19. Dang, T. D., Thiesing, N. C., Feld, W. A., Cerbus, C. A. and Arnold, F. E., Polymer Preprint, 2000, 41(2), 1213.
20. Gilman, J. W., Schlitzer, D. S., Lichtenhan, J. D., Journal of Applied Polymer Science, 1996, 60, (4), 591.
21. Gonzalez, R. I., Phillips, S. H. and Hoflund, G. B., J. Spacecraft Rock., 2000, 37, (4), 463.

22. Zimcik, D.G.; Wertheimer, M.R.; Balmain, K.B.; Tennyson, R.C. J. Spacecraft and Rockets, 1991, 28, 352.
23. Alexandre, M. and Dubois, P., Mater. Sci. Eng. 2000, 28, 1-63.
24. Grim, R.E. Clay Mineralogy, McGraw-Hill, New York, 1968. p 313-328, 353-387.
25. Kingery, W.D.; Bowen, H.K.; Uhlmann, D.R. Introduction to Ceramics, Wiley and Sons, New York, 1976. p. 298-315.
26. D'Alelio, G.F; Parker, J.A. eds. Ablative Plastics, Marcel Dekker, Inc. New York, 1971. p. 1-35
27. Hseih, D-T; Lloyd, T.B.; Rutledge, S.K. Int. SAMPE Symp. Proc., Conf. Proc. of 1998 meeting. (part 2 of 2) v 43 n2 SAMPE Covina, CA pp. 1170-1171.
28. Ar^+ etching rate of organic and inorganic is anticipated to be different, which results in preferential removal of organic and enrichment of inorganic with etch depth; thus compromising the absolute determination of elemental composition with depth.
29. XANES spectra were collected at the Canadian Synchrotron Radiation Facility located at the 1 GeV storage ring at the Synchrotron Radiation Facility in Stoughton, WI. The Si K-edge spectra were recorded using the double crystal monochromator beamline (DCM). The synchrotron radiation was monchromatized using InSb crystals with a photon resolution of ~0.9 eV.
30. Zerlaut, G. A.; Gilligan, J. E.; and Ashford, N. A., "Space Radiation Environmental Effects in Reactively Encapsulated Zinc Orthotitanates and Their Paints," AIAA 6th Thermophysics Conference, AIAA paper no. 71-449 (1971).
31. Wang, Z.; Lan, T.; Pinnavaia, T.J. Chem. Mater., **8**(9), 2200-2204 and Beneke, K.; Lagaly, G. Am. Mineral. **1977**, 62, 763.

Chapter 7

Nanocomposite Aerospace Resins for Carbon Fiber-Reinforced Composites

Chengang Chen[1], David Curliss[2], and Brian P. Rice[1]

[1]University of Dayton Research Institute, 300 College Park Avenue, Dayton, OH 45469–0168
[2]Materials and Manufacturing Directorate, Air Force Research Laboratory, Wright-Patterson Air Force Base, OH 45433–7750

Polymer-layered silicate nanocomposites have been the recent focus of a great deal of polymeric materials research due to enhancements of thermal, mechanical, and transport properties. In this research, thermoset polymer nanocomposites were successfully prepared and the polymer nanocomposite resins were also used to prepare carbon fiber-reinforced composites. Investigation of the rheological characteristics showed that the modified resin would still be suitable for resin transfer molding of fiber-reinforced composites. The data from differential scanning calorimetry (DSC) and *in situ* small-angle x-ray scattering (SAXS) are well related for the morphological development. Results from wide-angle x-ray diffraction (WAXD), SAXS, and transmission electron microscopy (TEM) of the polymer-silicate nanocomposites were used to characterize the morphology of the layered silicate in the epoxy resin matrix. The moduli of the polymer-silicate nanocomposites were found to be significantly greater. The diffusion coefficients of the solvent in the nanocomposites were found to be much lower and the oxygen plasma erosion rates were also reduced for the nanocomposites. The nanocomposite/IM7 carbon fiber composites were uniform with no "filtration" of nanoparticles from the resin during infusion. SAXS results indicate that the nanoclay planes are oriented parallel to the IM7 fiber axis. The thermal expansion coefficient of the composite for the nanocomposite as the matrix was reduced. There is some improvement for mechanical properties.

Introduction

Layered-silicate polymer nanocomposites are a new class of composite materials comprised of a thermoplastic or thermoset polymer and a nanoscale particle. Dispersed layered silicates are high aspect ratio nanoparticles with thickness of nanometer dimension and lateral dimension 100X–1000X greater. Nanocomposite polymers have been reported to exhibit increased modulus, increased thermal stability, improved flame retardancy, improved ablation performance, increased solvent resistance, reduced gas permeability, and decreased thermal expansion coefficients compared to the virgin polymers *(1-6)*. These increases in properties are attributed to the nanometer-size particle dispersion of the layered silicates with high aspect ratio, high surface area and high strength in the polymer matrix. Toyota researchers published some of the earliest comprehensive research on the polymer clay nanocomposites *(7)*, leading to increased interest in research on polymer layered-silicate nanocomposites *(1-6)*. The Toyota-developed Nylon-based nanocomposites have even been commercialized. New nanocomposite materials are being developed for demanding high-technology applications. Polymer nanocomposites have great potential to enhance the performance of polymers used for fiber-reinforced composite matrices. The reported performance improvements in polymers modified with low loading of organoclays offer promise of economically improving the performance of the fiber-reinforced composite. In this research, an epoxy resin suitable for use as a composite matrix (Epon 862 and Epi-cure curing Agent W) modified with organoclay (I.30E) was introduced into the carbon fiber IM7 to make an advanced composite.

The most widely used layered silicates for polymer nanocomposites are the smectite clays such as montmorillonite. They offer a high aspect ratio and a high surface area. Sodium montmorillonite is 2:1 phyllosilicate, constructed of repeating triple-layers with two silica tetrahedral sheets fused to an edge-shared octahedral sheet of alumina. The physical dimensions for these silicate sheets are approximately one hundred to several hundred nanometers in lateral dimension and 1 nanometer in thickness. The individual sheets in the silicates are generally stacked together and hydrophilic. Thus, montmorillonite is compatible with water-soluble polymers such as poly(vinyl alcohol) *(8)* and poly(ethylene oxide) *(9)*. However, in this state they are not compatible with most hydrophobic organic polymers. Fortunately, the cations in the gallery such as Na^+ can be easily exchanged with surfactants such as alkyl ammonium cations. The pendent organic group in the surface of the silicate sheets lowers the surface energy of silicate layers and improves the wettability and dispersibility of hydrophilic clay in hydrophobic thermoplastic polymers or thermoset resins. Generally, there are two types of nanocomposites defined by the morphology of the layered silicate in the nanocomposite: intercalated and exfoliated. Although most organoclay nanocomposites reported so far are intercalated, the exfoliated morphology is preferred, and expected to yield the greatest impact in property improvement of the polymeric materials. To date, a great deal of research has focused on preparation of the polymer nanocomposite *(1-6)* and the mechanism for exfoliation *(10-12)*. There have been very few reports on the use of a thermoset polymer nanocomposite as the matrix in a traditional fiber-reinforced composite. In this research work, the emphasis is placed on aerospace epoxy-silicate nanocomposites and their carbon fiber-reinforced composites. The aerospace epoxy resin used in this research effort is prepared from Shell Epon Resin 862 and Epi-cure curing agent W mixed at a stoichiometrically balanced ratio. This epoxy system has high T_g (~155°C), low viscosity, relatively slow cure kinetics, and desirable balanced thermomechanical properties. These characteristics make the resin an ideal candidate for infusion processing of composite materials using techniques such as RFI. The morphological development and exfoliation mechanism of the silicate in the thermoset resin during processing, mechanical properties, and transport properties in solvents of the nanocomposites and fiber-reinforced composites with a nanocomposite polymer matrix will be reported.

Experimental Details

Materials

Shell Epon Resin 862 (a bis-phenol F epoxy), Epi-cure curing agent W, *n*-dodecylamine (Aldrich), *n*-hexadecylamine (Aldrich), *n*-octadecylamine (Aldrich), hydrochloric acid (Aldrich), SNA (Southern Clay Products), I.30E (Nanocor), carbon fiber IM7G-12K tow (Hexcel) were the materials used in this study.

Preparation

Organoclays were prepared from Cloisite Na treated with hydrochloric acid and *n*-dodecylamine, *n*-hexadecylamine, and *n*-octadecylamine, respectively. SC18, as an example: 21.1 g of *n*-octadecylamine in the 750 mL of ethanol and water mixture solvent (v:v, 1:1) was added with aqueous hydrochloric acid (HCl, 1N, 67.5 mL). The mixture was stirred at ~70°C. When the solution was clear, 67.5 g of SNA were added to the above mixture solution, and the suspension was continuously stirred for 4 hours at ~70°C. The resultant mixture was filtered. The solid was washed with a mixture of warm ethanol and water, and dried.

Processing

Processing for Nanocomposites

The desired amount of Epon 862 and organoclay was mixed and stirred at ~70°C, followed by degassing in a vacuum oven. After resin degassing the corresponding amount of curing agent W was added to the mixture with continued stirring. The resulting nanocomposite resin was cast between release coated glass plates spaced at 0.635 cm and cured in a programmable convection oven. The following cure cycle was used for curing all of the Epon 862/W nanocomposite resins: heat the cast resin from RT to 121°C over 30 minutes; hold at 121°C for two hours; heat to 177°C over 30 minutes; hold for another two hours at 177°C; cool the cast nanocomposite resin in the oven to ambient temperature.

Processing for Carbon Fiber Composites

The carbon fiber-reinforced composites were fabricated using an RFI technique. A dry fiber preform was made by winding 16 plies of IM7 fiber onto a 25 x 25-cm Teflon-covered metal plate. Twenty-five percent excess matrix resin was placed inside a silicon dam and covered with a nonporous glass/Teflon film. A slit was cut down the center of the film to allow resin transport, and the fiber-covered plate was placed over the assembly. The entire dam containing the resin and the fiber-covered plate was then covered with a second layer of nonporous glass/Teflon film to prevent resin bleed. The part was then vacuum bagged and placed into an autoclave for cure.

Characterization

DSC was performed using a TA Instruments differential scanning calorimeter 2920 modulated DSC at 2°C/min with air sweep gas. Thermomechanical analysis was performed on a TA Instruments TMA 2940 thermomechanical analyzer at 10°C/min with nitrogen sweep gas. WAXD was performed in the Rigaku x-ray powder diffractometer. The generator power was 40 kV and 150 mA, the scan mode was continuous with a scan rate of 0.6°/min, and the scan 2θ range is from 2° to 10°. Some of the SAXS measurements were taken using a flat-film statton camera on a Rigaku RU-200 with Cu K_α as its radiation with a wavelength of 1.5418 Å. The power was 50 kV and 150 mA, and the exposure time was around 20 hours. Other SAXS experiments and *in situ* SAXS experiment were performed at the National Synchrotron Light Source at the Brookhaven National Laboratory (beamline X27C). In the *in situ* SAXS experiment, the mixture of the organoclay with epoxy and curing agent was mounted on the holder, and the sample was heated up at 2°C/min. The data were recorded every minute. The sample for TEM was microtomed in a Reichert-Jung Ultracut Microtome and mounted on 200-mesh copper grids. TEM was performed using a Philips CM200 transmission electron microscope with a LaB_6 filament operating at 200 kV. The dynamic mechanical analysis (DMA) was performed using a Rheometrics ARES dynamic spectrometer using torsion bar geometry at a frequency of 100 rad/sec, a strain of 0.1 percent, and a heating rate of 2°C/min; while the viscosity test was carried out on a Rheometrics ARES dynamic spectrometer using 50-mm-diameter parallel plates geometry at a frequency of 10 rad/sec, a strain of 3 percent and a heating rate of 2°C/min. Fracture toughness, Kq, was measured under ambient conditions using standard compact tension described in ASTM Standard E399. Ten specimens were tested for each sample.

The laminates of the carbon-fiber composite were first characterized by acoustic NDI (C-scan) and optical microscopy to evaluate their quality. Scanning electron microscopy (SEM) was used to evaluate laminate quality and nanoparticle distribution using a Philips XL30 ESEM TMP scanning electron microscope. Mechanical tests for the carbon-fiber composite were selected to measure resin-dominated properties. These tests were transverse four-point flexure with a span-to-depth ratio of 32:1 and longitudinal four-point flexure with a span-to-depth ratio of 16:1 designed to induce midplane shear failure. Ten specimens were tested for each material type and condition.

Oxygen plasma was generated in a GSC-200 plasma generator. The concentration in oxygen plasma was around 10^{18} radicals (or ions) per liter. Oxygen plasma erosion rates were determined by measuring the weight loss after treatment. The effect of the addition of clay on the solvent uptake properties of Epon 862/W was examined. Seven Epon 862/W panels were made with 0%

layer silicate, 1.0% SC16, 3.0% SC16, 6.0% SC16, 1.0% I.30E, 3.0% I.30E and 6.0% I.30E. The coupons were then weighed, and three coupons from each sample were placed in methanol at room temperature. The degree of solvent uptake was evident due to the increased mass of the coupons over time.

Results and Discussion

Processing

Rheology

The resin viscosity during cure for the nanocomposite 6% I.30E/Epon 862/W and the neat resin Epon 862/W was performed by DMA. The static viscosity during a constant heating rate scan is shown in Figure 1.

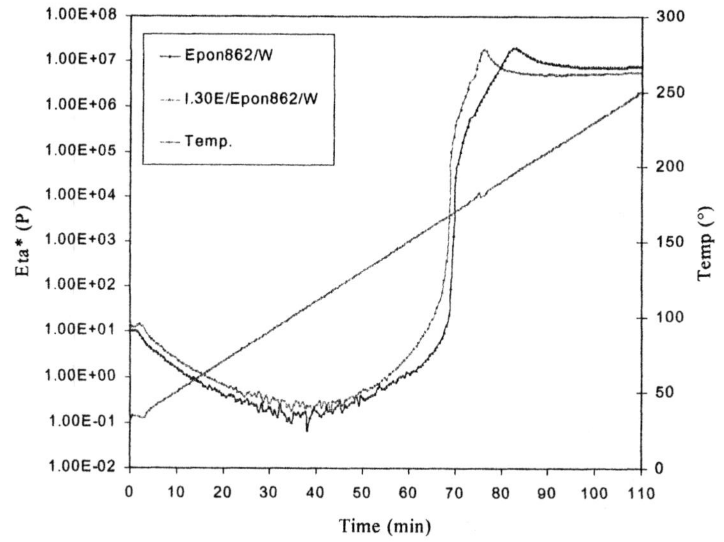

Figure 1. Viscosity of Epon 862/W and 6% I.30E/Epon 862/W at a heating rate of 2°C/mon. (See page 4 of color insert.)

At room temperature, the viscosity of the I.30E/Epon 862/W is slightly greater than that of Epon 862/W as expected. However, the increase is minimal and the processing characteristics are suitable for RFI. The viscosity vs. time and temperature behavior for these two systems is similar. Gelation occurs at a slightly lower temperature for the epoxy resin with organoclay. The temperature of the softening point for I.30E/Epon 862/W is similar to that of Epon 862/W, just at a slightly lower temperature (180 vs. 196°C). This demonstrates that the organoclay has some catalytic effect on the polymerization of Epon 862/W.

DSC

The cure behavior of the I.30E/Epon 862/W and Epon 862/W was evaluated using DSC. With the addition of the organoclay (I.30E) to the original Epon 862/W resin, the nanocomposite system exhibits a greater enthalpy change during cure (206.0 vs. 191.4 J/g) at lower curing onset temperature (102.5 vs. 122.5°C). This also demonstrates that the organoclay has some catalytic effect on the crosslinking polymerization of Epon resin 862 with Epi-cure curing agent W *(10)*.

In Situ Small-Angle X-ray Scattering

In situ small-angle x-ray scattering provides a powerful means of monitoring the morphological evolution of the nanocomposite during cure. When Epon resin 862 is mixed with I.30E, some of the Epon 862 monomer diffuses into the gallery (interplane region) of the organoclay. This intercalation is detected as a swelling of the interplane distances. In this case, the I.30E interplanar distance increased from 22.6 Å to ~35 Å. To illustrate the dynamic morphological development study, some key data are shown in Figure 2. When the I.30E/Epon 862/W nanocomposite resin was heated at 2°C/min, the peak gradually shifted and became a weaker, signifying that the interplanar spacings are increasing and become less uniform. As the temperature increased to ~102°C, the ordered structure collapsed and exfoliation took place as indicated by the almost disappearance of a coherent peak. At a slightly higher temperature (~126°C), a new clear peak appears at ~85 Å, and the intensity and interplanar spacing increases gradually with increasing temperature (and extent of cure) up to a maximum of ~130 Å. The temperature for the collapse of the ordered structure observed by *in situ* SAXS is the same as the curing onset temperature from DSC. The change in the SAXS data is related to the structural evolution of the nanocomposite during the curing. Before the onset temperature of curing, there is no cure taking place. The increase in the temperature just helps expand the gallery of the organoclay to a very limited extent. At the onset temperature of curing (~102°C), the curing takes place and provides extra energy to help the nanoclay sheets expand and the epoxy resin is migrated inside the gallery. While

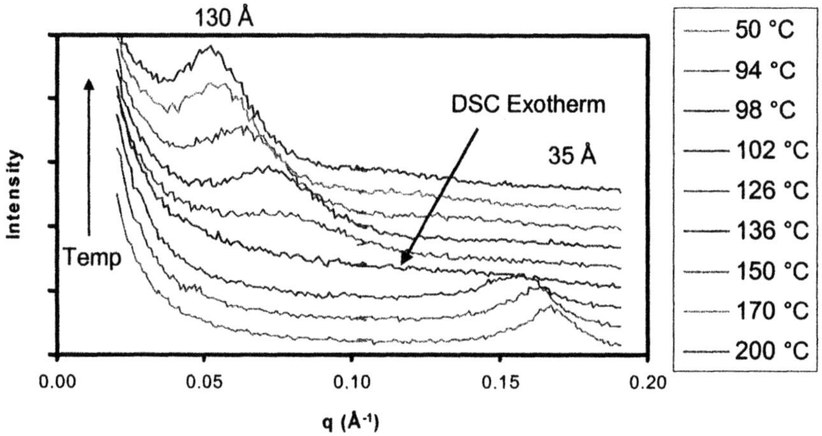

Figure 2. In-situ small-angle x-ray scattering of 6% I.30E/Epon 862/W at 2°C/min. (See page 4 of color insert.)

the epoxy system widens the clay layers from the edges of the particles (non uniform, curved layers), the loss of coherence between the nanoclay layers causes no clear peak in SAXS. With the increase in the temperature further, more and more epoxy resin outside the gallery migrates inside the gallery until the layers are widened completely through the entire particles. So the intensity of the peak in the SAXS is more intense and interplanar spacing is increased until the epoxy is almost fully cured. Then the epoxy surrounding the nanosheets of the layered silicate is rigid and a relatively ordered structure is formed in the nanocomposite.

Morphology

WAXD shows that there is no peak at 2 θ scans ranging from 2° to 10° for nanocomposites, which indicates that the interplanar spacing is larger than 44 Å and nanocomposites are generally considered to be exfoliated structure. In order to confirm the morphology of the nanocomposite, SAXS measurements were taken (Table I). SAXS measurements show that interplanar spacings for nanocomposites are larger than 100 Å at the organoclay loading of less than 8%. Even at the organoclay's loading of 10% for SC18, the nanosheets in the nanoclay are expanded to interplanar spacing of 90 Å. The weak peak in the SAXS measurements shows that there is some short-range ordered structure. The SAXS data show that the interplanar spacings for Epon 862/W nanocomposites with 3% loading of SC12, SC16, SC18 and I.30E are 146 Å,

Table I. Small-Angle X-ray Diffraction Data, Storage Moduli and Glass Transition Temperatures (Tg) of the Nanocomposites and their Pristine Polymer

Clay	*d-Spacing (Å)*		*Tg (°C)*	*G' (dyne/cm^{1})*	
	NSLS	*Rigaku*	*Tan δ*	*Glassy (30°C)*	*Rubber (180°C)*
NA			154	1.03E10	1.00E8
1.0% SC18	249				
3.0% SC18	135	135	145	1.23E10	1.39E8
6.0% SC18	129		154	1.30E10	2.14E8
10.0% SC18	90	89/44	144	1.56E10	2.82E8
3.0% SC12	146				
1.0% SC16		238	155	1.20E10	1.16E8
3.0% SC16	156	151	154	1.18E10	1.63E8
6.0% SC16	150	141	155	1.24E10	2.45E8
1.0% I.30E	172		153	1.20E10	1.24E8
3.0% I.30E	126	125			
6.0% I.30E	100/49	100/48			
9.0% I.30E		88/44	144	1.61E10	2.31E8
12.0% I.30E		68			

156 Å, 135 Å, 126 Å, respectively. The carbon length of the ammonium cation appears to have no direct effect on the dispersion and exfoliation of nanoclay in the epoxy resin. It appears that the acidic alkyl ammonium ions play a key role for the exfoliation of epoxy organoclay nanocomposite *(10)*.

Several images of TEM of these different nanocomposites were taken and the images are similar. Representative TEM images from low magnitude to high magnitude of 3% SC16/Epon 862/W nanocomposite are shown in Figure 3. The scale bar in Figure 3a is 1 μm. This image shows that the layered silicates are distributed very well in the polymer matrix even in the domain as large as 4 μm although there also exist some pure polymer matrix domain between the silicates. The TEM images with higher magnitude are shown in the Figures 3b, 3c and 3d. They show that the individual nanosheets are dispersed well in the polymer matrix. The dark lines are cross sections of the silicate sheets of nanometer thickness. The original aggregates of the silicate sheets are disrupted, and nanosheets are well dispersed in the epoxy resin. Some individual sheets are disordered, while some still preserve the parallel alignment of layers with ~15 nm separation. This is consistent with the SAXS results (~150 Å). Simultaneously, there also exist some relatively large pure polymer domains in

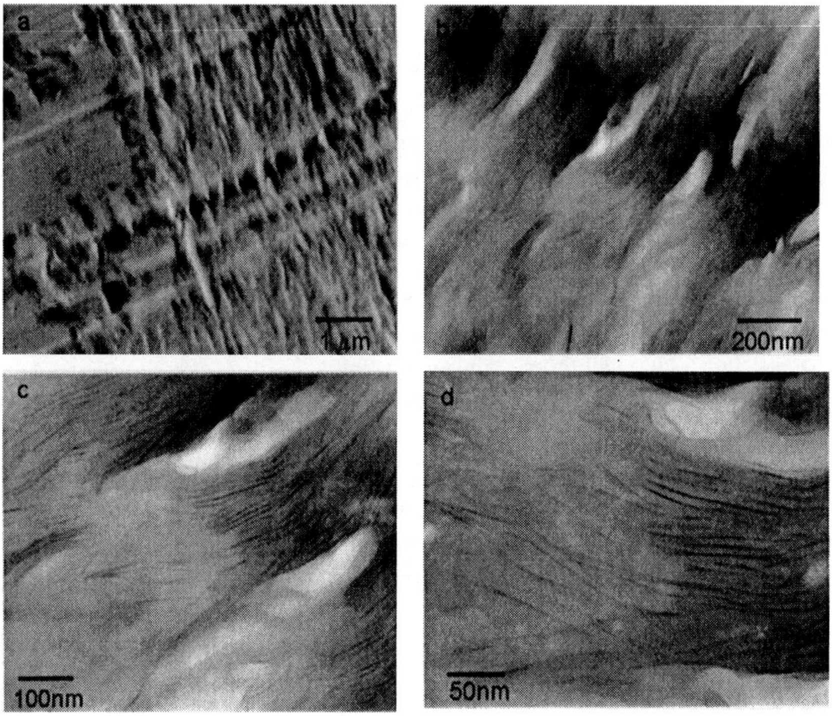

Figure 3. TEM images of the epoxy-nanocomposite of 3% SC16/Epon 862/W from low magnitude to high magnitude (a ,b, c and d).

the size of 50 nm to 100 nm between the nanosheets of silicate. So the nanocomposite contains some exfoliated nanostructure and some intercalated nanostructure with very large interplanar spacing. Compared with the original interplanar spacing (18 Å) of organoclay (SC16), the increase of the interplanar spacing for the nanocomposite (~150 Å) is extremely significant. There are large amount of the epoxy resin penetrating inside the gallery and expanding the gallery of the layered silicate. Although the ideal exfoliated morphology is that the individual silicate layers are completely separated and dispersed in a continuous polymer matrix, this nanocomposite can be considered as exfoliated nanocomposite because of the well-dispersed layered-silicate in the epoxy matrix. This observation is consistent with the morphology expected in a thermoset polymer nanocomposite system. The resin is clearly thermodynamically compatible with the organoclay, but dispersion of the individual silicate sheets is hindered by the rheology of the thermoset resin.

The processing of thermoset polymer nanocomposite resins into carbon fiber-reinforced composites presented concern about filtering of the organoclay particles by the carbon fiber preform. The thickness of the organoclay sheets is on the order of one nanometer, and the lateral dimensions are on the order of average fiber spacing of approximately one micron. It is conceivable that the nanoparticles would not be able to infiltrate the composite preform, or that the nanoparticle distribution would not be uniform. Ultrasonic evaluation (C-scan) of the carbon-fiber composite laminates indicated that the laminates were of good quality. SEM EDAX data were used to demonstrate that the compositions at the different locations of cross section in the composite are uniform, and therefore we believe there are no problems with particle filtering. A laminate cross-section photomicrograph, shown in Figure 4, of the IM7/I.30E (6% wt. epoxy)/Epon 862/W also demonstrates the composite is of high quality.

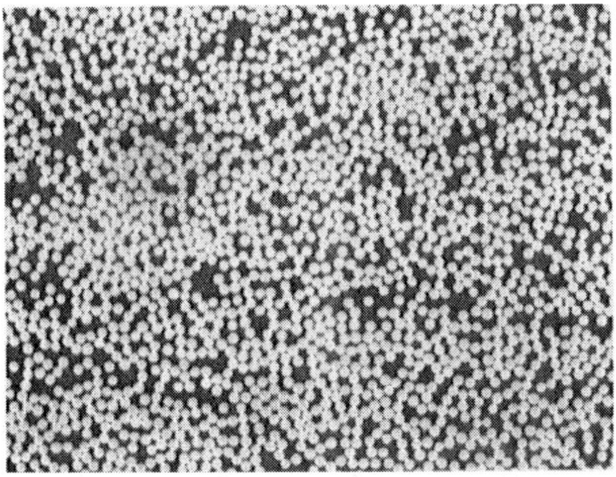

Figure 4. Photomicrograph of the composite of IM7/I.30E (6% wt. in epoxy)/Epon 862/W.

The SAXS of IM7 carbon fiber-reinforced I.30E (6% wt. epoxy)/Epon 862/W composite is shown in Figure 5. The SAXS data show that the layered silicates are preferentially oriented in the composite parallel to the carbon fiber axis. It is more likely that this orientation is induced by the resin flow around the fibers during infusion, leading to shear alignment of the silicate sheets. Interaction between the IM7 carbon fiber and layered silicate could include Van der Waals and/or acid-base interactions. The morphology and interaction should be important factors to the properties of the advanced composite.

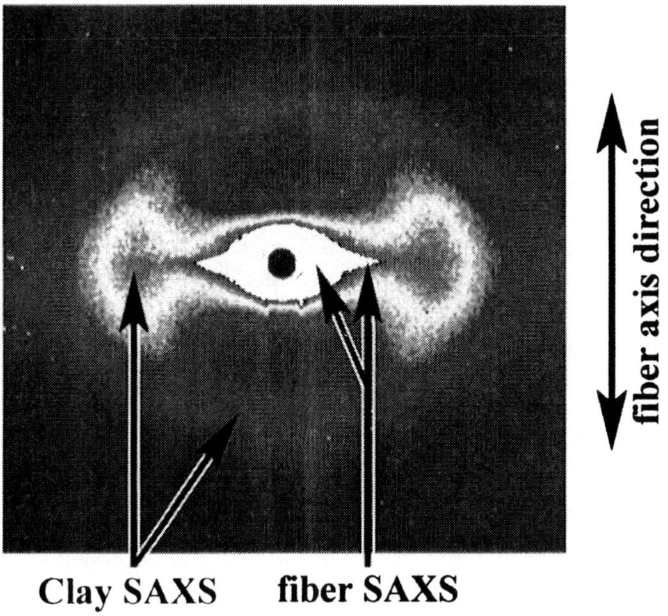

Figure 5. Small angle x-ray scattering of the composite of IM7/I.30E (6% wt. in epoxy)/Epon 862/W. (See page 5 of color insert.)

Properties

Mechanical

The dynamic mechanical analysis results (Table 1) show that the storage moduli of the nanocomposites are higher than those of the unmodified epoxy resin. Moduli increase with increased organoclay loading. In general, nanocomposites show a more significant increase of the storage modulus in the rubbery state than in the glassy state. In this research the dynamic storage modulus was increased more than 40% in the glassy state and 125% in the rubber state for 9% I.30E/Epon 862/W nanocomposites. This is attributed to the high aspect ratio and high stiffness of the organoclay filler.

In addition, for carbon fiber-reinforced composites, the 90° flex modulus of IM7/I.30E (6% wt. in epoxy)/Epon 862/W shows a 19% increase compared with IM7/Epon 862/W (control), while there is a 12% increase of the modulus for 6% I.30E/Epon 862/W nanocomposite compared with neat Epon 862/W polymer. The interlaminar shear strength measured using 0°, 4-point flex mechanical tests shows a 5% increase for IM7/I.30E (6% wt. in epoxy)/Epon 862/W compared with IM7/Epon 862/W. The fracture toughness for 6% I.30E/Epon 862/W nanocomposite, however, is reduced 18% compared with neat Epon 862/W polymer. The flexure modulus and interlaminar shear strength increases are believed to be due to the dispersion and preferential orientation of the silicate layers in the fiber-reinforced composite, but fracture toughness does not benefit from this effect. The interface between the nanoparticles and the polymer is poor, as there is no covalent bonding of the polymer with the surfactants used to treat the silicates. The particles are possibly acting as stress concentration defects that lead to reduced fracture toughness in the resin. The key to improving the mechanical performance (including toughness) of the nanocomposite is to enhance the interfacial interaction between the polymer matrix and layered silicates. Recently, we found that the fracture toughness of the polymer nanocomposite increased 30% when an aged mixture (instead of a freshly prepared mixture) of 6% I.30E/Epon 862 was used to prepare the thermoset nanocomposite polymer. This observation indicates that the dynamics of monomer/organoclay interaction and morphology development are more complex than currently understood. More research is underway to investigate this phenomenon.

Thermal Expansion Coefficient

Thermomechanical analysis shows that the thermal expansion coefficients of the nanocomposites are lower than those of the unmodified polymer in the glassy state. For example, the thermal expansion was found to be 76.6 μm/m°C for pure Epon 862/W polymer, 64.8 μm/m°C for 6% I.30E/Epon 862/W and 64.3μm/m°C for 6% SC16/Epon 862/W. It was also found to be 41.8 μm/m°C for IM7/I.30E (6% wt. in epoxy)/Epon 862/W and 46.9 μm/m°C for IM7/Epon 862/W. The reduction of the thermal expansion coefficient is due to the presence of low CTE silicate nanosheets and their constraining effect on the polymer. The reduction of thermal expansion is very important in the thermomechanical performance of the fiber-reinforced composite. The stresses induced by the thermal expansion mismatch between the polymer matrix and the fiber are significant and contribute to premature failure in composite laminates. The lower CTE of the polymer nanocomposite should improve the deterioration of fiber-reinforced composite laminates due to thermal residual stresses.

Solvent Transport

The solvent uptake was examined in methanol. The percentage of the sample mass gain vs. square root of immersion time of Epon 862/W, 1%, 3%, 6% SC16/Epon 862/W and 1%, 3%, 6% I.30E/Epon 862/W in methanol, is shown in Figure 6. The data show that the methanol uptake for the nanocomposite is significantly low compared with the pristine polymer. This is ascribed to the barrier effect of nanosheets of the nanoclay in the epoxy resin. High loading of organoclay is more effective for the solvent resistance, and SC16 appears more effective than I.30E. This is perhaps related to the better dispersion of organoclay in the epoxy matrix for SC16/Epon 862/W, as small-angle x-ray scattering studies indicate.

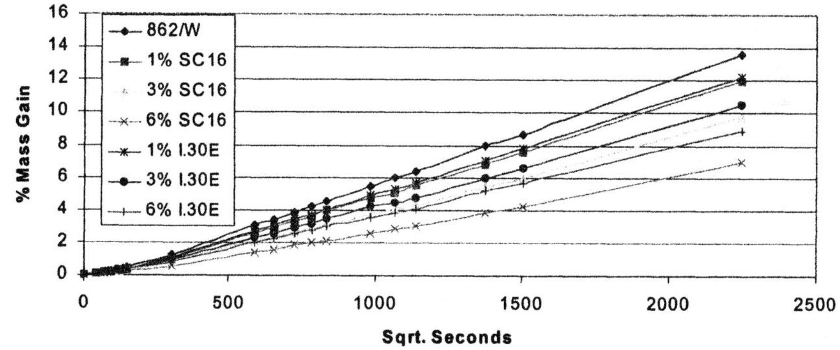

Figure 6. Methanol uptake of Epon 862/W; 1%, 3%, 6% SC16/Epon 862/W; and 1%, 3%, 6% I.30E/Epon 862/W nanocomposites. (See page 5 of color insert.)

From the percent mass gain-time data of the nanocomposite coupons, the one-dimensional binary diffusion coefficient was calculated using the following relationship *(13)*:

$$D = \pi \cdot ((2h/(4M_{max}))^2 \cdot (\Delta M/\Delta t^{0.5})^2)$$

where D: one-dimensional binary diffusion coefficient, h: half-thickness, M_{max}: maximum percent mass gain, and $(\Delta M/\Delta t^{0.5})$: initial slope of the percent mass gain vs. square root time plot. The derivation of this relationship was based on the assumption of one-dimensional diffusion. The initial slopes of the percent mass gain vs. the square root of time plots were calculated using linear regression. Values for the average one-dimensional binary diffusion coefficient of methanol in Epon 862/W with different I.30E loadings are listed as follows:

4.25x10^{-11} cm^2/sec (pure Epon 862/W resin), 3.54x10^{-11} cm^2/sec (1% I.30E/Epon 862/W nanocomposite), 2.79x10^{-11} cm^2/sec (3% I.30E/Epon 862/W nanocomposite), and 2.27x10^{-11} cm^2/sec (6% I.30E/Epon 862/W nanocomposite). The addition of I.30E had a dramatic effect on the ability of methanol to decrease the diffusion coefficient by 17% for 1% I.30E, 34% for 3% I.30E, and 47% for 6% I.30E loading, respectively. In all cases the SC16 caused a more significant decrease in the diffusion coefficient. The diffusion coefficients calculated for the samples loaded with 6.0% SC16 were typically 20% lower than the diffusion coefficients calculated for the samples loaded with 6.0% I.30E. Again, this is perhaps related to the better dispersion of nanosheets in SC16/Epon 862/W nanocomposites.

Survivability in Oxygen Plasma

In order to test the survivability of the nanocomposites in aggressive oxidizing environments, two nanocomposites (3% I.30E/Epon 862/W and 6% I.30E/Epon 862/W) and pure epoxy resin (Epon 862/W) were exposed to an oxygen plasma. Oxygen plasma contains many species such as atomic oxygen, radicals, ozone molecules, cations, anions, electrons, etc., which are extremely strong oxidants. Up to now, polymers were one of the most vulnerable materials under oxygen plasma. The surface erosion thicknesses after a 5-hour exposure to the oxygen plasma are ~34, 25, and 22 microns for pure epoxy resin, 3% I.30E/Epon 862/W, and 6% I.30E/Epon 862/W, respectively. The surface erosion rates for epoxy nanocomposites are significantly retarded compared with pure epoxy resin. The retardation effect of 6%-clay nanocomposite is greater than that of 3%-clay nanocomposites. The SAXS showed that the interplanar spacing of nanocomposites after exposure to the oxygen plasma was reduced, indicating some epoxy resin inside the gallery was eroded. The possible mechanism for the retardation is that the nanocomposite materials can form intact inorganic layers when they are exposed to the oxygen plasma *(14)*. The preferential oxidation of polymers and corresponding enrichment of the nanoscale-layered silicate on the surface make the formation of the ceramic-like inorganic layer. This inorganic layer retards the penetration of the oxygen plasma and thus prevents further degradation. The nanocomposite may enhance the survivability of the polymeric materials in aggressive environments such as low-earth orbit (LEO) against atomic oxygen. Similar materials are being exposed in LEO under the Materials on International Space Station Experiment (MISSE) project.

Conclusions

Exfoliated aerospace epoxy organoclay nanocomposites with organoclays (SC12, SC16, SC18 and I.30E) and carbon fiber-reinforced composites with these nanocomposites as a matrix were successfully made. The characterization from WAXD, SAXS and TEM confirms the exfoliated nanostructure. The carbon length of the ammonium cation has no direct effect on the dispersion and exfoliation of nanoclay in the epoxy resin, and the acidic alkyl ammonium ions play an important role in the exfoliation of epoxy nanocomposite. Rheological characteristics showed that the addition of clay to the resin did not significantly alter the viscosity. *In situ* SAXS clearly shows the morphological development of the nanocomposite during cure. DSC and *in situ* SAXS studies show that the organoclay has some catalytic effect for polymerization and that the exothermal heat at the onset curing temperature is an important factor for nanosheet exfoliation. The dynamic storage of the nanocomposites was increased. The solvent uptake for the nanocomposites is significantly low compared with the pristine polymer, and diffusion coefficients of methanol in nanocomposites are much reduced. Aerospace epoxy-organoclay nanocomposites can also form a passivating inorganic layer, which will enhance the polymeric survivability under oxygen plasma. In addition, the nanocomposite, composed of organoclay (I.30E), Epon 862 and curing agent W, was successfully introduced into carbon fiber IM7 to make a high quality advanced composite. The SEM EDAX data show that the composition of the composite is uniform without a filtering problem. The advanced composite shows some extra improvement in mechanical properties with the addition of clay into the matrix. SAXS studies indicate that the nanoclays are oriented and parallel to the fiber axis. The orientation and interaction between the nanoclay and carbon fiber may play an important role in the extra improvement in the mechanical properties. The thermal expansion coefficient of the composite for the nanocomposite as the matrix was measured to be lower than that of neat epoxy resin as the matrix.

Acknowledgements

This work is supported by the Air Force Office of Scientific Research and Air Force Research Laboratory (F33615-00-D-5006). The authors would also like to thank Dr. Charles Lee, Dr. Richard A. Vaia, Dr. David P. Anderson and Mr. Larry Cloos for their support and help.

References

1. E. P. Giannelis, *Adv.Mater.*, **1996**, *8*, 29.
2. R. A. Vaia and E. P. Giannelis, *MRS Bulletin,* **2001**, *26* (5), 394.
3. P. C. LeBaron, Z. Wang and T. J. Pinnavaia, *Applied Clay Science*, **1999**, *15*, 11.
4. R. Krishnamoorti and R. A. Vaia, eds., *Polymer Nanocomposites,* American Chemical Society, Washington DC, **2001**.
5. M. Alexandre and P. Dubois *Mater. Sci. Eng.,* **2000**, *28*, 1.
6. J. W. Gilman, *Applied Clay Science*, **1999**, *15,* 31.
7. Usuki, Y. Kojima, A. Okada, Y. Fukushima, T. Kurauchi and O. Kamigiato, *J. Mater. Res.,* **1993**, *8,* 1179.
8. K. A. Carrado, P. Thiyagarajan and D. L. Elder, *Clays Clay Mineral.,* **1996**, *44,* 506.
9. R. A. Vaia, S. Vasudevan, W. Krawiec, L. Scanlon and E. P. Giannelis, *Adv. Mater.,* **1995**, *7*, 154.
10. T. Lan, P. D. Kavirayna and T. J. Pinnavaia, *J. Phys. Chem. Solids,* **1996**, *57,* 1005.
11. T. Lan, P. D. Kavirayna and T. J. Pinnavaia , *Chem. Mater.,* **1995**, *7*, 2144.
12. I. J. Chin, T. Thurn-Albrecht, H. C. Kim, T. P. Russell and J. Wang, *Polymer,* **2001**, *42*, 5947.
13. C. S. Kresse, *University of Dayton, Masters Degree Thesis,* **1991**.
14. H. Fong, R. A. Vaia, J. H. Sanders, D. Lincoln, A. J. Vreugdenhil, W. Liu, J. Bultman and C. Chen, *Chem. Mater.,* **2001**, *13,* 4123.

Novel Nanomaterials, Processes, and Nanomaterial Applications

Chapter 8

Millimeter-Wave Beam Processing and Production of Ceramic and Metal Materials

D. Lewis[1,*], L. K. Kurihara[2], R. W. Bruce[3], A W. Fliflet[3], and A. M. Jung[3]

[1]Materials Science Division, Code 6304, [2] Materials Physics Branch, Code 6340, and [3]Beam Physics Branch, Plasma Physics Division, Code 6793, Naval Research Laboratory, 4555 Overlook Avenue, SW, Washington, DC 20375
[*]Corresponding author: lewis@anvil.nrl.navy.mil

An 83 GHz millimeter-wave beam system has been used in several areas of material processing, including sintering, joining, coating and production of nanophase materials via the polyol process. This system has a number of unique features relative to material processing and production. These features have been used to advantage in rapid sintering of ceramics, joining of ceramics and polyol production of nanophase metal and metal oxide materials. With the latter we have been able to prepare powders of single elements, alloys, metastable alloys, composites and coatings. Examples of a few of the metals processed in this study include Fe, Co, Ni, Cu, Ru, Rh, Pt, Au, FePt, Fe_xCo_{100-x}, NiAg and Cu-Ni. The millimeter-wave driven polyol process has been operated in both a batch process, with a reflux system driven in superheated mode and in a continuous system, where residences times at temperature are only a few seconds. Recently, a continuous, 2.45 GHz system has been used for the polyol process with significant advantages and is described briefly here.

The high intensity millimeter-wave beams (10^2–10^5 W/cm^2) that can be generated by powerful gyrotron oscillators have unique capabilities for rapid, selective heating of many materials to high temperatures. Previous work by the authors has demonstrated the efficacy of such rapid heating for both production of nanophase materials and processing of such materials into useful components (*1-3*). The advantages of using a millimeter wave beam as the heat source derive from localization of energy deposition, and the ability to deposit energy within the volume of material. The resultant advantages include rapid heating and cooling, volumetric heating, elimination of thermal inertia effects, spatial control of heating, and use of low cost tooling, process vessels and instrumentation. These features of the millimeter-wave beam system, described subsequently, have been used in several types of material processing and have been the subject of four recent patent applications *(4-7)*.

In the case of joining of functional and structural ceramics, the system permits rapid localized heating (and cooling) of joint regions. This permits rapid thermal processing, with controlled temperature profiles and minimizes heating effects on the materials adjacent to the joint. With piezoelectric ceramics, such as lead zirconate titanate, this has permitted production of high strength, electrically conductive braze joints in materials which cannot be heated above 300°C. In the case of structural ceramics, such as aluminum oxide, this has permitted production of joints with high strength and high temperature capability, using low cost, base metal tooling for component alignment and application of pressure during joining. Some of the joining results are described elsewhere (*3*). The ability to provide rapid thermal cycles and in-depth heating has been used in rapid sintering (total cycle time less than 10 min.) of ferrite materials. In the particular case of polyol processing, the system permits: very short processing times, high heating rates due to bulk heating of the solution that results from direct coupling of the beam energy to the solution elements, the ability to selectively coat substrates where the beam has been focused, and working in a superheated liquid region. By working in this regime, it is possible to suppress the growth process and to promote nucleation. This is a particular feature of microwave heating of liquids, where the interior of a volume may be heated well beyond the boiling point, provided the liquid is reasonably free of boiling nuclei. Conventional heating from an external source is limited by boiling nucleated at the surface of the container. Polyol processing has been done both in a batch system, using the millimeter-wave beam system to heat polyol solutions in a reflux condenser, and in a continuous mode, where the polyol solution if heated by the millimeter wave beam as it flows through a silica reaction tube. Millimeter wave driven polyol processing has been used to prepare nanocrystalline powders of Fe, Fe_xCo_{100-x}, FePt, Co, Ni, NiAg, Cu, Cu_xNi_{100-x}, Ru, Rh, Ag, Pt and Au. Nanostructured coatings of copper have been also been deposited on AlN substrates.

The remainder of this article will focus on the millimeter-wave polyol production of nanophase materials, which are of great technical and commercial interest. Fine metal particles, as produced in the polyol process, have many uses–in pigments, magnetic materials, catalysts, electromagnetic shielding, ferrofluids, sensors, biomedical, electronics and advanced-engineered materials. Among the various preparative techniques used to prepare nanoscale particles, the chemical routes provide the most practical route for preparing fine and ultrafine powders. The polyol method, in which the polyol acts as a solvent and as the reducing agent, and also acts as a surfactant, is a suitable method for preparing nanophase and micron size particles with well-defined shapes and controlled particle sizes (*8-18*). This process also provides intrinsic protection, for the high surface area particles formed, from agglomeration and reaction with the atmosphere, particularly with regard to oxidation of ultrafine metal particles.

EXPERIMENTAL

A continuous-wave (CW) gyrotron-based material processing system has been established at the Naval Research Laboratory (NRL). The system is comprised of a Gycom™, Ltd. industrial 15-kW CW gyrotron and associated superconducting magnet, power supplies, cooling system, control system, and a material processing chamber. The complete system is shown in Fig. 1. The gyrotron operates near 83 GHz, and the output is produced in the form of a free-space quasi-Gaussian beam, which is directed into the work chamber. The gyrotron can be operated over a wide range of output powers (0.1–15 kW) by varying the cathode voltage from 22 to 28 kV, the cathode current from 1 to 2 A, and by making small changes in the applied magnetic field to vary the interaction efficiency. At optimum conditions, the gyrotron is the most efficient beam power source available, reaching about 50% overall efficiency between power input and power output in the millimeter-wave beam. The gyrotron can be operated in the CW mode or for pulses as short as 1 second.

After exiting the gyrotron boron nitride output window, the microwave beam is injected into a cylindrical, stainless steel work chamber (length = 100 cm, diameter = 60 cm) via a 60-cm conical transition, which contains a secondary, Teflon™-coated BN window to protect the primary gyrotron window from processing environment and process by-products. This processing chamber is gas and vacuum tight, will withstand overpressures of several atmospheres, and is constructed entirely of stainless steel allowing capability for working in a wide range of processing environments. In the work chamber, the beam is directed toward the workpieces by reflective optics–typically gold-plated copper mirrors. The optics can be used to both direct and focus (or defocus) the millimeter-wave beam, and with motorized drives could be used to

Fig. 1. NRL 83 GHz Millimeter-Wave Beam Material Processing Facility: a) gyrotron and superconducting magnet, b) transition and secondary BN window, c) processing chamber, d) access doors, e) exhaust system, f) 3-axis optics manipulator, g) 2 color optical pyrometer

raster the beam in a controlled manner if desirable. In the batch polyol processing, a 15 cm diameter concave mirror was used to focus the beam down onto a flask containing the polyol solution. The beam was approximately 3 cm in diameter at the flask. The reaction flask was placed in the work chamber with an attached, air-cooled reflux condenser in initial experiments, with the beam centered on the flask. The reflux condenser was not in the path of the beam. In subsequent experiments, water cooling was added to the reflux condenser with feedthroughs into the chamber. Additionally, a fine wire, type K thermocouple was placed into the center of the reaction flask, with feedthroughs to readouts via the Labview™ control system for the millimeter-wave system. This setup is shown in Figure 2. The above illustrates one of the key advantages of the millimeter-wave beam system over conventional heating and low frequency cavity microwave heating. With the beam system, it is possible to have apparatus (water-cooled reflux condenser) and instrumentation (K thermocouple) in the processing system without these being affected by the process heat source. Basically, anything not in the direct path of the beam is affected very little, and the primary requirement is for electromagnetic shielding of electrical leads. In addition to these advantages, this process also has significant advantages relative to process scaling over both conventional polyol processing with external heat sources and microwave heating of polyol solutions in closed vessels (Parr bombs) in conventional microwave ovens. (*16*) In the latter case, scaling the process vessel size dramatically alters the nature of the process, with much longer process times and greater process heterogeneity due to thermal gradients and larger convection current cells. These should result in greater heterogeneity of the product with increases in scale of reaction vessels. In the latter case, the process volume is severely constrained by power available and by the requirements for the process vessel to withstand the pressure generated during the process. Published work on the latter process is based on reacting polyol solution of very small volumes (50-100 ml of ca. 0.1 M solutions), which would result in extremely low production quantities (ca. 0.5

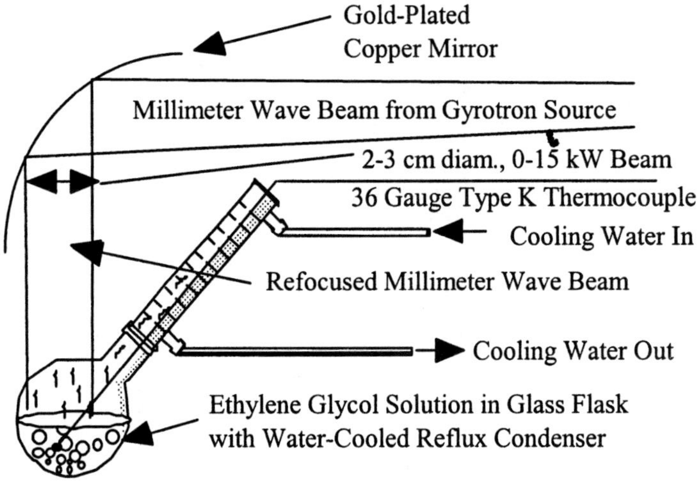

Fig. 2. Schematic of Millimeter Wave-Driven Polyol Process for Nanophase Material Production and Coating with Rapid Direct Heating with Millimeter Wave Beam Source

g/day) and extremely high product cost. In the millimeter wave process, scaling up the operation simply involves using larger reaction vessels, and higher power and larger beam diameters in the millimeter wave beam. With the current batch system, reaction vessels as large as 1000-2000 ml could be used, with power inputs of ca. 10 kW and reaction times of 10-20 minutes, resulting in a substantial production rate (50-100 g per day).

The general procedure for the synthesis of different metallic powders and coatings involved preparing a mixture of the metal precursor in ethylene glycol or 1,2- propanediol, typically about 0.01-0.1M, and bringing the mixture to reflux in the glass flask with the solution heated by the millimeter-wave beam. The chemistry here is thought to be fairly well understood in general, with reduction of metal salts and metal oxides occurring with the ethylene glycol acting as a reducing agent, though the specific products and intermediates (acetaldehydes, glycoaldehydes, gyoxal, glycolic acid, etc.), depend on the specific process *(19)*. For example, for a dilute solution of copper acetate monohydrate (CuAc) in ethylene glycol (EG):

$$\text{CuAc} + \text{EG} + \text{Heat} \Rightarrow \text{CuO} + \text{Organic Intermediates} + \text{EG}$$

$$\text{CuO} + \text{EG} + \text{Heat} \Rightarrow \text{Cu} + \text{Organic Products} + H_2O + CO_2 + \text{EG}$$

where we observe both the copper oxide and copper in various stages of the continuous microwave and millimeter wave processes, as is also seen in the batch processes.

The initial batch experiments were conducted on relatively small volumes of solution (25-100 ml) with very low power levels (100-300W). In most cases, refluxing of the glycol solution was observed within a minute of turning on the millimeter-wave beam.*(16)*. Reaction times were varied from about 1 minute to 60 minutes in experiments, but results indicate that copper acetate could be reduced to nanophase Cu in 5-10 minutes. During the processing, the gyrotron power was controlled to maintain continuous refluxing, based on visual observations of the process through a viewport and the temperature readouts from the K thermocouple placed in the solution. After the reaction was completed, the metal–glycol mixture was cooled to room temperature. For film deposition experiments, an AlN substrate was placed in the metal-glycol solution . The crystal structure of the powders and coatings were studied using X-ray diffraction (XRD). Line broadening of XRD peaks was used to estimate the average crystallite size. The morphology of the particles was investigated using scanning electron microscopy (SEM) and transmission electron microscopy (TEM).

Following these batch experiments which showed the feasibility of millimeter-wave polyol processing, analysis indicated that a continuous polyol process might be feasible, and might have significant advantages over the batch process. Conceptually, a continuous process should have much higher production rates, in this case with little increase in capital cost of equipment, thus producing product at much lower cost. In addition, the continuous process should produce materials of consistent quality, a possible issue with batch processing. Finally, analysis of the polyol process suggested that a continuous process, with very short processing times, might produce much finer particles, if desired, than the batch process. Based on these considerations, limited experiments were done in the gyrotron system with the setup shown in Fig. 3 below. In this system, the polyol solution of a metal salt was pumped into the chamber through feedthroughs, using silicone rubber tubing which couples very little with microwaves, into a silica reaction tube which was placed under the beam. The products were fed out of the chamber into a collection vessel. Reaction temperatures at the outlet of the reaction tube and at the inlet to the gyrotron chamber were monitored with K thermocouples. The solutions were pumped through the system with a peristaltic pump that allowed precise control over flow rates and permitted some overpressure in the system (ca. 1 atm.).

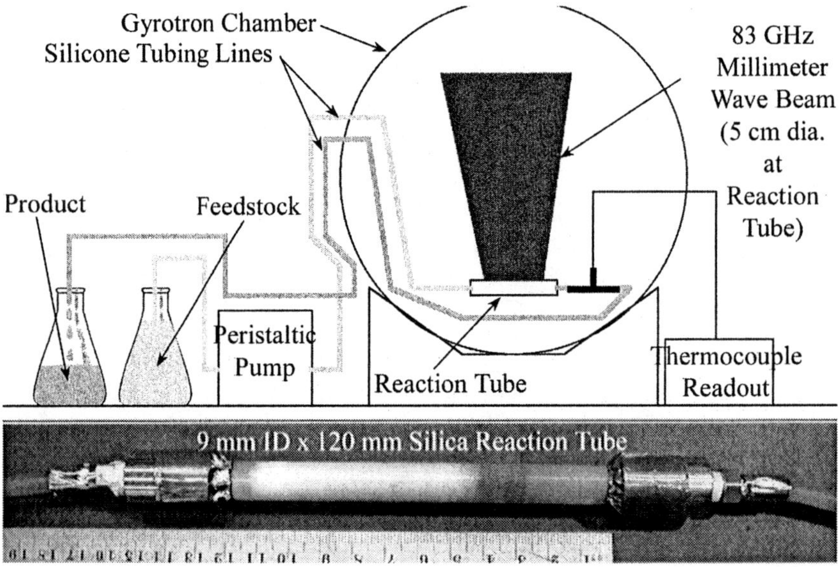

Fig. 3. Schematic of continuous polyol process using 83 GHz millimeter-wave beam system and silica reaction tube

With pumping rates that corresponded to residence times in the reaction tube of about 10 seconds and power levels of 1-2 kW in the beam, reaction temperatures of ca. 205°C at the outlet of the reaction tube were achieved. The copper acetate in ethylene glycol was reduced to nanophase copper of 10-100 nm crystallite size (by XRD line broadening).

Based on these results, the process was transitioned to another specialized microwave system of much lower capital cost ($20k vs $400k for the millimeter-wave system) and simpler configuration, further experiments were done, and a patent application has been filed on the process *(5)*. In this newer system, the precursor solutions, e.g., glycol solutions of metals salts, are pumped through a silica glass tube along the centerline of an S-Band waveguide fed by a 2.45 GHz source. The solutions are rapidly heated to high temperatures and the metal salts rapidly converted to metal oxides and metals. In this process reduction to metals is achieved in a process time of 20-30 seconds as the liquid flows through the waveguide. The system is pressurized to about 200 kPa, permitting operation of the polyol process at temperatures of 230-240°C, without boiling. The present experimental system is capable of processing several liters of polyol solution per hour in continuous operation at a power level of 2-3 kW. Operating at its maximum CW power output and the temperatures cited above, this system should be able to processing several hundred liters of polyol solution per day. Processes that proceed at lower temperatures could be performed at even higher production rates.

RESULTS AND DISCUSSION

Powders of a variety of transition metals have been synthesized by reducing metal precursors in a millimeter wave driven polyol solution. The precursors used in our study included: acetates, oxides, hydroxides, chlorides and carbonates. Results for some of these experiments are summarized below.
Table I below shows the relationship between the processing parameters of power and time for copper. Table II below shows some results for copper and other metals with particle sizes as determined by light scattering and XRD line broadening. Some of the latter observations were confirmed by TEM as well. With power at 100 W, and processing time of 9 minutes, the predominant phase of copper by XRD is Cu_2O, although there was a small amount of copper metal present (~20%). Increasing the average power led to formation of copper metal. However, after 60 minutes with the beam on, copper was present, but now agglomerated. The average agglomerate size was 1-3 microns, with an average crystallite size of 500 nm. Further increasing the power output, while decreasing the time and concentration, also produced copper metal. In this case, (power = 220 W, time =10 minutes) the average crystallite size dropped to 150-200 nm. Increasing the power further to 330 W produced copper with an average crystallite size of 80-90 nm. Further increases in power and shorter

Table I. Conventional and Millimeter-Wave Polyol Processes for Copper

Copper Acetate concentration (M)	*Time (min)*	*Gyrotron Power (W)*	*Results*
.215	9	100	Cu_2O, Cu
.215	60	120*	Cu
.2	8	186	Cu
.1	7	220	Cu
.1	10	220	Cu
.1	15	330	Cu
.1	20	688	Cu
.1	0.15	2000#	Cu
		Conventional	
.2	60	Heating Mantle	Cu_2O
0.1	15	Heating Mantle	Copper Acetate

*average power #continuous process

Table II. Nanophase Metals Made by Batch Millimeter-Wave Polyol

Metal Precursor	Process Parameters	Observations
Copper Acetate	60 min., 120W	1-3 μm agglomerates, 200 nm crystallites
Copper Acetate	10 min., 220W	80-100 nm particles
Copper Acetate	20 min., 688W	10-20 nm particles
Copper Acetate	8 min., 186W	20-60 nm particles
Copper Acetate	1.5 min., 186W	1-5 nm crystallites
Cobalt Acetate	40 min., 750W	50-60 nm particles
Gold Chloride	10 min., 750W	30-40 nm particles
Nickel Acetate	20 min., 688W	10-20 nm particles

processing times led to even small particle sizes. At the highest powers and shortest times in the batch process, copper particles on the order of 10 nm were produced. These are shown in Fig. 4 below. These results and the comparison with conventional polyol batch processing with a heating mantle agree with the usual model for the reduction process, in that shorter process times should produce smaller particle size. An interesting comparison here is that copper acetate reduced using a conventional heating mantle requires more than an hour processing time under reflux conditions, while the process in the millimeter-wave batch system can be done in a few minutes (in the continuous process, this

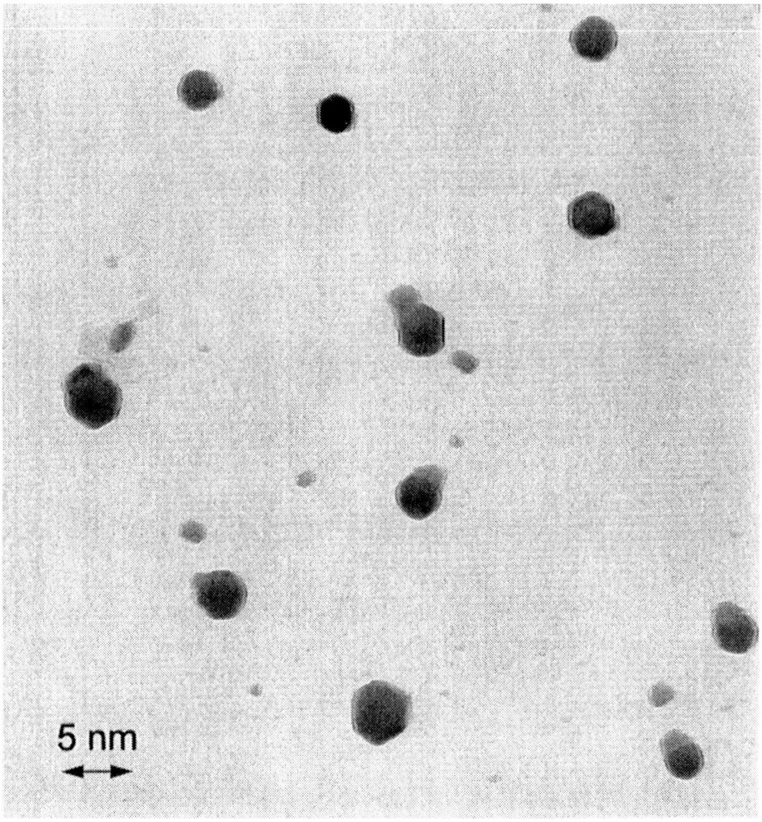

Fig. 4. Nanophase copper produced by millimeter-wave polyol process; particles show faceting consistent with growth of crystalline particles

only take about 10 seconds). One hour of refluxing under conventional conditions leads to Cu_2O (14) and when the solution is refluxed for only 15 minutes, just copper acetate is present (9). As noted above the shorter times should result in smaller crystallite and particle size, and, in the case of the continuous process, probably a reduction in crystallite and particle size variability. In general, it would be expected that a steady-state continuous process should generate products of more consistent quality (more uniform particle size and particle size distributions) than batch processes.

CONCLUSIONS AND SUMMARY

The millimeter wave and microwave driven polyol approaches have been used to synthesize metallic nanocrystalline powders and coatings. The results of this study show that there are two promising trends in processing with millimeter waves and microwaves. First, is that an increase in power decreases the time needed to reduce the metallic precursor, and also leads to a decrease in particle size. Secondly, that the millimeter wave process produce materials with smaller crystallite sizes than those prepared by the conventional thermal method, at the same processing times. Preliminary experiments with an S-Band (2.45 GHz) continuous microwave-driven polyol process have demonstrated that materials comparable to those of the conventional polyol, batch and continuous millimeter-wave processes can be produced in much larger quantities in the continuous S-Band system. This latter system also has very great advantages in terms of system capital cost and scaleability. The present laboratory-size system S-Band system should be capable of producing nanophase metal powders in quantities approaching 1 kg/day, for a capital cost of only $20k. These results suggest that a continuous microwave system should be capable of producing sizeable quantities of nanophase metal and metal oxide powders at reasonable cost, potentially with finer particle size, if desirable, and with better particle size control than current conventional batch polyol processes. A millimeter-wave system, though of higher capital cost, could achieve even higher powers and shorter reactions times, and might produce the finest particle size.

ACKNOWLEDGEMENTS

This work was supported by an Office of Naval Research Phase II SBIR grant and by Naval Research Laboratory Plasma Physics and Materials Science Divisions 6.1 research programs.

REFERENCES

1. L. K. Kurihara, G. M. Chow, D. Lewis, B. Bender, M. I. Baraton, and P. E. Schoen, in press *J. Am. Ceram. Soc.*, 2003
2. D. Lewis, M. A. Imam, L. K. Kurihara, A. W. Fliflet, A. Kinkead, S. Miserendino, S. Egorov, R. W. Bruce, S. Gold and A. M. Jung, Mater. and Mfg. Proc. 18 [2] 151-167 (203).
3. D. Lewis, R. W. Bruce, M. Kahn and A. W. Fliflet, Proc. 13th Int. Fed. for Heat Treatment and Surface Engineering Congress, ASM, April 2003.
4. G. M. Chow, L. K. Kurihara, and P. E. Schoen, U.S. Patent 5,759,230(1998).

5. D. Lewis, R. W. Bruce, S. Gold, A. W. Fliflet, "Microwave-Assisted Continous Synthesis of Nanocrystalline Powders and Coatings Using the Polyol Process," Patent Application, Navy Case No. 83,975, 2003.
6. D. Lewis, A. W. Fliflet, and R. W. Bruce, "Development of a High Frequency Microwave-Based Method and System for the Environmentally Benign Removal of Coatings," Patent Application, Navy Case No. 83,976, 2002.
7. R. W. Bruce, D. Lewis, A. W. Fliflet and M. Kahn, "Microwave-Assisted Reactive Brazing of Ceramics, Patent Application, Navy Case No. 83977, 2003.
8. P. Tonguzzo, G. Viau, O. Archer, F. Guillet, E. Bruneton. *J. Mater. Sci.*, **35**(2000).
9. P. Tonguzzo, G. Viau, O. Archer, F. Fievet-Vincent, and F. Fievet, *Adv. Mater.* **10**, 1032 (1998)
10. H. -O. Jungk and C. Feldmann, *J. Mater. Res.* **15**, 2244(2000).
11. J. Merikhi, H. -O. Jungk, and C. Feldmann, *J. Mater. Chem.* **10**, 1311(2000).
12. P. Tonguzzo, O. Archer, G. Viau, A. Pierrard, F. Fievet-Vincent, F. Fievet and I. Rosenman, *IEEE Trans. Magn.* **35**, 3469(1999).
13. L. K. Kurihara, G. M. Chow and P. E. Schoen, *Nanostruct. Mater.* **5**, 607(1995).
14. G. M. Chow, L. K. Kurihara, K. M. Kemner, P. E. Schoen, W. T. Elam, A. Ervin, S. Keller, Y. D. Zhang, J. Budnick, and T. Ambrose, *J. Mater. Res.* 10, 1546(1995).
15. L. K. Kurihara, D. Lewis, A. M. Jung, A. W. Fliflet, unpublished results 2000.
16. D. Lewis, R. J. Rayne, B. A. Bender, L. K. Kurihara et al. *Nanostruct Mater,* **9** , 97(1997)
17. L. K. Kurihara, G. M. Chow, S. H. Lawrence, and P. E. Schoen, Processing and Properties of Nanocrystalline Materials, C. Suryanarayana, J. Singh and F. H. Froes, Eds., TMS Minerals, Metals, Materials, Warrendale PA **49**,235 (1996).
18. S. Komarneni, D. Li, B. Newalkar, H. Katsuki and A. S. Bhalla, *Langmuir* **2002**, 18, 5959-5962
19. D. Larcher and R. Patrice, J. Solid State Chem. 154, 405-411, 2000.

Chapter 9

Bioinspired Organic–Inorganic Hybrid Devices

Lawrence L. Brott[1], Rajesh R. Naik[1], Sean M. Kirkpatrick[1], Patrick W. Whitlock[2], Stephen J. Clarson[2], and Morley O. Stone[1]

[1]Materials and Manufacturing Directorate, Air Force Research Laboratory, 3005 P Street, Wright-Patterson Air Force Base, OH 45433–7702
[2]Department of Materials Science and Engineering, University of Cincinnati, 497 Rhodes Hall, Cincinnati, OH 45221–0012

In order to achieve the high amount of index of refraction mismatch necessary for the fabrication of a photonic bandgap device, a highly ordered hybrid material of organic and inorganic compounds must be developed. Through the identification of peptides from the diatom *Cylindrotheca fusiformis*, simple silica nanospheres can now be synthesized from silanes under physiological conditions. By incorporating these peptides into a monomer formulation, peptide-rich regions can be created on the polymer surface using a holographic two-photon induced photopolymerization process. After exposing the cured polymer to a silane precursor, silica nanospheres are embedded in the peptide-rich regions resulting in a highly ordered two-dimensional array of silica spheres on the polymer backing. The diffraction efficiency of these devices increases nearly fifty-fold when compared to a polymer hologram without the silica spheres.

With the recent growth in interest in the area of nanotechnology, a great deal of interdisciplinary research has focused upon the ability to create nanometer size-scale devices for potential applications in biomedicine, electronics and optics. Numerous cases of nanopatterning and nanostructure are commonly found in nature, with some notable examples appearing in the marine diatoms and sponges (*1,2*). These simple organisms are able to form nano- and micro-structured components with precise control using proteins. It is humbling to realize that what these single-cell organisms accomplish so elegantly, requires extreme laboratory conditions to duplicate even the simplest of structures (*3,4*). Kröger *et al.* identified a set of polycationic peptides (referred to as silaffins) from the diatom *Cylindrotheca fusiformis* (*5*). The silaffins, when added to a hydrolyzed silane precursor, catalyze the formation of simple silica nanospheres at neutral pH and ambient temperatures and pressures (*6*). The R5 peptide, a short 19 amino acid (SSKKSGSYSGSKGSKRRIL) repeat unit of the silaffin protein, is also able to catalyze silica precipitation (*5*). Here we describe our work to understand the process of biosilification through a study of the reaction conditions by varying the reaction time and pH. This knowledge was then applied to the fabrication of a practical optical device by incorporating the peptide into a polymer hologram to produce a novel hybrid organic/inorganic ordered nanostructure. Continuing work is also presented on potential ways to enhance the properties of this device by modifying the shape of the silica nanostructure from a sphere to a more complex morphology by replacing the static reaction conditions to a more dynamic one.

The R5 peptide based on the published sequence was chemically synthesized and used in the research presented here. Using the reaction conditions previously described (*5*), the R5 peptide (100 μg/mL)was dissolved in a sodium phosphate-citrate buffer 7.0 pH and added to a 0.1M tetrahydroxysilane solution to form silica spheres with a diameter of 400-700 nm within 10 minutes (Figure 1). In the absence of R5 peptide, the tetrahydroxysilane solution remains stable for hours before it slowly converts to a clear amorphous gel. We determined the relevant kinetic parameters (time and pH) in order to maximize the rate of formation of silica spheres. The UV-Vis spectrum obtained using the silica spheres synthesized by R5 peptide has been previously shown experimentally to absorb strongly at around 290 nm. In the kinetic experiments, we monitored the formation of silica spheres by measuring the absorbance at 290 nm. From the results in Figure 2, it is evident that the formation of the silica spheres was essentially complete in 10 minutes and the activity of the R5 peptide is maximal at pH 8.0. Little or no silica precipitation is observed at pH below 7. This is consistent with the finding of Kröger *et al.* requiring lysine modification for activity at acidic pH (*5*).

The ability to use peptides to catalyze the formation of inorganic material offers the possibility of spatially controlling the deposition of the

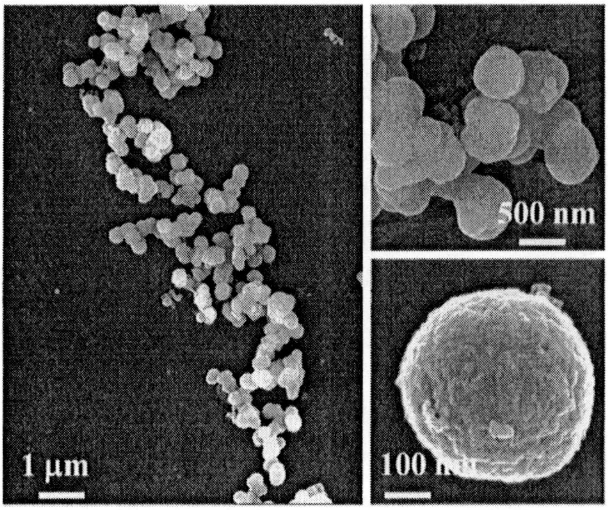

Figure 1. Scanning electron micrographs (SEM) of spherical biosilica structures obtained by reacting the R5 peptide with tetrahydroxysilane in solution.

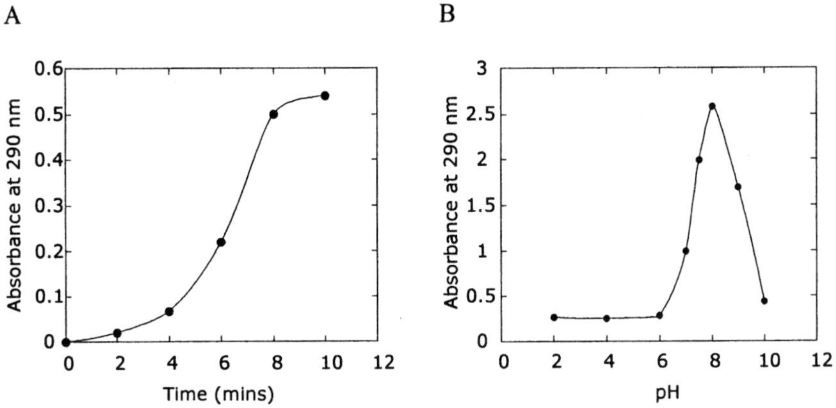

Figure 2. The rate of silica precipitation using the R5 peptide as a function of (A) time and (B) pH.

inorganic material. In the recent past, a number of techniques have been exploited for the deposition and patterning of peptides (*7,8*). Using a polymer hologram created by a holographic two-photon induced polymerization (H-TPIP) process to spatially control the deposition within a polymer matrix, we were able to create a hybrid organic/inorganic ordered nanostructure (*9,10*). The advantage of using a two-photon polymerization process to cure a peptide-containing monomer matrix is that the ultrafast infrared laser used in this process typically does not alter the biological activity of the incorporated protein. It was also expected that during the polymerization process, the peptide would be segregated into regions of low crosslinking density. The approach of using ultraviolet lasers to phase separate small liquid crystal molecules in a polymer-based hologram has been used extensively (*11*) and it was reasoned that this technique was also applicable to the H-TPIP process as well. Furthermore, by exposing the peptide-containing hologram to the liquid silicic acid, it was theorized that silica would form in the holographic nanopattern. A water miscible monomer formulation was created by combining two poly(ethylene glycol)-based tri- and penta-acrylates with triethanol amine and isopropylthioxanthone as initiators (*12,13*). The peptide was added to this formulation, which was then spin coated onto a glass slide. The sample was cured under nitrogen in a two-beam transmission holographic arrangement using a 790 nm titanium-sapphire laser. Because certain areas of the sample cure more rapidly than others, the smaller molecules (namely water and the peptide) phase separate from the areas of higher crosslink density and migrate into areas of lower density. As a result, peptide rich domains are created in the polymer sample with the periodicity of the hologram. After the curing process, the sample was briefly rinsed with water to remove any uncured monomer.

Freshly prepared hydrolyzed silane was applied to the hologram and allowed to react with the R5 peptide embedded in the hologram for 10 minutes before being rinsed with water to remove any unreacted silane. A study of this hologram by scanning electron microscopy revealed that silica spheres formed a regular two-dimensional array with the periodicity of the hologram (see Figure 3). A study of the size distribution of the silica spheres reveals that the average nanosphere diameter is 452 nm (± 81 nm) and the periodicity of the hologram is 1.60 μm. As a result, this photonic device exhibits a nearly fifty-fold increase in diffraction efficiency over a comparable polymer hologram without silica. The untreated grating exhibited a diffraction efficiency of approximately 0.02%, while the grating with the silica spheres showed an efficiency of approximately 0.95%. This large increase can be attributed to the fact that the spheres form an almost continuous line of silica in the hologram, achieving a high fill factor.

While these results are promising, further improvements upon this photonic device would include the replacement of the individual silica spheres with a long, continuous region of silica. Work has begun to realize this goal

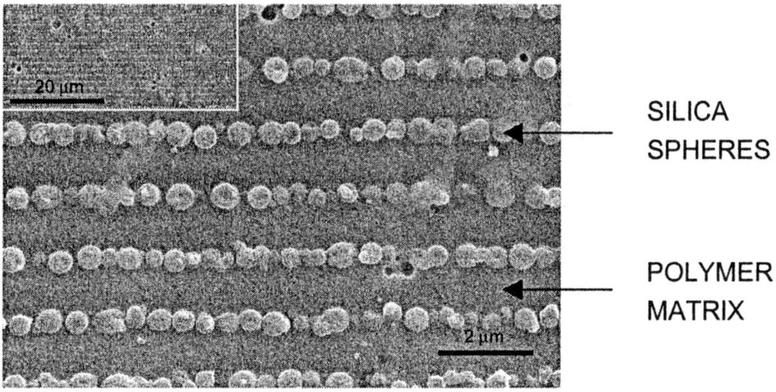

Figure 3. Two-dimensional array of ordered silica nanospheres formed by reacting the R5-embedded hologram to the silane.

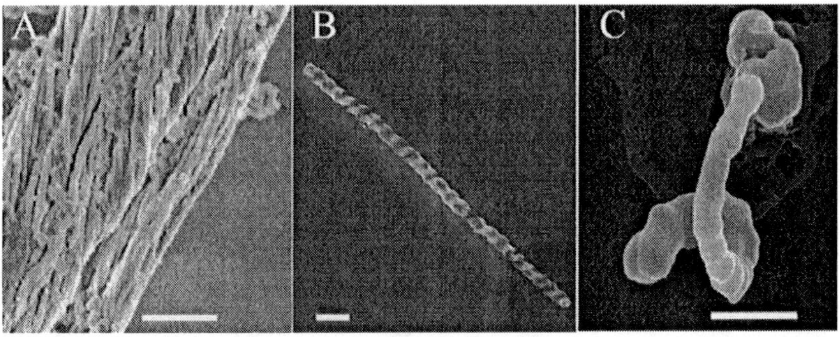

Figure 4. Scanning electron micrographs of fibrillar arch-shaped morphologies obtained by (A,B) applying a mechanical shear or (C) bubbling nitrogen gas through a solution of the R5 peptide and tetrahydroxysilane. Scale bar equal 1 µm.

through the careful manipulation of the physical reaction environment, in particular the microfluidic environment, during the R5/silicic acid reaction. Preliminary results indicate that this can be achieved by either bubbling nitrogen gas through the reaction vessel or applying mechanical shear to the nucleating sites. Silica produced in the presence of bubbling nitrogen results in the formation of arched, elongated structures that resembles rods (see Figure 4c). However, biosilification in the presence of a shear flow in a linear environment results in fibrillar-like structures present throughout the sample with a complete absence of the spherical structures (see Figure 4a).

These preliminary results are encouraging and research continues on applying these new post-processing techniques to create rod-shaped silica nanostructures on the surface of the peptide-containing hologram. Such materials will have considerable potential for use in photonic devices since these techniques combine the ease of processability of an organic polymer with the improved mechanical and optical properties of an inorganic to produce a hybrid system with properties not possible from each structure alone. Additionally, the holographic phase separation technique is universally applicable for any catalyst or binding agent that can be incorporated into a polymer. For example, as different catalysts are identified, a wide variety of unique hybrid structures will be possible with differing shapes and mechanical properties. Consequently, the promise of this technique is that it allows a simple yet general and easily modifiable method for nanopatterning.

Acknowledgements

This research was supported by the Air Force Office of Scientific Research.

References

1. Round, R.E.; Crawford, R.M.; Mann, D.G. *The Diatoms: Biology and Morphology of the Genera*; Cambridge University Press: Cambridge, 1990.
2. Volcani, B.E. "Cell Wall Formation in Diatoms: Morphogenesis and Biochemistry," in *Silicon and Siliceous Structures in Biologic Systems*; Springer-Verlag: New York, 1981.
3. Parkinson, J.; Gordon, R. *Trends in Biotechnology*, **1999**, *17*, 190-196.
4. Morse, D.E. *Trends in Biotechnology*, **1999**, *17*, 230-232.
5. Kröger, N.; Deutzmann, R.; Sumper, M. *Science*, **1999**, *286*, 1129-1132.

6. Cha, J.N.; Stucky, G.D.; Morse, D.E.; Deming, T.J. *Nature*, **2000**, *403*, 289-292.
7. Berggren, K.K.; Bard, A.; Wilbur, J.L.; Gallaspy, J.D.; Helg, A.G.; McClelland, J.J.; Rolston, S.L.; Phillips, W.D.; Prentiss, M.; Whitesides, G. *Science*, **2000**, *260*, 1255-1257
8. Wilson, D.L.; Martin, R.; Hong, S.; Cronin-Golomb, M.; Mirkin, C.A.; Kaplan, D.L. *Proc. Natl. Acad. Sci. USA*, **2001**, *98*, 13660-13664.
9. Brott, L.L.; Naik, R.R.; Pikas, D.J.; Kirkpatrick, S.M.; Tomlin, D.W.; Whitlock, P.W.; Clarson, S.J.; Stone, M.O. *Nature*, **2001**, *413*, 291-293.
10. Kirkpatrick, S.M.; Baur, J.W.; Clark, C.M.; Denny, L.R.; Tomlin, D.W.; Reinhardt, B.A.; Kannan, R.; Stone, M.O. *Appl. Phys.*, **1999**, *69*, 461-464.
11. Bunning, T.J.; Natarajan, L.V.; Tondiglia, V.; Sutherland, R.L.; Vezie, D.L.; Adams, W.W. *Polymer*, **1995**, *36(14)*, 2699-2708.
12. Belfield, K.D.; Schafer, K.J.; Liu, Y.; Liu, J.; Ren, X.; Van Stryland, E.W. *J. Phys. Org. Chem.*, **2000**, *13*, 837-849.
13. Brott, L.L.; Naik, R.R.; Kirkpatrick, S.M.; Pikas, D.J.; Stone, M.O. *Polymer Preprints*, **2001**, *42(1)*, 675-676.

Chapter 10

Decontamination of Chemical Warfare Agents with Nanosize Metal Oxides

George W. Wagner[1,*], Lawrence R. Procell[1], Olga B. Koper[2], and Kenneth J. Klabunde[2]

[1]Research and Technology Directorate, U.S. Army Edgewood Chemical Biological Center, Aberdeen Proving Ground, MD 21010
[2]Department of Chemistry, Kansas State University, and Nanoscale Materials, Inc., 1500 Hayes Drive, Manhattan, KS 66502

Chemical warfare agents VX, GB, GD, and HD undergo room temperature reactions with nanosize MgO, CaO, and Al_2O_3. For HD on CaO, a rather fast autocatalytic dehydrohalogenation is observed yielding up to 80% divinyl sulfide (DVS) and 20% thiodiglycol (TG). On MgO, no catalytic reaction is observed for HD, and a slower reaction yields about 50% DVS and 50% TG. A quite fast hydrolysis is observed for HD at lower loadings on Al_2O_3 to yield TG (71%) and a minor amount of the CH-TG sulfonium ion (12%); a small amount of DVS is also observed (17%). At high loadings, the reactive capacity of Al_2O_3 for HD is exceeded where complete conversion to TG is hindered. However, addition of excess water results in the quantitative conversion of sorbed HD to CH-TG. On Al_2O_3 dried to remove physisorbed water a surface alkoxide consistent with that of TG is observed. VX, GB and GD simply hydrolyze on nanosize oxides to form surface-bound phosphonates. At sufficiently high loadings metal phosphonates form.

Introduction

Inorganic oxides are currently being considered as reactive sorbents for the decontaminantion of chemical warfare agents (CWA).[1] These versatile materials, well-known for their industrial uses as adsorbents, catalysts and catalyst supports, have many potential decontamination applications such as environmentally-friendly "hasty" decontamination on the battlefield, protective filtration systems for vehicles, aircraft and buildings, and the demilitarization of CWA munitions and stockpiles. To stress the enormity of the latter problem, the US and Russia have declared CWA stockpiles of 25,000 and 42,000 tons, respectively.[1b]

Although conventional MgO,[2] and other inorganic oxides such as alumina,[3] possess reactivity towards CWA simulants, such reactivity is anticipated to be enhanced in nanosize particles.[4] The increased reactivity is due not only to the larger surface area of smaller particles, but also to the greater amount of highly-reactive edge and corner "defect" sites. Additionally, unusual lattice planes, also possessing greater reactivity, are stabilized and exposed in nanosize particles.

In the work described herein, solid state Magic Angle Spinning (MAS) NMR is employed to examine the room temperature reactions of neat VX, GB, GD, and HD liquid with nanosize MgO (AP-MgO), CaO (AP-CaO) and Al_2O_3 (AP-Al_2O_3). More complete accounts of this work have been published previously.[5-7] VX, GB (Sarin),and GD (Soman) are "nerve" agents, whereas HD (mustard) is a "blister" agent.[1a] Multinuclear MAS NMR is employed to comprehensively characterize the reactions. In particular, ^{27}Al MAS NMR is utilized to confirm the formation of discreet aluminophosphonate compounds on AP-Al_2O_3. Indeed, ^{31}P MAS NMR suggests the formation of corresponding metal phosphonates on all three metal oxides.[5-7]

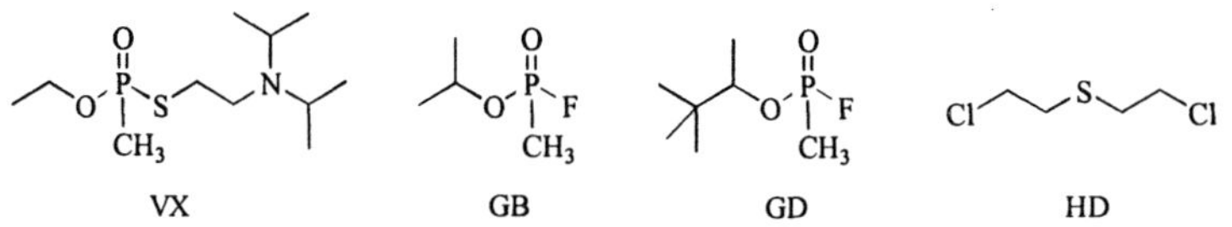

Figure 1. Structures of chemical warfare agents examined in this study.

Surface Bound Products

Using IR, Kuiper et al.[8] and Yates et al.[9] have detected alumina surface-bound isopropyl methylphosphonate (the hydrolysis product of GB) and 2-

hydroxyethyl ethyl sulfide (the hydrolysis product of 2-chloroethyl ethyl sulfide, CEES, a common HD simulant), respectively. As discussed below, MAS NMR also detects these species. For VX, GB, and GD, broadened ^{31}P MAS NMR peaks, indicative of restricted mobility, are detected for the phosphonate products. Moreover, spinning sidebands consistent with the expected large chemical shift anisotropy (CSA)[10] of these species, which is not averaged by motion,[11] are also evident. For HD, broadened ^{13}C MAS NMR peaks are also found for the TG product, again indicative of surface immobilization. ^{27}Al, in addition to ^{31}P and ^{13}C, MAS NMR is employed to better characterize the surface complexes of these products on AP-Al_2O_3. For the nerve agents, spectra of the surface-bound phosphonate products are compared to those of synthesized aluminophosphate model compounds, which are prepared by simple precipitation reactions.[12] Hypothetical structures for these latter compounds are shown in Figure 2. For comparison, proposed structures[13] of five types of surface hydroxyl groups are also given in Figure 2. Type Ia is the most basic and type III is the most acidic. ^{27}Al NMR[14] readily distinguishes tetrahedral sites and octahedral sites which possess shifts of ca. 50 and 0 ppm, respectively; thus the coordination of various aluminum species is readily apparent. Additionally, ^{27}Al Cross-Polarization (CP) MAS NMR[15] selectively probes aluminum species in proximity to protons, i.e. surface aluminum sites and aluminophosphonate species.

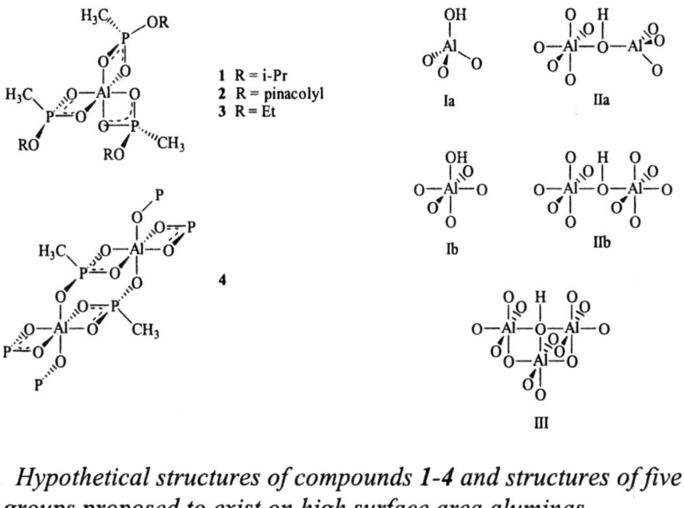

Figure 2. Hypothetical structures of compounds 1-4 and structures of five hydroxyl groups proposed to exist on high surface area aluminas.

Experimental

Materials

AP-MgO, AP-CaO, and AP-Al_2O_3 were synthesized from alkoxide precursors as previously described.[5-7] The nanosize oxides were in agglomerated form. The porosity of these materials has not been thoroughly investigated. Surface areas for AP-MgO and AP-Al_2O_3 were 344 m^2/g and 273.6 m^2/g, respectively. AP-CaO possesses typical surface areas of 100 – 160 m^2/g.[4a] The commercial γ-alumina is Selexsorb CDX (Alcoa), and has a surface area of 400 m^2/g. AP-MgO and AP-CaO were used as received. The aluminas were used as received, dried in air at 100 °C to remove physisorbed water, and dried in air at 400 °C to yield partially-dehydroxylated aluminas (PDA). ^{13}C labeled HD was used for low-loading reactions followed by ^{13}C MAS NMR. Unlabeled HD was used for high-loading reactions.

Aluminophosphonate Syntheses

Aluminophosphonate model compounds were synthesized by mixing appropriate molar ratios of phosphonic acid and $Al(NO_3)_3$•$9H_2O$ (all from Aldrich) in water. $Al[OP(O)(CH_3)(OiPr)]_3$ (**1**) and $Al[OP(O)(CH_3)$-(OPinacolyl)$]_3$ (**2**) spontaneously precipitated and were filtered and dried under vacuum at room temperature to yield dry, white powders. $Al[OP(O)(CH_3)$-(OEt)$]_3$ (**3**) remained soluble, and attempts to precipitate it by raising the pH (NaOH) resulted in the predominant precipitation of aluminum hydroxide. Thus this compound, which is extremely hygroscopic, was isolated by drying under vacuum at 80°C to yield a dry, white powder. $Al_2[OP(O)(CH_3)(O)]_3$ (**4**), which also did not spontaneously precipitate, was isolated by adjusting the pH to 3 with NaOH to effect precipitation, followed by drying in air at 70°C to yield a dry, white powder. Compounds **1**, **2**, and **4** are not hygroscopic. The compounds were analyzed by ^{31}P and ^{27}Al MAS NMR, which showed minor impurities of IMPA, PMPA, and MPA in **1**, **2**, and **4** respectively.

NMR

^{31}P, ^{13}C, and ^{27}Al MAS NMR spectra were obtained at 7 T using a Varian Unityplus 300 NMR spectrometer equipped with a Doty Scientific 7 mm high-speed VT-MAS probe. Double O-ring sealed rotors (Doty Scientific) were used. Both direct polarization (DP) and cross-polarization (CP) spectra were obtained. For ^{27}Al spectra, the 90° pulse width was determined using 0.1M $Al(NO_3)_3·9H_2O$, which also served as the external shift reference. Quantitative spectra were obtained using 0.2 μsec (3°) pulses.[14a] For ^{1}H, ^{13}C and, ^{31}P, external TMS and 85% H_3PO_4 were used as shift references.

Reaction Procedure

Caution: These experiments should only be performed by trained personnel using applicable safety procedures. In a typical kinetic run, 5 wt % neat, liquid agent (3-9 μL) was added via syringe to the center of a column of metal oxide (80-200 mg) contained in the NMR rotor. The rotor was then sealed with the double O-ring cap. MAS NMR spectra were obtained periodically to monitor the reaction in situ. Heavily-loaded samples of AP-Al_2O_3 were prepared by adding 250 μmol neat, liquid agent (30-70 μL) to 100 mg contained in a glass vial. The vial was capped and allowed to stand at room temperature for an extended period.

Results and Discussion

Low-Loading Reactions

Selected ^{31}P and ^{13}C MAS NMR spectra obtained for the reactions of 5 wt % VX, GD, and HD (^{13}C labeled) with "as received" AP-Al_2O_3 are shown in Figure 2. Quite similar spectra are obtained for these agents on AP-MgO and AP-CaO. Discussions of the spectral features observed for each agent follow.

VX

On all three oxides, VX (sharp peak at 52.0 ppm in Figure 3) exhibits selective hydrolysis (see Scheme 1) for ethyl methylphosphic acid (EMPA, broad peak at 23 ppm). As mentioned in the Introduction, the EMPA product binds to the oxide surface as shown in Scheme 1 for the alumina surface. In the reaction with the nanosize oxides, no toxic EA-2192 is observed, which would yield a peak near 40 ppm.[16] This selectivity for EMPA has been previously observed on numerous inorganic oxides,[2] including zeolites.[17] Such selectivity for VX hydrolysis is also shown in the slow reaction of VX with an equimolar amount of water.[18] As shown in Scheme 1, EMPA is also prone to secondary hydrolysis to methylphosphonic acid (MPA) and this has been observed at high VX loadings on AP-Al_2O_3 (see below). Also at high loadings on AP-Al_2O_3, and on AP-MgO[5] and AP-CaO,[6] multiple phosphonate species are present consistent with the secondary formation of metal phosphonates (discussed below).

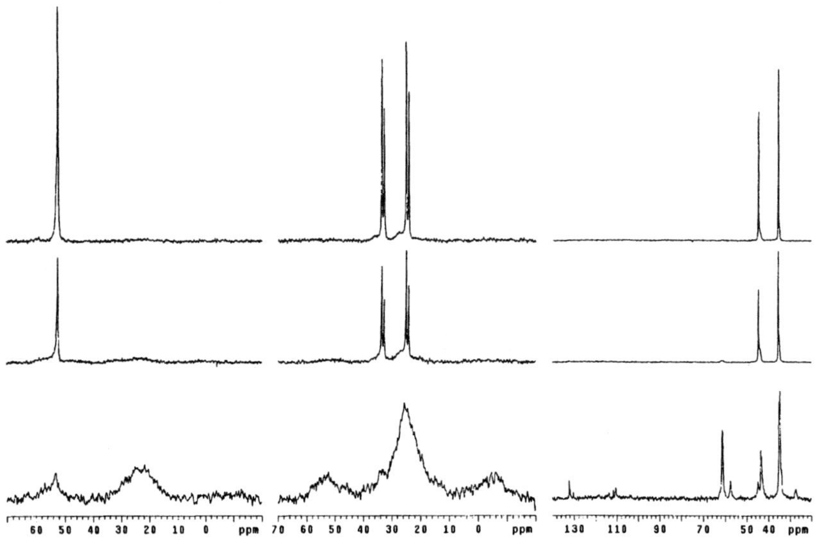

Figure 3. Selected MAS NMR spectra obtained for the reactions of 5 wt % VX, GD, and HD with "as received" Al_2O_3.

GD

GD (sharp doublets centered near 29 ppm, Figure 3) readily hydrolyzes (see Scheme 1) to pinacolyl methylphosphonic acid (PMPA, broad peak at 24 ppm) which binds to the oxide surface as shown in Scheme 1. Additionally, on AP-MgO[5] and AP-CaO[6] a second, minor product peak appears near 20 ppm. Based on the results for high-loading GD on AP-Al_2O_3 (see below) this second product is assigned to the corresponding metal pinacolyl methylphosphonates rather than MPA since PMPA is extremely resistant to hydrolysis.

HD

HD (sharp peaks at 44.6 and 35.4 ppm, Figure 3) exhibits both hydrolysis to thiodiglycol (TG, peaks at 61.2 and 34.5 ppm) and the CH-TG sulfonium ion (62.1, 57.6, 44.6, 34.5, and 27.2 ppm)[19] and elimination of HCl to yield 2-chloroethyl vinyl sulfide (CEVS), and divinyl sulfide (DVS). These reactions are shown in Scheme 1. For AP-Al_2O_3 and AP-MgO, the ratio of hydrolysis to elimination is 83:17 and 50:50, respectively. However, on AP-CaO the ratio is 20:80. For this reaction, an induction period is observed during which slow hydrolysis primarily occurs. This is then followed by a rather fast elimination reaction. Thus the reaction appears autocatalytic, with the elimination mechanism operative during the fast reaction.

Scheme 1

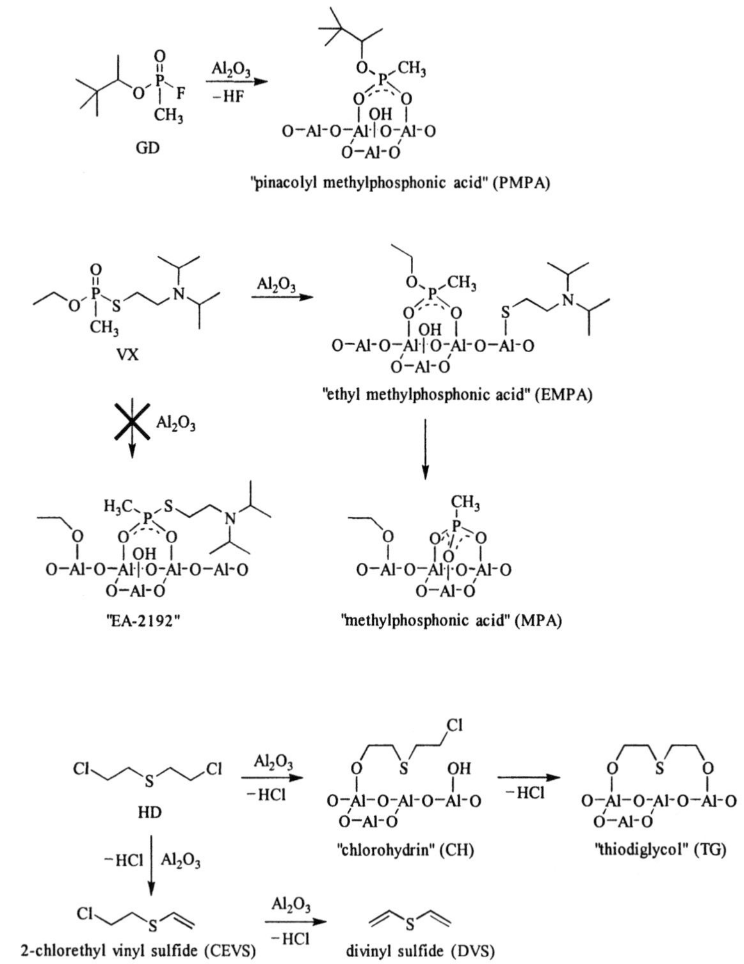

Low-Loading Reaction Kinetics

The reaction profiles for the low-loading reactions are shown in Figure 4. Generally, the kinetics are characterized by an initial fast reaction followed by slower, diffusion-limited reactions exhibiting first-order behavior. Half-lives for the steady-state reactions are indicated in Figure 4. For VX the half-lives were quite similar after a few days, but the reaction continued to slow and the true steady-state reaction was not reached even after 1 week. The fast reactions are attributed to facile liquid spreading through the pores across fresh, unreacted surface. Spreading stops once the liquid achieves its volume in the pores; the reaction ceases, too, once the surface is consumed with product. The mechanism by which fresh surface is reached is via evaporation and gas-phase diffusion. Consistent with this latter mechanism, the diffusion-limited half-lives observed for GD, HD, and VX are in relative agreement with their vapor pressures: 0.4, 0.072, and 0.0007 mmHg, respectively. However, for HD on AP-CaO, after a brief induction period, the reaction rate actually *increases* with time, suggesting the reaction is autocatalytic. A reasonable explanation for this behavior is as follows. Slow hydrolysis occurs during the induction period to generate HCl. Liberated HCl then reacts with CaO to generate $CaCl_2$ "islands". $CaCl_2$ is a more active elimination catalyst than CaO;[20] thus facile elimination of the remaining HD may ensue.

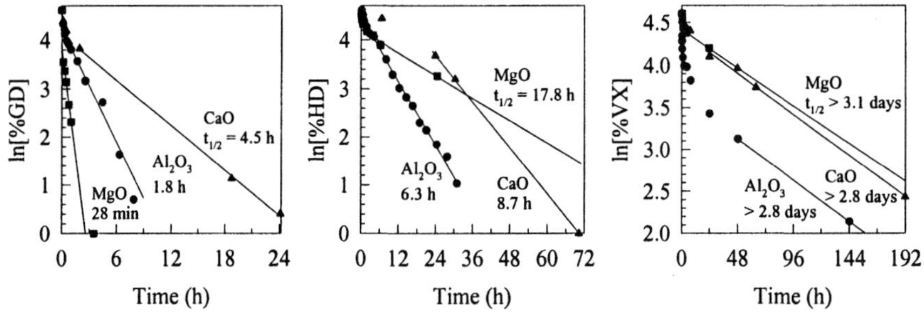

Figure 4. Reaction profiles for 5 wt % GD, HD, and VX with nanosize metal oxides.

Comparing the various metal oxides, the reaction of GD on AP-MgO is remarkable as a quite short half-life of 28 min is observed. However, AP-MgO also yields the slowest reaction for HD. The reaction profiles for VX are nearly identical on AP-MgO and AP-CaO. AP-CaO possesses the slowest reaction for GD, whereas the rate of autocatalytic HD elimination is comparable to the rate of HD hydrolysis on AP-Al_2O_3. It is important to note that the divinyl sulfide elimination product is problematic in that it retains irritant properties;[21] thus hydrolysis is the preferred HD decontamination mechanism and AP-Al_2O_3 is the oxide most selective for this reaction. Although AP-Al_2O_3 yields a somewhat slower reaction for GD than AP-MgO, it provides for relatively fast HD hydrolysis and, furthermore, significantly more VX reacts during the initial fast reaction than on the other nanosize oxides. Thus AP-Al_2O_3 is judged to exhibit the best overall reactivity for the agents.

High-Loading Agent Reactions with AP-Al_2O_3

High loadings of GB, GD, VX and HD were employed to facilitate the use of ^{27}Al MAS NMR to characterize the potential surface complexes of these products on AP-Al_2O_3. "As received" material was employed for these reactions as water is required for complete reaction (see below). For each sample 250 μmol agent was reacted with 100 mg AP-Al_2O_3, nominally 980 μmol (ignoring water weight); thus the molar loading was kept constant. Model compounds of the phosphonic acid surface complexes, i.e. aluminophosphonates $Al[OP(O)(CH_3)(OR)]_3$ [R = i-Pr (**1**), pinacolyl (**2**), and Et (**3**)] and $Al_2[OP(O)(CH_3)(O)]_3$ (**4**) were also prepared and characterized by MAS NMR to compare with the surface complexes.

HD

^{13}C and ^{27}Al MAS and ^{27}Al CP-MAS spectra were obtained for 28 wt % HD (unlabeled) reacted with "as received" AP-Al_2O_3 (spectra not shown). Such a high loading apparently overwhelms the reactive capacity of AP-Al_2O_3. After 51 days, little overall conversion of HD to products occurred, and no unusual features were present in the ^{27}Al MAS NMR spectra. Thus, no evidence of a surface alkoxide (Scheme 1) was found. Such species yield resonances between 70 to 80 ppm in ^{13}C MAS NMR spectra.[22] If formed, alkoxides would presumably undergo facile hydrolysis and thus are not stable in the presence of water. Addition of 33 wt % H_2O to the sample effected hydrolysis of the HD to the H-2TG and CH-TG sulfonium ions[19] within 17 days. The H-2TG slowly decomposed to yield CH-TG as the exclusive product after 2.5 months (this behavior is remarkably similar to that of bulk HD droplets in water[19]). At this point, the ^{27}Al MAS NMR spectrum showed an extremely sharp peak at 1.3 ppm which is assigned to free, solvated $Al(H_2O)_6^{3+}$. This species was not observable by ^{27}Al CP-MAS NMR, consistent with its substantial motion. Formation of this species is important for the reaction of CW agents as their acidic products will react to the core of alumina particles, consuming them in stoichiometric fashion. Moreover, the ability of acid to generate free $Al(H_2O)_6^{3+}$ is a crucial step in the formation of aluminophosphonates from phosphonic acid products (see below).

A second sample was prepared containing 30 wt % HD on AP-Al_2O_3 dried at 100 °C to remove physisorbed water. ^{13}C and ^{27}Al DP and CP-MAS spectra were obtained for this sample. Similar to the "as received" material, the extent of reaction was not great, even after an extended period. After 4 months, ^{13}C MAS revealed mostly unreacted HD. However, ^{13}C CP-MAS NMR did detect some minor products: a broad peak at 62.1 ppm assignable to TG hydrogen bonded to the surface,[22] and a broad shoulder extending to 80 ppm consistent with the anticipated surface TG alkoxide.[9,22] ^{27}Al DP and CP-MAS spectra did not reveal any evidence of the surface alkoxide, however. Possible reasons for this include 1) too low a concentration to detect, 2) overlap of this signal with the surface peak, and/or 3) a large quadrupolar coupling constant (Q_{cc})[14a] for this species.

GB

In Figure 5, ^{31}P, ^{13}C, and ^{27}Al CP-MAS spectra obtained after 2 weeks for 26 wt % GB reacted with "as received" AP-Al_2O_3 are shown along with ^{31}P MAS and ^{13}C and ^{27}Al CP-MAS NMR spectra of **1**. In the ^{31}P CP-MAS NMR spectra of GB/AP-Al_2O_3, two broad peaks are observed at 22.5 and 14.4 ppm which possess intense spinning side bands, indicative of two distinct, immobilized phosphonates. Extraction of the surface species and analysis by ^{31}P

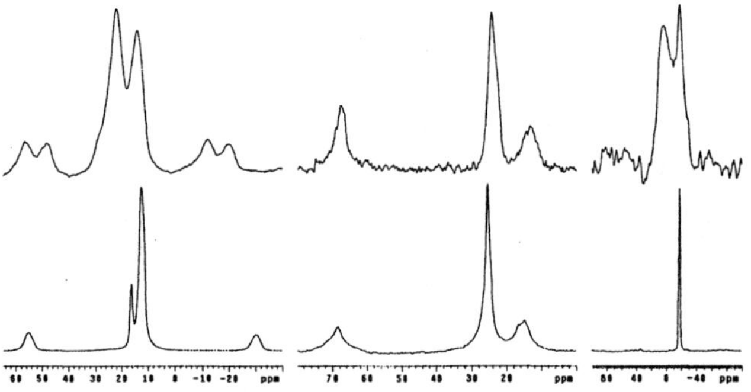

Figure 5. MAS NMR spectra obtained after 2 weeks for the reaction of 26 wt % GB with "as received" AP-Al_2O_3, top row (left to right): ^{31}P CP, ^{13}C CP, and ^{27}Al CP. MAS NMR spectra obtained for model compound 1, bottom row (left to right): ^{31}P, ^{13}C CP, and ^{27}Al CP.

NMR found only isopropyl methylphosphonic acid (IMPA), but no methyl phosphonic acid (MPA); a potential secondary hydrolysis product (Scheme 1). Thus the two resonances detected in the ^{31}P CP-MAS spectrum reflect IMPA in two different environments. The ^{13}C CP-MAS spectrum does not differentiate the two IMPA's, detecting only single P-OiPr and P-CH_3 groups. The ^{27}Al CP-MAS NMR spectrum reveals a new, sharp peak at –17.9 ppm not present in the pristine AP-Al_2O_3. Comparison of these spectra with those of model compound **1** reveals that **1** itself is forming on the AP-Al_2O_3 surface as the ^{31}P MAS and ^{27}Al CP-MAS NMR peaks of **1**, 12.6 and –17.6 ppm, respectively, are in close agreement with the peaks detected for one of the surface species (14.4 and –17.9 ppm). The IMPA surface complex giving rise to the ^{31}P CP-MAS NMR peak at 22.5 ppm is consistent with the types of surface complexes shown in Scheme 1, and its shift is in agreement with those of the EMPA and PMPA surface complexes observed for the low-loading VX and GD samples (23.0 and 24.0 ppm, respectively). No distinct ^{27}Al CP-MAS NMR peak is detected for this surface complex, and this is attributed to the potentially-large Q_{cc}[14a] for this species and/or its overlapping with the alumina surface O_h peak.

GD

^{31}P MAS and ^{13}C, and ^{27}Al CP-MAS spectra were obtained for 32 wt % GD reacted with "as received" AP-Al_2O_3 after 2 weeks, and compared to ^{31}P and ^{27}Al MAS and ^{13}C CP-MAS NMR spectra obtained for model compound **2**. Like GB, two broad peaks possessing intense spinning sidebands are observed for two phosphonate species in the ^{31}P MAS NMR spectrum at 23.4 and 14.3 ppm, and a new peak is observed in the ^{27}Al CP-MAS NMR spectrum at –12.2 ppm. Extraction of the surface species and analysis by ^{31}P NMR found only pinacolyl methylphosphonic acid (PMPA), but no MPA. Thus the resonances found for the two surface phosphonates are similarly assigned to surface-bound PMPA and the discreet aluminophosphonate complex **2**. The shifts of the GD-derived aluminophosphonate species are consistent with those of **2** (^{31}P: 14.3 ppm; ^{27}Al: –16.7 ppm), and the ^{31}P MAS NMR shift of the surface-bound PMPA is in agreement with the same species observed in the low-loading sample (24 ppm). As seen for GB, the ^{13}C CP-MAS NMR spectrum does not distinguish the two PMPA species, yielding only peaks for single O-pinacolyl and P-CH_3 groups. The surface-bound PMPA complex also does not yield a distinct ^{27}Al CP-MAS peak (see above).

VX

^{31}P MAS and ^{13}C, and ^{27}Al CP-MAS spectra were obtained after 7 weeks for 40 wt % VX reacted with "as received" AP-Al_2O_3. For VX, the ^{31}P MAS NMR spectrum is quite different from those of GB and GD in that the main peak at 21.0 ppm is sharp with diminutive spinning sidebands, indicating a rather mobile species. The second ^{31}P MAS NMR peak at 17.3 ppm is broad and possesses the spinning sidebands expected for a rigid species. Consistent with the low-loading reaction, no detectable resonance (ca. 40 ppm) for toxic EA-2192 is evident. Extraction and analysis by ^{31}P NMR also did not detect any EA-2192, but did find both EMPA (32.2 ppm) and MPA (22.6 ppm) in the same proportion as the two phosphonate species evident in the ^{31}P MAS spectrum. Thus unlike the GB and GD reactions, secondary hydrolysis has occurred to yield MPA. The shift of the sharp ^{31}P MAS NMR peak at 21.0 ppm is in closest agreement with the 19.7 ppm peak of model compound **3**; thus this peak is assigned to **3** which is quite mobile on the surface and presumably heavily hydrated. Consistent with the evident mobility of hydrated, surface **3**, wet, "gummy" samples of authentic **3** also yielded ^{31}P MAS NMR spectra possessing small spinning sidebands. The MPA species yielding the broad peak at 17.3 ppm is consistent with either a surface bound MPA species (Scheme 1) or aluminophosphonate complex **4**.

Outlook for CWA Decontamination

The room temperature reactivity exhibited by the nanosize oxides (especially AP-Al_2O_3) for VX, GB, GD, and HD is promising for the use of such a material for "hasty" battlefield decontamination, where CWA deposited on personnel, equipment, vehicles, and perhaps even strategic locations, could be quickly removed, adsorbed, and subsequently detoxified. Such reactive materials would also be useful in building, vehicle, and aircraft filtration systems designed to protect occupants against CWA attacks, perhaps extending the life of conventional carbon-based filters. Finally, the fact that such materials can consume large quantities of CWA is promising for their use in the demilitarization of CWA stockpiles and/or munitions.

Conclusions

Nanosize metal oxides undergo room temperature reactions with VX, GB, GD, and HD. The reactive capacity is exceedingly large for the nerve agents as their phosphonic acid hydrolysis products effect facile erosion of the oxide surface, thus allowing reactions to proceed to the particle core. Aluminophosphonate products have been detected and identified by comparison to authentic compounds. In particular, aluminophosphonates **1**, **2**, and **4** have been identified as potential forensic markers for GB, GD, and VX, respectively, owing to their insolubility at neutral pH.

For HD on AP-Al_2O_3 hydrolysis is the primary reaction mechanism; but the reactive capacity is limited, with higher loadings requiring copious amounts of water to effect facile hydrolysis. TG is the preferred product at low loadings, but sulfonium ions H2-TG and CH-TG exclusively form under high loading/wet conditions. HD on AP-MgO reacts via both hydrolysis and elimination and these reactions occur at similar rates. The elimination products, i.e. divinyl sulfide, still retain irritant properties and are thus less desirable than the TG hydrolysis product. HD on AP-CaO exhibits autocatalytic behavior: a slow induction period involving primarily hydrolysis, is followed by a much faster reaction involving elimination.

Acknowlegements

We thank Messrs. Philip W. Bartram, David C. Sorrick, and Richard J. O'Connor and Ms. Vikki D. Henderson for assistance with the agent operations, and Dr. Shekar Munavalli for the synthesis of compounds **1** – **3**.

References

1. (a) Yang, Y.-C.; Baker, J. A.; Ward, J. R. *Chem. Rev.* **1992**, *92*, 1729-1743. (b) Yang, Y.-C. *Acc. Chem. Res.* **1999**, *32*, 109-115, and references therein. (c) Yang, Y.-C.; Szafraniec, L. L.; Beaudry, W. T. *J. Org. Chem.* **1993**, *58*, 6964-6965. (d) Bartram, P. W.; Wagner, G. W. *US Patent Number 5,689,038*, Nov. 18, 1997.
2. Wagner, G. W.; Bartram, P. W. Unpublished results.
3. (a) Wagner, G. W.; Bartram, P. W. *Interaction of VX-, G- and HD-Simulants with Self-Decontaminating Sorbents: A Solid State MAS NMR Study*, ERDEC-TR-375; Aberdeen Proving Ground, MD, Nov. 1996. (b) Wagner, G. W.; Bartram, P. W. *J. Mol. Catal. A: Chem.* **1996**, *111*, 175-180. (c) Wagner, G. W.; Bartram, P. W. *J. Mol. Catal. A: Chem.* **1995**, *99*, 175-181. (d) Wagner, G. W.; Bartram, P. W. *J. Mol. Catal. A: Chem.* **1999**, *144*, 419-424.
4. (a) Klabunde, K. J; Stark, J; Koper, O.; Mohs, C.; Park, D. G.; Decker, S.; Jiang, Y.; Lagadic, I.; Zhang, D. *J. Phys. Chem.*, **1996**, *100*, 12142-12153.

(b) Stark, J. V.; Park, D. G.; Lagadic, I.; Klabunde, K. J. *Chem. Mater.* **1996**, *8*, 1904-1912.

5. Wagner, G. W.; Bartram, P. W.; Koper, O.; Klabunde, K. J. *J. Phys. Chem. B* **1999**, *103*, 3225-3228.
6. Wagner, G. W.; Koper, O.; Lucas, E.; Decker, S.; Klabunde, K. J. *J. Phys. Chem. B* **2000**, *104*, 5118-5123.
7. Wagner, G. W.; Procell, L. R.; O'Connor, R. J.; Munavalli, S.; Carnes, C. L.; Kapoor, P. N.; Klabunde, K. J. *J. Am. Chem. Soc.* **2001**, *123*, 1636-1644.
8. Kuiper, A. E. T.; van Bokhoven, J. J. G. M.; Medema, J. *J. Catal.* **1976**, *43*, 154-167.
9. Mawhinney, D. B.; Rossin, J. A.; Gerhart, K.; Yates, J. T., Jr. *Langmuir* **1999**, *15*, 4789-4795.
10. Duncan, T. M.; Douglass, D. C. *Chem. Phys.* **1984**, *87*, 339-349.
11. (a) Vila, A. J.; Lagier, C. M.; Wagner, G.; Olivieri, A. C. *J. Chem. Soc., Chem. Commun.* **1991**, 683-685. (b) Olivieri, A. C. *J. Magn. Reson.* **1990**, *88*, 1-8.
12. Cao, G.; Lee, H.; Lynch, V. M.; Mallouk, T. E. *Inorg. Chem.* **1988**, *27*, 2781-2785.
13. (a) Knozinger, H.; Ratnasamy, P. *Catal. Rev. Sci. Eng.* **1978**, *17*, 31. (b)Peri, J. B. *J. Phys. Chem.* **1975**, *79*, 1582.
14. (a) Huggins, B. A.; Ellis, P. D. *J. Am. Chem. Soc.* **1992**, *114*, 2098-2108. (b) Fitzgerald, John J.; Piedra, G.; Dec, S. F.; Seger, M.; Maciel, G. E. *J. Am. Chem. Soc.* **1997**, *119*, 7832-7842.
15. Coster, D.; Blumenfeld, A. L.; Fripiat, J. J. *J. Phys. Chem.* **1994**, *98*, 6201-6211.
16. Yang, Y.-C.; Szafraniec, L. L.; Beaudry, W. T.; Rohrbaugh, D. K. *J. Am. Chem. Soc.* **1990**, *112*, 6621.
17. Wagner, G. W.; Bartram, P. W. *Langmuir* **1999**, *15*, 8113-8118.
18. Yang, Y.-C.; Szafraniec, L. L.; Beaudry, W. T.; Rohrbaugh, D. K.; Procell, L. R.; Samuel, J. B. *J. Org. Chem.* **1996**, *61*, 8407-8413.
19. Yang, Y.-C.; Szafraniec, L. L.; Beaudry, W. T.; Ward, J. R. *J. Org. Chem.* **1988**, *53*, 3293-3297.
20. Noller, H.; Hantsche, H.; Andreu, P. *J. Catal.* **1965**, *4*, 354-362.
21. Davis, G. T.; Block, F.; Sommer, H. Z.; Epstein, J. *Studies on the Destruction of Toxic Chemical Agents VX and HD by the All PurposeDecontaminants DS-2 and CD-1;* EC-TR-75024, May, 1975.
22. Pilkenton, S.; Hwang, S.-J.; Raftery, D. *J. Phys. Chem. B* **1999**, *103*, 11152-11160.

Chapter 11

Nanomaterials as Active Components in Chemical Warfare Agent Barrier Creams

E. H. Braue, Jr.[1], J. D. Boecker[1,2], B. F. Doxzon[1], R. L. Hall[1], R. T. Simons[1], T. L. Nohe[1], R. L. Stoemer[1,3], and S. T. Hobson[1]

[1]Drug Assessment Division, U.S. Army Medical Research Institute of Chemical Defense, Aberdeen Proving Ground, MD 21010–5400
[2]Current address: Northwestern University, Evanston, IL 60208
[3]Current address: Department of Chemistry, The Pennsylvania State University, University Park, PA 16802

A material that acts as a physical barrier and an active destructive matrix against chemical warfare agents is vital to military operations in chemically contaminated environments. Our research into active topical skin protectants (aTSPs) identified nanomaterials as a promising class of reactive moieties. We determined the efficacy of aTSPs against sulfur mustard (HD) and soman (GD). Using a penetration cell, we determined the cumulative amount of agent that penetrates the aTSP. Proof of neutralization of these materials was obtained using two assays: headspace GC/MS and one- and two-dimensional NMR. Both organic and inorganic nanomaterials showed efficacy against GD and/or HD. Relative to the inactive base cream, incorporation of nanomaterials into the aTSPs resulted in a 99.6% reduction of both GD and HD vapor after 20 hours. None of these nanomaterials, however, passed the initial *in vivo* testing with HD vapor.

Historical Summary

Chemical warfare agents (CWAs) represent a real and growing threat to both United States Armed Forces and civilians. The United States Army classifies CWAs into seven categories (*1*), but in this chapter we will focus on protection against two classes: nerve agents and blister agents. The use of CWAs within the last three decades includes Soviet use in Cambodia (yellow rain, tricothecene mycotoxins) (*2*), the Iraqi use against Iran (sulfur mustard [HD] and tabun [GA]) (*3*), and Iraqi use against its own dissident Kurdish population at Halabja (HD, $HCN_{(g)}$) (*4*). In World War I, almost one-third of Allied casualties were hospitalized as a result of CWA injuries (*5*). In 1995, a Japanese religious cult terrorized the civilian population by releasing sarin (GB) in a Tokyo subway. This attack resulted in over 1000 casualties and 12 deaths (*6*). Most recently, a plan by Al Qaeda terrorists to use sarin on the European Parliament Building in Strausberg was prevented by German police (*7*). These examples demonstrate that the civilian population is no longer immune from the threat of CWAs.

The current protection scheme against these agents in the United States Army consists of a chemically resistant outer layer of clothing (BDO), protective mask (M40), butyl rubber gloves, and rubber overboots (*8*). This ensemble does allow continued operation in a chemically contaminated area, but results in decreased performance and increased heat retention. In a continuing effort to develop a barrier cream to CWAs that will increase protection without degrading a warfighter's performance, the U.S. Army has investigated materials that serve as a physical barrier to these agents *and* contain an active moiety to neutralize any chemicals that may come in contact with the material. These active topical skin protectants (aTSPs) would be used in conjunction with other protective procedures.

Applying a topical protectant to vulnerable skin surfaces prior to entry into a chemical combat arena was proposed as a protective measure against percutaneous CWA toxicity soon after the use of HD by Germany at Ypres, Belgium in 1917 (*9*). As early as the summer of 1917, the United States Army began examining various soaps and ointments. Although several simple formulations were found to be effective in reducing 'skin redness' produced by agents such as hydrogen sulfide, no product was available before the end of World War I (*10*). Research in the area of protective ointments continued after the war, but this effort did not produce a fielded product before the beginning of World War II. During World War II, a concentrated effort to develop ointments for protection against HD took place at the Chemical Warfare Service in Edgewood Arsenal, Maryland. The Army produced the M-5 protective ointment, which was manufactured in 1943 and 1944. However, because of limited effectiveness, odor and other cosmetic characteristics, by the mid 1950s the M-5 ointment was no longer issued to soldiers (*11*).

Between 1950 and the early 1980s, the focus on research shifted to medical countermeasures rather than protective creams. However, research started in the early 1980s resulted in two non-active barrier skin cream formulations based on a blend of various perfluorinated polymers. These two formulations developed at the United States Army Medical Research Institute of Chemical Defense (USAMRICD), Aberdeen Proving Ground, Maryland were transferred to advanced development in October 1990 (*12*). The best formulation was selected and progressed through development with an Investigational New Drug (IND) filed with the Food and Drug Adminstration (FDA) in 1994, and approval of a New Drug Application (NDA) in 2000. This new product is now called Skin Exposure Reduction Paste Against Chemical Warfare Agents (SERPACWA). SERPACWA consists of fine particles of polytetrafluoroethylene (PTFE) solid dispersed in a fluorinated polyether. The excellent barrier properties of this polymer blend are related to the low solubility of most materials in it. Only highly fluorinated solvents like Freon® have been observed to show appreciable solubility. SERPACWA will be available to augment the protection of the current protective scheme for warfighters in 2002 (*13*).

SERPACWA extends the protection afforded by the current protective garments and allows a longer window for decontamination, but it does not neutralize CWAs into less toxic products. Furthermore, although the SERPACWA formulation provides excellent protection against liquid challenges of soman (GD, pinacolyl methyl phosphonofluoridate), VX (o-ethyl-s-(2-iisopropylaminoethyl) methyl phosphonothiolate), and HD (bis(2-chloroethyl) sulfide), its protection against HD vapor is less than optimal.

To overcome these limitations, we have developed an improved SERPACWA that contains an active component. The aprotic non-polar environment of SERPACWA provides a unique but challenging medium for active moieties to neutralize CWA. Reaction mechanisms that do not involve charged transition states should be favored in this medium. Ideally, this aTSP will meet the following criteria. First, the aTSP will neutralize CWAs including HD, GD, and VX. Second, the barrier properties of SERPACWA will be maintained or increased. Third, the protection against HD vapor will increase. And fourth, the cosmetic characteristics (e.g., odor, texture) of the TSP will be maintained.

Formulation and Analysis of aTSPs

Materials for the formulation of the aTSPs and the active moieties are as previously described (*14*). As seen in Figure 1, formulation of aTSPs followed

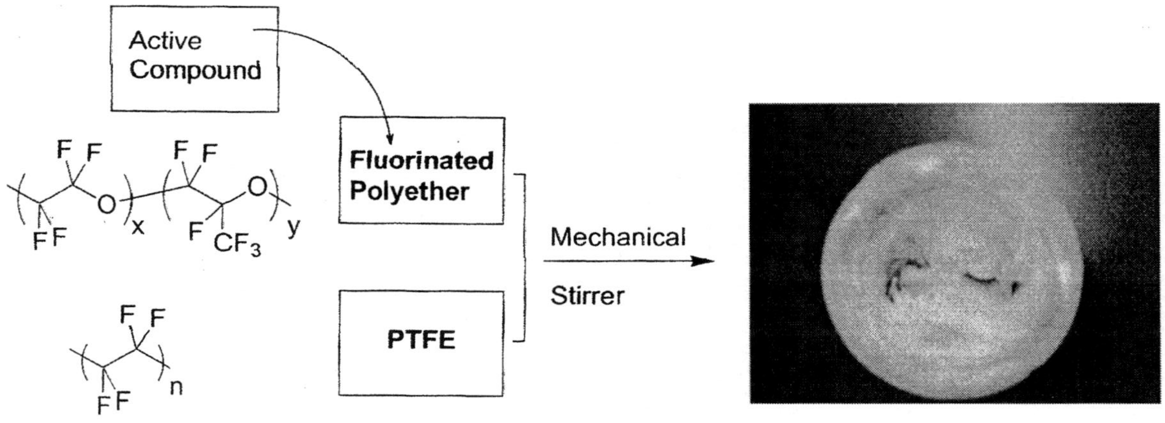

Figure 1. Formulation of Active Topical Skin Protectants by adding active compounds from different classes of active moieties into the base cream.

as closely as possible the technique used for the production of a *non*-active TSP base cream (SERPACWA) (*12*). HD, GD, and VX were obtained from the United States Army Soldier and Biological Chemical Command at Aberdeen Proving Ground, Maryland. NMR spectra for proof of neutralization were recorded on a Varian Unity INOVA NMR at the appropriate frequency (^{1}H: 600 MHz, ^{13}C: 150 MHz, ^{31}P 242 MHz). Evaluation of formulations was conducted with a decision tree network (DTN) that describes the path that aTSPs follow during evaluation (Figure 2). The DTN is divided into two pathways: one for vesicants (HD) and the other for nerve agents (GD). Within these pathways there are three blocks each with a decision point. The first block consists of a series of three *in vitro* modules used to determine the initial efficacy of candidate formulations and to eliminate non-effective candidates before animal testing. The second block consists of *in vivo* modules and the third block consists of an advanced animal module to determine the influence of time, water/sweat and interactions with other products. Experimental details for most of the DTN modules have been previously described (*15*, *16*).

The proof of neutralization tests are used to verify that aTSP formulations actually react with CWAs to produce less toxic materials. A headspace gas chromatography mass spectroscopy (HS-GC/MS) test uses the solid phase microextraction (SPME) technique for the collection of CWAs. Samples collected on the extraction filament are analyzed by gas chromatography/mass spectroscopy (*17*). A 100 mg sample of aTSP is challenged with 0.1μl of undiluted CWA (HD, GD, or VX) in a small vial. The headspace above the mixture is sampled periodically to determine the amount of CWA remaining in the flask. Efficacy is determined by calculating the percent of SERPACWA. Percent of SERPACWA is the ratio (expressed as percentage) of the headspace concentration of the CWA determined for each formulation divided by the headspace concentration determined for SERPACWA. Thus, percent of SERPACWA represents the amount of agent remaining in the headspace for aTSP formulations following challenge compared to the amount of agent remaining in the headspace for the SERPACWA control. Other analytical techniques such as Nuclear Magnetic Resonance (NMR) and Fourier-Transform Infrared Spectrometry (FTIR) have also been used in this module.

Efficacy of Nanomaterials in aTSPs.

Two criteria constrain our selection of active components. First, the barrier properties of the base cream must not be degraded by the incorporation of the active compound(s). Second, the active moiety must neutralize CWAs in the environment of the base cream (SERPACWA, *vide supra*). We have investigated over 150 active moieties that are divided into four classes (Table I).

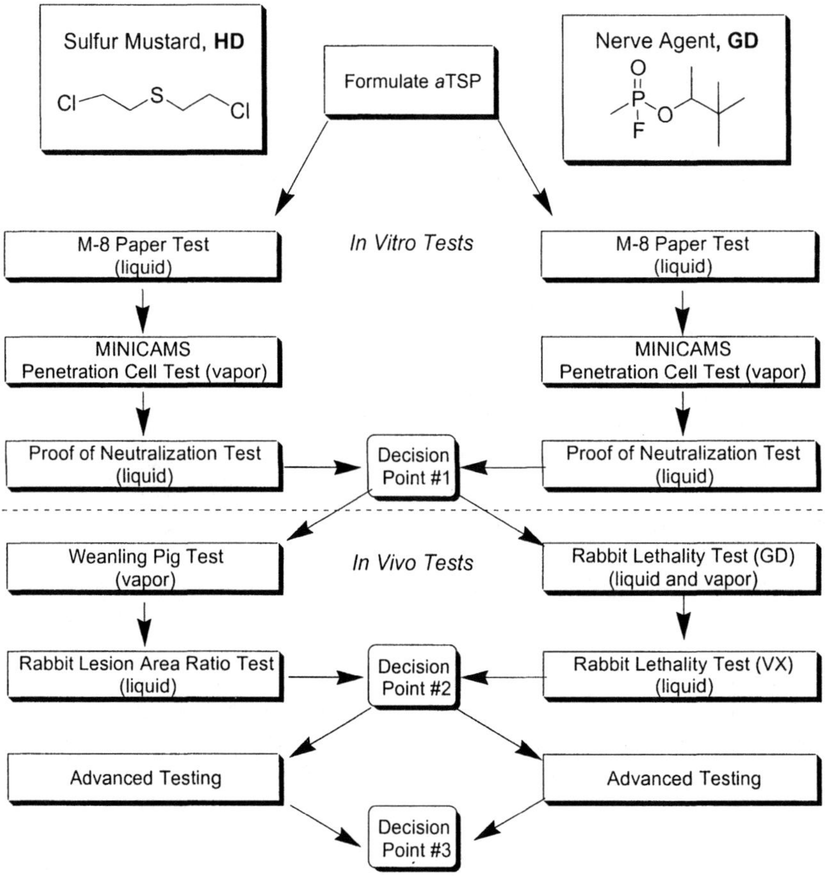

Figure 2. Decision Tree Network used in the evaluation of aTSPs.

The 150 active materials were formulated into over 350 mixtures with perfluorinated polyether, polytetrafluoroethylene, surfactants, and other additives. The prepared formulations were evaluated for efficacy in the DTN (Figure 2). Whereas several classes of active moieties demonstrated significantly ($p<0.05$) improved efficacy compared with SERPACWA, the focus of this chapter will be on only nanomaterial active moieties.

Against HD, nanomaterials that demonstrated significantly improved efficacy compared with SERPACWA in at least one DTN module were nanoreactors (NR), micro precipitates, and several reactive nanoparticles (RNP), including MgO, TiO_2, CaO, and ZnO.

Table I. Classes of Active Moieties

Class	*Examples*
Organic Molecules	S-330, iodobenzenediacetate, DNA, chloramines, 2,3-butanedione monoxime, phosphodithiates
Inorganic Compounds	polymer coated metal alloys (TiFeMn, MgNi, CaNi), micro precipitates, reactive nanoparticles (MgO, CaO, TiO_2, ZnO), polyoxometalates ($M_5PV_2W_{10}O_{40}$)
Organic Polymers	Nanoreactors, XE-555 resin
Enzymes	organophosphorous acid anhydride hydrolase (OPAA)

Against GD, nanomaterials that demonstrated significantly improved efficacy compared with SERPACWA in at least one DTN module included nanoreactors, micro precipitates, MgO RNP, ZnO RNP, CaO RNP, and TiO_2 RNP. Against VX, nanomaterials that demonstrated significantly improved efficacy compared with SERPACWA in at least one DTN module included nanoreactors, MgO RNP, CaO RNP, and TiO_2 RNP.

Figure 3 summarizes the results of nanomaterial aTSPs tested in the penetration cell modules. Four nanomaterials passed the penetration cell test for HD vapor with less than 1000 ng breaking through in 20 hr. Nanoreactors were the most effective nanomaterials, reducing the amount of HD vapor by 99.6% relative to SERPACWA. These were followed by TiO_2 (92.8 % reduction), micro precipitates (87.4 % reduction), and MgO RNP (84.8 % reduction). Five nanomaterials passed the penetration cell test for GD vapor with less than 1000 ng breaking through in 20 hr. TiO_2 was most effective, reducing the amount of GD vapor by 99.6 % relative to SERPACWA, followed by ZnO (98.7 % reduction), micro precipitates (94.9 % reduction), nanoreactors (91.3 % reduction), and MgO (85.5 % reduction).

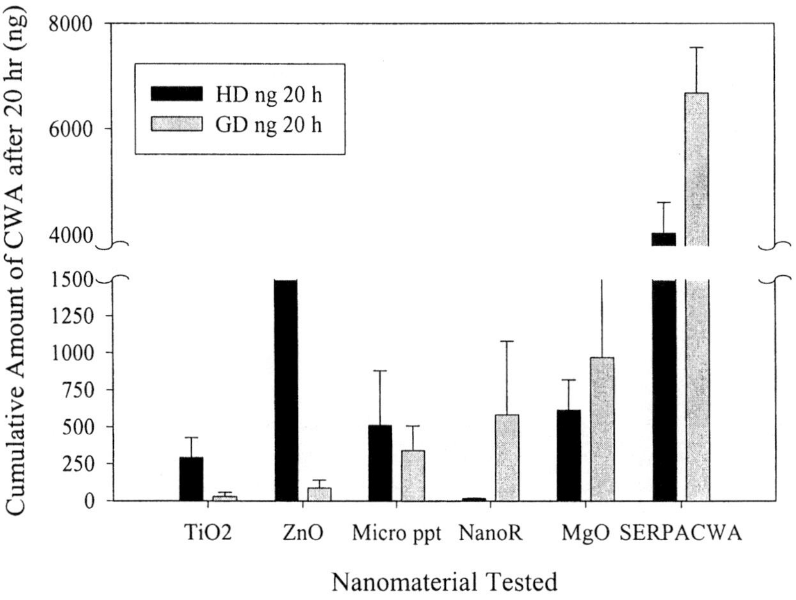

Figure 3. Nanomaterials passing the penetration cell HD and GD Vapor Test. Total ng represents the amount of CWA that penetrates a 0.15 mm thick aTSP barrier in 20 hr.

Note that the penetration cell test only measures the increase in the barrier properties. The Miniature Continuous Air Monitoring System (MINICAMS) (*18*) is designed to selectively monitor the amount of challenge agent and does not respond to the neutralization products. Thus, we can infer that the increase in the barrier properties is due to neutralization of the CWA by the nanomaterials incorporated into the aTSP. Other assays must be run to directly determine the fate of the CWA.

We have used two assays to determine the extent of neutralization with these materials: HS-GC/MS and one- and two-dimensional NMR. Using the HS-GC/MS test, we determined the concentration of HD, GD, or VX and calculated the percent of SERPACWA (Figure 4). Against HD liquid, all materials tested in the penetration cell test passed the HS-GC/MS test. Four out of five of the nanomaterials passing the penetration cell test for GD vapor also passed the HD HS-GC/MS test for GD liquid. A formulation that provides significantly ($p < 0.05$) better protection than SERPACWA passes each test.

Although the MINICAM system can detect VX, we were unable to develop a quantitative penetration cell method for VX. The very low equilibrium vapor pressure of VX (0.00063 mm Hg @ 25°C) (*19*) prevented reproducible transfer of VX from the penetration cell to the MINICAMS detection system. Thus, the only quantitative mechanical module in the DTN for VX is the HS-GC/MS test for VX liquid. Four nanomaterials passed this test (Figure 4). The high correlation of the nanomaterials that passed the HS-GC/MS tests for all three CWAs (HD, GD, and VX), as well as the penetration cell test for HD and GD, indicates the universal ability of these materials to neutralize both vesicants and nerve agents.

To further probe the reaction products of nanomaterials with CWAs, we have used one- and two-dimensional NMR. For example, we evaluated the reaction products of a HD simulant, 2-chloroethylethylsulfide (CEES), with TiO_2 RNP in an environment similar to the base cream (Figure 5). Although reaction products were not isolated and characterized, the spectra clearly demonstrate that CEES has chemically reacted and is no longer present. As seen in Figure 6, adventitious water on the surface of the nanomaterial TiO_2 RNP may neutralize CEES, and presumably HD, by hydrolysis. Another possibility is that CEES may be adsorbed on the surface of the material followed by elimination (*20*). We have also examined the neutralization of a GD simulant, diisopropylfluorophosphate (DFP), in similar conditions using ^{31}P NMR (Figure 7). It is likely that DFP, and presumably GD, is adsorbed by the TiO_2 RNP followed by hydrolysis. This was previously reported for other nanoscale metal oxides (*20*).

Having completed the top portion of the DTN, a decision point on continuing through the *in vivo* tests is reached. A diverse selection of nanomaterials was able to pass multiple modules in the *in vitro* tests (Table II).

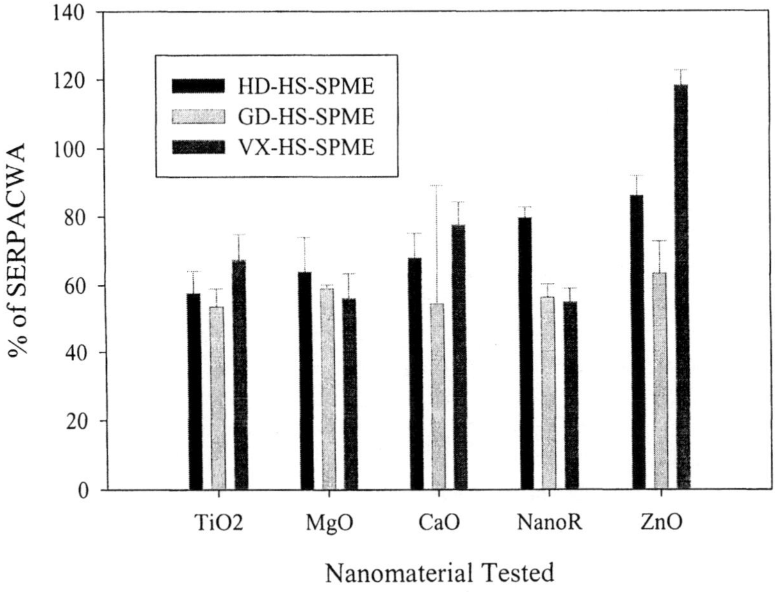

Figure 4. liquid proof of neutralization test. Error bars display SD. Percent of SERPACWA is the ratio (expressed as %) of the headspace concentration of the CWA determined for each formulation divided by the headspace concentration determined for SERPACWA.

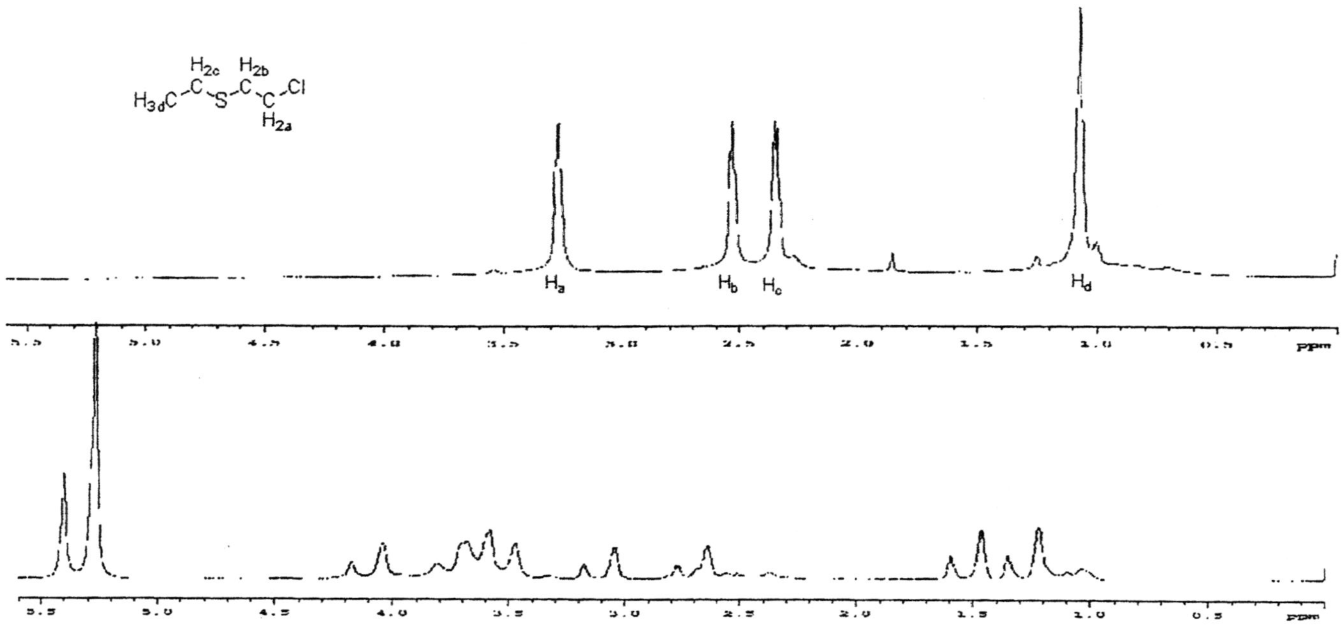

Figure 5. *^{1}H NMR of 2-chloroethylethylsulfide (CEES top), and reaction products of CEES with TiO_2 (bottom). Spectra were obtained in perfluorinated oil with a coaxial insert of D_2O as an external lock.*

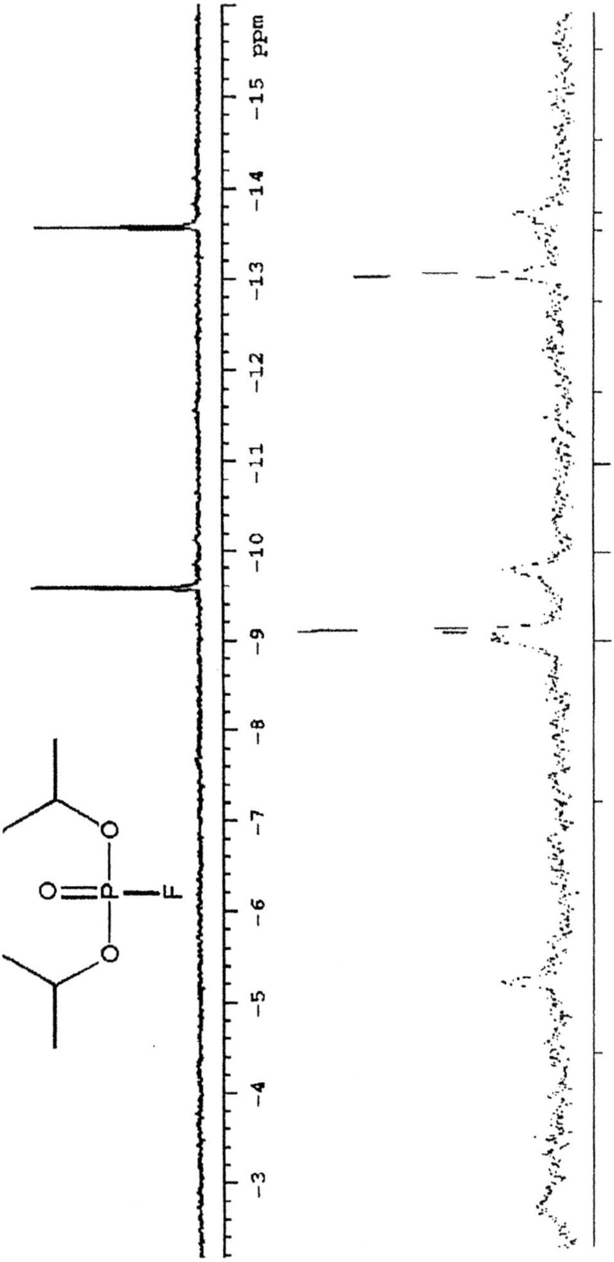

Figure 6. Possible neutralization reactions of CEES on TiO_2.

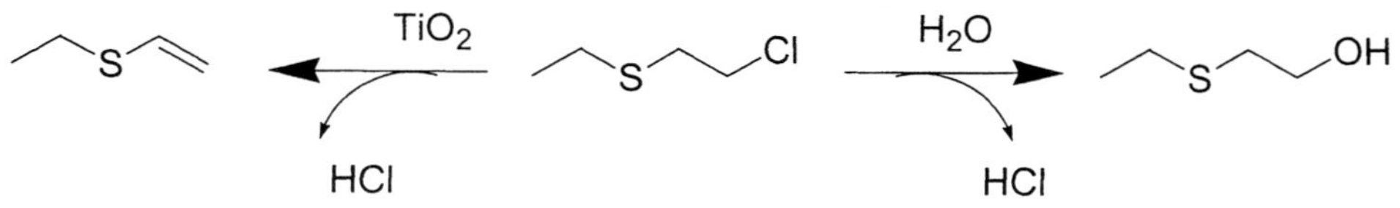

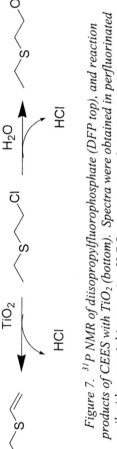

Figure 7. ^{31}P *NMR of diisopropylfluorophosphate (DFP top), and reaction products of CEES with* TiO_2 *(bottom). Spectra were obtained in perfluorinated oil with a coaxial insert using* H_3PO_4 *as an external standard and* D_2O *as an external lock.*

Table II. *In vitro* results of aTSPs containing nanomaterials.

Nanomaterial	*$HD_{(v)}$ pen cell*	*$GD_{(v)}$ pen cell*	*$HD_{(l)}$ HS-SPME GC/MS*	*$GD_{(l)}$ HS-SPME GC/MS*	*$VX_{(l)}$ HS-SPME GC/MS*	*NMR CEES and DFP*
Nanoreactor	Pass	Pass	Pass	Pass	Pass	Pass
TiO_2	Pass	Pass	Pass	Pass	Pass	Pass
ZnO	Fail	Pass	Pass	Pass	Fail	NT
MgO	Pass	Pass	Pass	Pass	Pass	NT
CaO	Fail	Fail	Pass	Fail	Pass	NT
Micro Ppt	Pass	Pass	NT	NT	NT	NT

NOTE: NT means material has not yet been tested in the module. Pass is defined as a significant ($p < 0.05$) increase in protection compared to SERPACWA.

Formulations that pass one or more of the *in vitro* tests are then evaluated in the *in vivo* DTN modules. The first animal model in the testing sequence is the weanling pig test for HD vapor. None of the nanomaterials that passed the *in vitro* tests passed the weanling pig test for HD vapor. We have seen similar trends with other compounds and have five possible explanations.

First, the skin is occluded by the aTSP, increasing agent penetration and thus the observed erythema. It is well known that increasing the moisture of skin (*21*, *22*) magnifies the effects of HD exposure. Applying a 0.15mm layer of SERPACWA reduces the transepidermal water loss from human skin by about 50 % (*23*). These data indicate that aTSP applied to the skin is partially occlusive and results in increased skin moisture.

Second, the skin may be sensitized by the aTSP, and thus the small amount of HD vapor that penetrates into the skin may result in greater erythema. We observed in the haired guinea pig model (*24*) that hair removal by chemical depilation resulted in sensitized skin that produced more severe lesions following HD vapor exposure compared with the hairless guinea pig that did not require chemical depilation (*25*). Although not depilation agents, the active moieties in the aTSP formulations may interact in other ways with the skin and cause an increased response from HD. In control experiments with no HD exposure, however, none of the nanomaterial formulations evaluated caused any erythema when applied to pig skin for 24 hours.

Third, the reaction products may irritate the skin. Active moieties that neutralize harmful CWAs use three main modes of action: oxidation, reduction, or hydrolysis. The modes of action for each of the nanomaterials discussed are still under investigation. Nevertheless, it is reasonable to suggest that reactive intermediates or the final reaction products (HCl from hydrolysis; sulfur mustard sulfone from oxidation) may cause skin irritation and erythema.

Fourth, the agent may dissolve into the aTSP during exposure and not be completely removed by the cleaning procedure. The cleaning procedure following HD vapor exposure involves wiping each site with two dry foam-covered cotton swabs. This process removes most of the aTSP but a small film still remains on the skin surface (*26*). If HD dissolved into the aTSP and was not completely neutralized, a small amount could have been in contact with the skin surface until the postexposure readings were made (24 hr).

Fifth, there may be a fundamental interaction between the skin surface and the layer of aTSP that interferes with the neutralizing ability of the active moieties. The penetration cell test results indicate that when the aTSP is supported on a nitrocellulose membrane, small amounts (i.e., 15 – 20 ng, *vide supra*) of HD penetrate through the barrier. Since it is believed that μg quantities of HD are required to cause noticeable erythema, this is clearly not enough HD to cause the response observed in the weanling pig model (*22,27*). The nature of this potential interaction is unknown at present. We plan to modify the penetration cell experiments by replacing the nitrocellulose membrane with excised split-thickness pig skin. We are hopeful that these experiments will help us to understand the discrepancies observed between the penetration cell testing and weanling pig experiments.

The nanomaterials have not yet been tested in the other *in vivo* models. We are hopeful that better correlation between the *in vitro* and *in vivo* experiments will be observed.

Summary

We have reported the preparation and evaluation of nanomaterials as active components in chemical warfare agent barrier creams. We developed these formulations and optimized their efficacy by adjusting the amounts of active nanomaterials, perfluorinated polyether oil, polytetrafluoroethylene resin, and other additives. Thus far, the optimum formulations display excellent resistance against HD and GD (99.6 % reduction in vapor break-through after 20 hr compared with SERPACWA). Against VX the best formulation reduced the VX vapor concentration by 55 % compared with SERPACWA. These materials continue to move towards advanced development with the ultimate goal of complete protection for U.S. warfighters and civilians against CWAs.

References

1. Takafuji, E.T.; Kok, A.B., in *Textbook of Military Medicine, Medical Aspects of Chemical and Biological Warfare*, Sidell, F. R.; Takafuji, E. T.;.

Franz, D. R., Eds. Office of the Surgeon General at TMM Publications, Washington, D. C. 1997; pp 118-119.

2. Bartley, R. L.; Kucewicz, W. P., *Foreign Affairs*, **1993**, *61*, 805-826.
3. UN Security Council, *Report of the specialists appointed by the Secretary-General to investigate allegations by the Islamic Republic of Iran concerning the use of chemical weapons*, S/16433, 26 March 1984, pp. 11-12.
4. (a) Over 10,000 casualties were reported. Spiers, E. M. *Chemical and biological weapons: A study in proliferation.* St. Martin's Press: New York, NY, 1994; p 18; (b) Kirkham, N. Cyanide bombers lay waste a town. *The Daily Telegraph*, 22 March 1988, p. 1.
5. Heller, C. E. *Leavenworth Papers. Chemical Warfare in World War I: The American Experience* Combat Studies Institute: Ft. Leavenworth, KS, 1984; pp 91-92.
6. Woodall, J., *Lancet*, **1997**, *350*, 296.
7. Bamber, D.; Hastings, C.; Syal, R.. Bin Laden British Cell Planned Gas Attack On European Parliament. *London Sunday Telegraph*, 16 September 2001. http://www.telegraph.co.uk (accessed January 2002).
8. O'Hern, M.R.; Dashiell, T.R.; Tracey, M.F. in *Textbook of military medicine, Part I: Medical aspects of chemical and biological warfare*, Sidell, F. R.; Takafuji, E. T.;. Franz, D. R., Eds. Office of the Surgeon General at TMM Publications, Washington, D. C. 1997; pp 371-372.
9. Papirmeister B.; Feister, A. J.; Robinson, S. I; Ford, R. D. *Medical Defense Against Mustard Gas: Toxic Mechanisms and Pharmacological Implications.* CRC Press: Boston, MA, 1991; p. 2.
10. Papirmeister B.; Feister, A. J.; Robinson, S. I; Ford, R. D. *Medical Defense Against Mustard Gas: Toxic Mechanisms and Pharmacological Implications.* CRC Press: Boston, MA, 1991; p. 3.
11. Romano, J. R. United States Army Medical Research Institute of Chemical Defense, Aberdeen Proving Ground, MD. Personal communication, 2001.
12. McCreery, M. J. U.S. Patent 5,607,979 March 4, 1997.
13. Liu, D.K.; Wannemacher, R.W.; Snider, T.H.; Hayes, T.L. *J. Appl Toxicol.*, **1999**, *19*, S41-S45.
14. Hobson, S. T.; Lehnert, E. K.; Braue, E. H. Jr. *Mat. Res. Soc. Symp. Proc.* **2000**, *628*, 10.8.1-10.8.8.
15. Braue, E. H. Jr. *J. Appl. Toxicol.,* **1999**, *19*, S47-S53.
16. Hobson, S. T.; Braue, E. H., Jr., *Polymeric Materials: Science & Engineering.* **2001**, 84, 80..
17. Stuff, J. R.; Cheicante, R. L.; Morrissey, K. M.; Durst, H. D. *J. Microcolumn Separations*, **2000**, *12*(2), 87-92.

18. Miniture Continuous Air Monitors, CMS Field Products, Division of OI Analytical, 2148 Pelham Parkway - Building 400 - Pelham, AL 35124-1131
19. SBCCOM ONLINE: Material Safety Data Sheet, Lethal Nerve Agent (VX). http://www.sbccom.apgea.army.mil/RDA/msds/vx.htm (accessed January 2002).
20. Similar reactivity has been reported for Nanosize ZnO and CaO. (a) Wagner, G. W.; Bartram, P. W.; Koper, O.; Klabunde, K. J. *J. Phys. Chem B.* **1999**, *103*, 3225-3228. (b) Wagner, G. W.; Koper, O. B.; Lukas, E.; Decker, S.; Klabunde, K. J. *J. Phys. Chem B.* **2000**, *104*, 5118-5123.
21. Renshaw, B. *J. Invest. Dermatol.,* **1947**, *9*, 75-85.
22. Papirmeister B.; Feister, A. J.; Robinson, S. I; Ford, R. D. *Medical Defense Against Mustard Gas: Toxic Mechanisms and Pharmacological Implications.* CRC Press: Boston, MA, 1991; pp. 14-21.
23. Braue, E.H. Jr. The U.S. Army Medical Research Institute of Chemical Defense, APG, MD. Personal communication, 2001.
24. Braue, E. H Jr.; Nalls, C.R.; Way, R.A.; Zallnick, J. E.; Rieder, R. G.; Lee, R. B.; Mitcheltree, L. W. *Skin Research and Technology.* **1988**, *4*, 99-108.
25. Braue, E. H.; Koplovitz, I.; Mitcheltree, L. W.; Clayson, E. T.; Litchfield, M. R; Bangledorf, C. R.; *Toxicology Methods*, **1992**, *2(4)*, 242-254.
26. Kurt, E. M.; Braue, E. H. Jr. *Skin Adsorption Pharmacology of Topical Skin Protectant ICD #2289;* USAMRICD-TR-97-03, The U.S. Army Medical Research Institute of Chemical Defense, APG, MD, June 1997.
27. Minimal amount of HD for vesication = 1000 ng, see Sidell, F. R.; Urbanetti, J. S.; Smith, W. J.; Hurst, C. G. in *Textbook of Military Medicine, Medical Aspects of Chemical and Biological Warfare*, Sidell, F. R.; Takafuji, E. T.;. Franz, D. R., Eds. Office of the Surgeon General at TMM Publications, Washington, D. C. 1997; p 201.

Chapter 12

Carbon Black Dispersions: Minimizing Particle Sizes by Control of Processing Conditions

D. Forryan[1], D. Rasmussen[2], and R. Partch[1,*]

[1]Department of Chemistry and Center for Advanced Material Processing, Clarkson University, 8 Clarkson Avenue, Potsdam, NY 13699–5814
[2]Department of Chemical Engineering, Clarkson University, 8 Clarkson Avenue, Potsdam, NY 13699–5705

The objective of the this work was to determine a technique for reducing the particle size of commercial carbon black aggregates for their possible use as laser absorbing fillers in optical lenses. Aggregate size in the range of 40-80nm has been achieved using one commercial carbon black sample in both water and a chlorinated organic solvent, using an ultrasonic horn and argon purge to create the dispersion. Addition of anionic, cationic or nonionic surfactant to the solvent before processing did not result in further size reduction. Cabot Monarch 1100 large carbon black aggregates composed of primary particles having 15-20nm diameter gave the best results. After processing the dispersions were stable for over twenty-four hours and after precipitation occurred, redispersion to the 40-80nm aggregates was achieved using only an ultrasonic bath.

Introduction

Complete and uniform dispersibility of fillers in a matrix is a prerequisite for a composite to have optimum properties. Regardless of composition, shape or size of the particles, less than optimum distribution in, for example, ceramic, metal or polymer material can result in lower mechanical strength, random discoloration or decreased electrical or thermal conductivity. For these and other reasons much effort has been and continues to be devoted to understanding fundamental reasons why some powders readily disperse in a medium and others do not. It is clear from many historic studies (*1-4*) that the surface chemistry of a particle, which dictates relative hydrophilicity-hydrophobicity and zeta potential, is the dominant factor. Benefits of perfected filler dispersibility are found in dental resins (*5*), personal body armor (*6*), cosmetics and sunscreens (*7*), rubber products (*8*), latex paint (*9*), metal matrix composites (*10*), inks and gels (*11*), many foods, and in abrasive slurries used for chemical mechanical planarization (CMP of wafers during computer chip manufacture (*12*).

For decades colloid scientists have prepared and characterized multi- and sub-micron sized particles to the extent that processing equipment and analytical instrumentation allowed. But the routinely handled sub-micron species, now called nanoparticles, were produced only in small quantities and usually in dilute dispersions (*13*). The development of scaled-up methods for synthesizing and collecting nanoparticles of single or mixed composition has caused a revolution in composite engineering, which has already resulted in material properties previously unavailable (*14,15*). But the price to pay as filler particle sizes are reduced to nano scale is their greater tendency to form soft or hard (fused) aggregates of the primary units, especially when in the dry state. Their surface areas are so large that Van der Waals forces of attraction inhibit their separation when mixed into a matrix. Large sized aggregates of nanoparticles dispersed in a matrix are not capable of enhancing the properties of a composite to the same degree as are smaller aggregates or unaggregated primary species.

The recent availability of many types of dry nanoparticle fillers requires that techniques be devised to break apart the aggregates and to process the as-generated smaller units before they re-assemble. The objective of the current study was to determine conditions by which carbon black powder aggregate sizes can be minimized below previously reported values and stabilized in various liquid media. The incentive was to provide filler theoretically (*16*) capable of absorbing laser radiation for use in optical lenses.

Discussion

Starting Materials

Powdered carbon formed by natural or artificial combustion or pyrolysis of organic material has existed for centuries and been employed for a wide variety of uses (17-20). The dry powders are generally in the form of aggregates greater than 100nm in size and do not de-aggregate into primary nanoparticles when placed in a liquid even with application of high energy mixing or addition of a surfactant. The latter, like many other organic chemicals such as colorants, flavors and odorants, are well known to adsorb strongly to the surfaces of the carbon aggregates.

Four representative types of commercial carbon black samples were selected for use in the present study. As shown in Table I they were all of the Cabot Monarch variety and exhibited a range of primary nano-particle diameters from 13-16nm, surface areas and oxygen content. All as-received dry samples yielded only aggregates larger than 300nm when manually shaken in water. It was established (21) that the larger aggregates were weakly associated soft assemblies of smaller 40-50nm hard aggregates composed of fused primary particles. Prior to this study the larger aggregates were acceptable as fillers in most composites (22,23). Our goal was to achieve size reduction of particles to less than 50nm in a liquid, which, along with the dispersed carbon nanoparticles, may absorb and dissipate a range of laser photon energies.

Table I. Properties of Cabot Carbon Black Samples[1]

Carbon Sample	*Particle Size (nm)*	*DBP Oil Adsorption (cc/100 grams)*	*Nitrogen Surface Area (M2/gram)*	*Density (lbs/ft^3)*	*Iodine Number (mg/gram)*
Monarch 880	16	112	220	8	258±20
Monarch 1000	16	110	343	13	500±30
Monarch 1100	14	65	240	15	258±20
Monarch 1300	13	100	560	20	665±50

[1]All data taken from Cabot product literature.

Dispersion Procedures and Results

Initially, 0.01-0.1wt% amounts of the four Monarch types of carbon black were mixed separately with water and treated as described in Table II. Particle sizes were obtained using both a Brookhaven BI-200SM instrument operating at 514nm and interfaced with a BI-9K digital auto correlator, and an ALV goniometer operating at 647nm. All particle size numbers are weight average with 8-10% deviations. As can be seen, Monarch 1100 carbon black yielded the smallest aggregates. The ultrasonic bath used was of the standard cleaning bath variety common to most laboratories. It's level of energy did not break apart the large soft aggregates any more efficiently than when the mixtures were shaken by hand.

Contrary to the above observation, the use of an ACE 600W ultrasonic horn operating at 20 kHz frequency and 2.5s pulses with 1.0s intervals, as well as a Microfluidics M-110 microfluidizer, caused Monarch 1100 aggregates to be reduced in size down to 60-80nm clusters of partially fused primary particles. A slow purge of argon gas into the dispersion during sonication facilitated the process (*24-28*). To the naked eye the samples obtained after microfluidization appeared less dark in color. This could have been due to retention of some amount of the carbon inside the apparatus even though none was evident in wash fluids. An alternative explanation for the slight decolorization is that the fluidization process produces a higher percentage of less than 30nm particles, which is supported by light scattering data. Under the best conditions of treatment none of the other Monarch type carbon black samples yielded desirable results and were therefore not used in further studies.

A review of the data in Table I reveals that Monarch 880 carbon black is possibly most similar in chemical and physical property to Monarch 1100 except for density. The iodine number, which indicates reactivity of vinyl and other functional groups, does not correlate with the results in Table II. Monarch 1100 is twice as dense as Monarch 880, which means greater mass per volume. This feature should correlate with tighter packing, possibly due to more interparticle fusion, of the primary carbon black nanoparticles in the aggregates. If this postulate is valid, the observation that the more dense aggregates yield to breakup more efficiently cannot be explained by the current results.

Solvent effects on Monarch 1100 dispersions were also investigated. Table III shows the results when 0.1wt% solids were mixed with water, toluene and hexachlorobutadiene under identical conditions of ultrasound treatment. It is evident that toluene is not a good medium into which any of the Monarch type samples should be dispersed. Furthermore, by comparing with Table II data, water has been superceded by the chlorinated organic as the best of the solvents evaluated to achieve carbon black size reduction. It is not possible to explain the observed trend of solvent effects on particle size reduction by shear or sonic

Table II. Conditions for Dispersion of Cabot Carbon Black Samples in Water[1]

Carbon black	*Media*	*Dispersion method*	*Particle size*
Monarch 1100	Water	Shaken by hand	>450nm
Monarch 1100	Water	Ultrasonic bath	450nm
Monarch 1100	Water	Sonic horn 0.5 hr	120nm
Monarch 1100	Water/ Argon	Sonic horn 0.5 hr	80nm
Monarch 1100	Water/Argon	Sonic horn 1.0 hr	80nm
Monarch 1100	Water	Microfluidiser 5 passes	60nm
Monarch 880	Water/Argon	Sonic horn 0.5 hr	100nm
Monarch 1000	Water/Argon	Sonic horn 0.5hr	115nm
Monarch 1300	Water	Sonic horn 0.5 hr	214nm
Monarch 1300	Water/Argon	Sonic horn 0.5 hr	128nm

[1]Particle sizes are given as wt. ave. and have 8-10% deviation.

Table III. Solvent Effects on Monarch 1100 Particle Size Reduction[1]

Carbon Black	Water	Toluene	Hexachloro-1,3-butadiene
Monarch 880	100	360	200
Monarch 1000	115	440	190
Monarch 1100	80	4300	40
Monarch 1300	130	2070	190

[1]Dispersions treated with sonic horn/argon for 0.5 hr. Particle sizes are given as wt. ave. and have 8-10% deviation.

Table IV. Effect of Surfactants on Size of Monarch Carbon Black Particles in Water[1]

Carbon black	*Surfactant level (on wt carbon)*	*Particle size (nm)*
Monarch 880	0.1% PVSA	100
Monarch 880	0.1% Triton	120
Monarch 1000	0.1% PVSA	100
Monarch 1000	0.1% PDDA	130
Monarch 1000	0.1% Triton	160
Monarch 1100	0.1% PVSA	90
Monarch 1100	2.0% PVSA	80
Monarch 1100	0.1% PDDA	80
Monarch 1100	1.0% PDDA	90
Monarch 1300	0.1% PVSA	110
Monarch 1300	0.1% Triton	220

[1]Dispersions treated with sonic horn/argon for 0.5 hr. Particle sizes are given as wt. ave. and have 8-10% deviation.

PDDA = poly (diallyldimethylammonium hydrochloride salt)
PVSA = poly (vinylsulfonic acid sodium salt)
Triton = poly (ethylene oxide) (X-100)

energy. Solvent polarity cannot contribute to the difference because toluene and hexachlorobutadiane have similar low values compared to water. The symmetry of the perchlorinated diene prevents it from having much different polarity than toluene. No data is available to compare the energy due to cavitation in the dispersions in the different solvents employed in this study. Other chlorocarbon or chlorohydrocarbon solvents were not investigated in this study because the butadiene had the chemical and physical properties to act synergistically with the finely dispersed carbon to absorb and dissipate laser photon energy (*16*). The long-term objective is to transfer such effectiveness into optical lens composites for eye protection.

Time dependency of the stability of the 40nm and 80nm aggregate dispersions in hexachlorobutadiene and water, respectively, was measured qualitatively. In each case black solid settled to the bottom of the container after 48 hr. The redispersing-settling-redispersing sequence was cycled several times by use of an ultrasonic bath (not horn) and each time the redispersed aggregates returned to the 40-80nm size range.

The effect of various surfactants on the above-described procedures in water is shown in Table IV(*29*).

Comparing particle sizes in water for the four Cabot carbon types in Table III with the data in Table IV reveals that for Monarch 880 neither anionic nor nonionic surfactant helps reduce particle size during sonication. In the case of Monarch 1000 and 1300 particles, the nonionic surfactant has a negative effect (115 to 160nm ; 130 to 220nm), while the anionic surfactant appears to facilitate slight reduction in particle size (130 to 100nm). The data in Table IV show that for Monarch 1100 none of the surfactants added to the water during sonication effect a size difference. However, the time dependent stability of Monarch 1100 dispersions in water containing surfactant appeared to be somewhat greater after sonication than without surfactant.

Conclusions

Large aggregates of dry, commercial Cabot Monarch 1100 carbon black can be reduced in size to 80nm or 40nm particles in water and hexachlorobutadiene, respectively., by employing sonication. Other types of Cabot carbon black do not yield the same results. Toluene is a poor solvent in which to attempt size reduction. Added surfactants do not cause aggregate size to decrease further during processing but purging the dispersion with argon during sonic horn treatment does. Surfactants increase dispersion stability to some extent. Future research will employ a laser to input energy into carbon black dispersions to determine if further aggregate breakup can be achieved.

Acknowledgments

The authors gratefully acknowledge discussions with Andrew Clements and Roger Becker, both associated with US Army TACOM, Michael Sennett with US Army Natick Lab, and with Barbara Severin of AMPTIAC. Financial support came from IITRI and the US Army Contract Office. Facilities were provided by the Center for Advanced Materials Processing at Clarkson University, sponsored by the New York State Science and Technology Foundation.

References

1. *Hydrophobic Surfaces*; Fowkes, F.; Ed.; Kendall Award Symposium; Academic Press: New York, NY, 1969.
2. Sato, T.; Ruch, R. *Stabilization of Colloidal Dispersions by Polymer Adsorption*; Surfactant Science Series; Marcel Dekker: New York, NY, 1980.
3. *Colloidal Dispersions*; Goodwin, J.; Ed.; Royal Society of Chemistry Special Publication No. 43; Dorset Press: London, 1981.
4. *Colloidal Dispersions*; Ottewill., R.; Ed.; Faraday Discussions of the Chemical Society No. 90; Arrowsmith Ltd.: London, 1990.
5. Xu, H.; Eichmiller, F.; Antonucci, J.; Schumacher, G.; Ives, L. *Dental Materials* 2000, *16, 356-63.*
6. Medvedovski, E. *Bull. Am. Ceram. Soc.* 2002, *81, 27-32.*
7. Reisch, M. *Chem. & Eng. News.* June 24, 2002, p 38.
8. Olmstead, H. *Proc. Intertech Conference on Functional Tire Fillers*; Intertech Co.: Portland, ME, 2001.
9. Duivenvoorde, F.; van Nostrum, C.; Laven, J.; van der Linde, R. *J. Coatings Tech.* 2000, *72,145052.*
10. *Proceedings of the MRS Meeting*, Singh, R.; Hofmann, H.; Senna, M.; Partch, R.; Muhammed, M., Eds.; Materials Research Society: Warrendale, PA, 2002, Vol. 704.
11. Smay, J.; Cesarano III, J.; Lewis, J. *Langmuir* 2002, *18, 5429-37.*
12. Steigerwald, J.; Murarka, S.; Gutman, R. *Chemical Mechanical Planarization of Microelectronic Materials*; Wiley Interscience: New York, NY, 1997.
13. Matijevic, E. *Langmuir* 1994, *10, 8-16.*
14. Rittner, M. *Bull. Am. Ceram. Soc.* 2002, *81, 33-36.*
15. Thayer, A. *Chem. & Eng. News.* May 6, 2002, p 23-30.
16. Goedert, R.; Becker, R.; Clements, A.; Whittaker,T., III. *J. Opt. Soc. Am. B.* 1998, *15, 1442-62.*
17. Powell, R. *Chemical Process Review* No. 21; Noyes Development Corp.: Park Ridge, NY, 1968.
18. Smisek, M; Cerny, S. *Active Carbon: Manufacture, Properties and Applications*; Elsevier: Amsterdam, 1970.
19. Mattson, J.; Mark, H., Jr. *Activated Carbon: Surface Chemistry and Adsorption from Solution;* Marcel Dekker: New York, NY, 1971.
20. Cooney, D. *Activated Charcoal: Antidotal and other Medial Uses,* Drugs and Pharmaceutical Science; Marcel Dekker: 1980, Vol. 9.
21. private communication from Cabot Corp.
22. Zlaczower, I. In reference 8; Chapter 11.
23. Gerspacher, M. In reference 8; Chapter 8.

24. Didenko, Y.; Suslick, K. *Nature* 2002, *418, 394.*
25. Bonard, J.; Stora, T.; Salvetat, J.; Maier, F.; Stockli, T.; Duschl, C.; Forro, L.; de Heer, W.; Chatelain, A. *Adv. Mater.* 1997, *9, 827-31.*
26. Coughlin, R.; Azra, F.; Tan, R. In reference 1, p 44-54.
27. Steele, W.; Karl, R. In reference 1, p 55-60.
28. Walker, P.; Janov, J. In reference 1, p 107-115.
29. Kiselev, A. In reference 1, p 88-100.

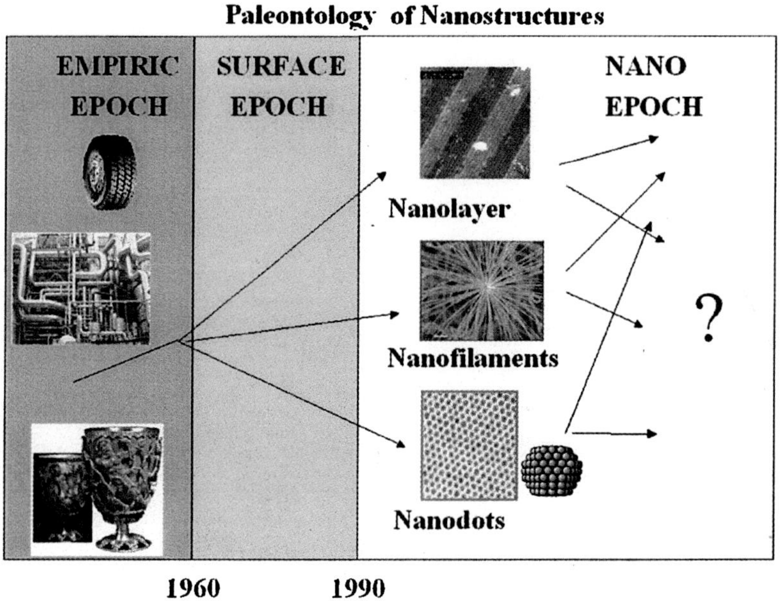

Plate 1.1. Paleontology of Nanostructures

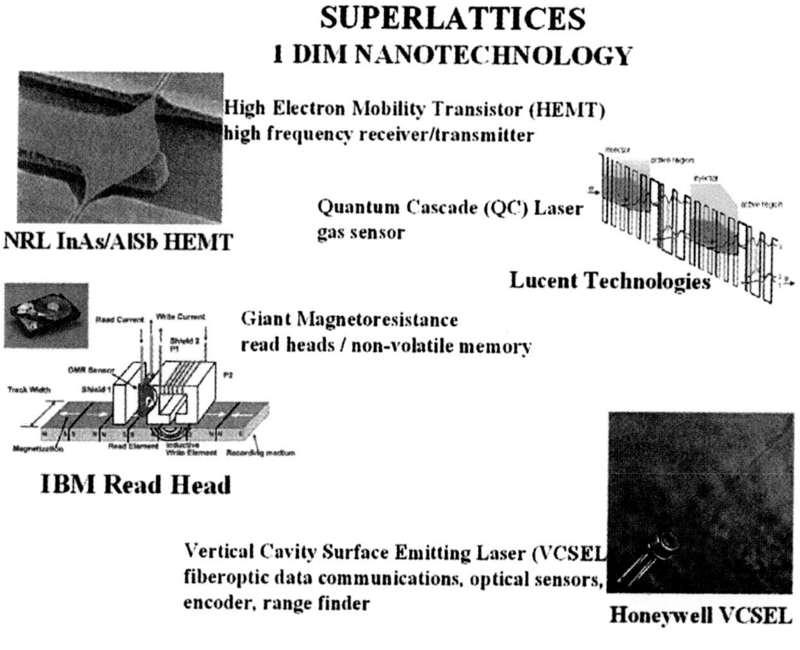

Plate 1.2. Superlattice-based Commercial Devices

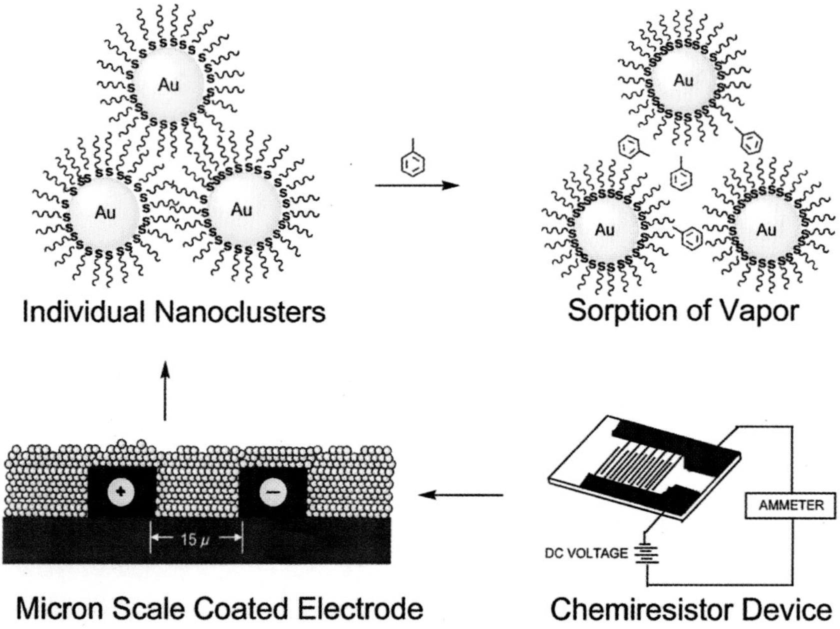

Plate 3.1. Schematic of the Metal-Insulator-Metal ensemble (MIME) sensor concept. A micron-scale interdigital electrode is coated with a film of alkanethiol stabilized gold nanoclusters and exposed to toluene vapor. The toluene adsorbs into the alkanethiol monolayer shell, and the consequent swelling causes an increase in the separation distance between gold cores and a reduction of electron tunneling between them. (Reproduced from NRL Review. U.S. Government work in the public domain.)

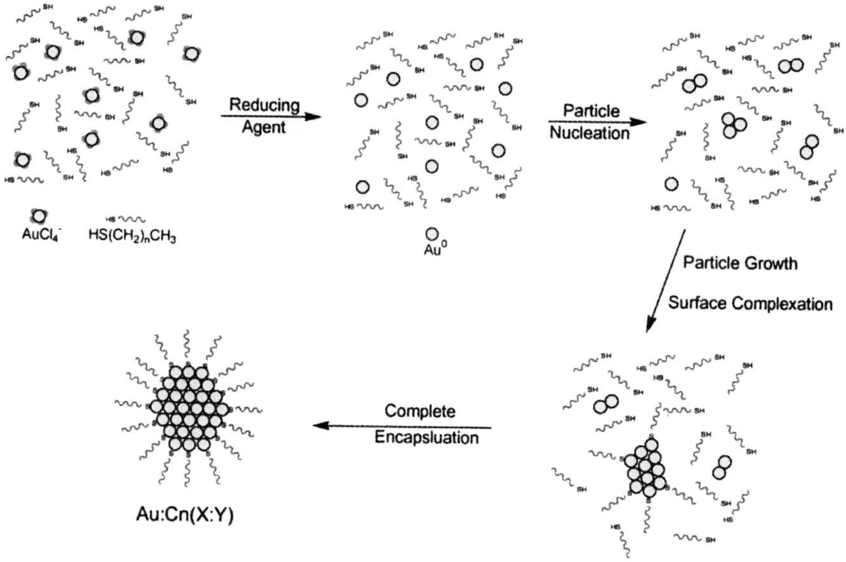

Plate 3.2. Schematic of the alkanethiol stabilized cluster synthesis. After reduction of the gold ions, the competitive processes of gold particle growth and alkanethiol surface complexation determine the size of the gold nanocluster.

Table 1. Au:Cn(X:Y) Cluster Series Core and Shell Dimensions and Bulk Electrical Conductivity ($(ohm\ cm)^{-1}$)

Cn/(X:Y)	(1:3)	(1:1)	(3:1)	(5:1)	(8:1)	R_{Shell} (nm)
C_{16}		8×10^{-11}		1×10^{-8}	4×10^{-7}	1.0
C_{12}	2×10^{-9}	1×10^{-8}	5×10^{-8}	1×10^{-6}	3×10^{-6}	0.80
C_8	5×10^{-9}	1×10^{-6}	7×10^{-6}	2×10^{-5}		0.60
C_6	8×10^{-8}	6×10^{-6}	2×10^{-4}	6×10^{-4}		0.50
C_4	2×10^{-6}	2×10^{-4}				0.40
R_{Core}(nm)	0.65	0.88	1.2	1.8	2.4	

SOURCE: NRL Review. U.S. Government work in the public domain. See page 3 of the color insert.

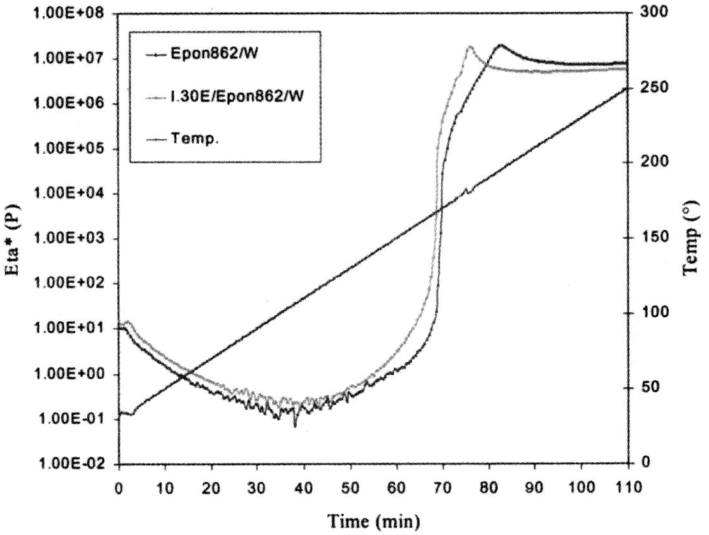

Plate 7.1. Viscosity of Epon 862/W and 6% I.30E/Epon 862/W at a heating rate of 2°C/mon.

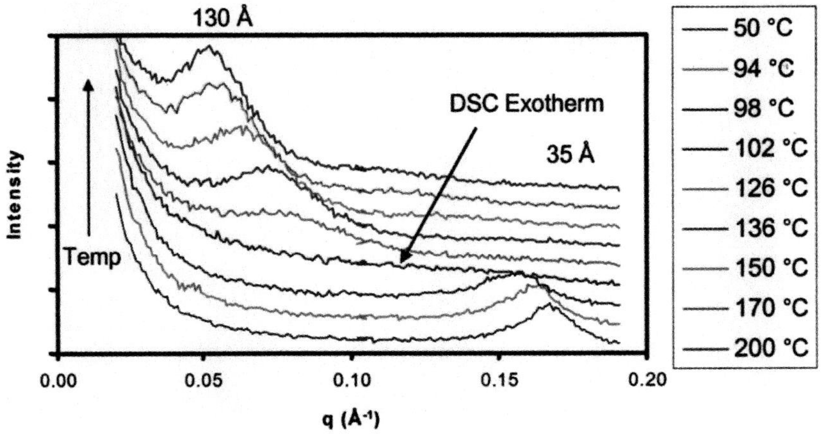

Plate 7.2. In-situ small-angle x-ray scattering of 6% I.30E/Epon 862/W at 2°C/min.

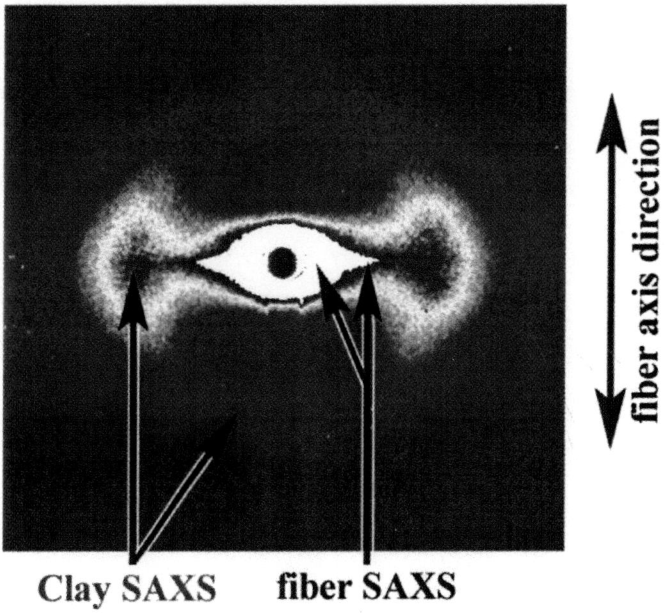

Plate 7.5. Small angle x-ray scattering of the composite of IM7/I.30E (6% wt. in epoxy)/Epon 862/W.

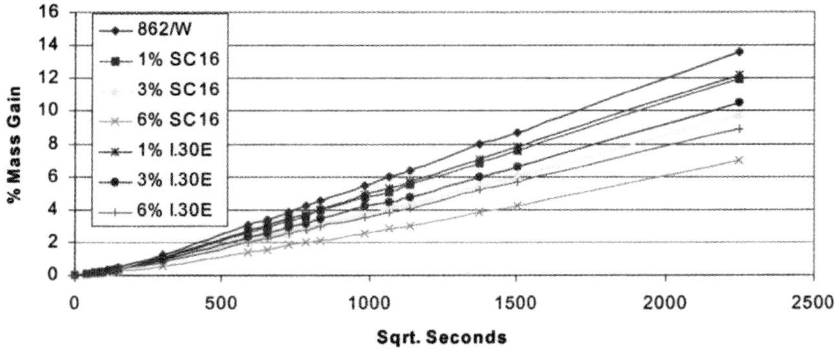

Plate 7.6. Methanol uptake of Epon 862/W; 1%, 3%, 6% SC16/Epon 862/W; and 1%, 3%, 6% I.30E/Epon 862/W nanocomposites.

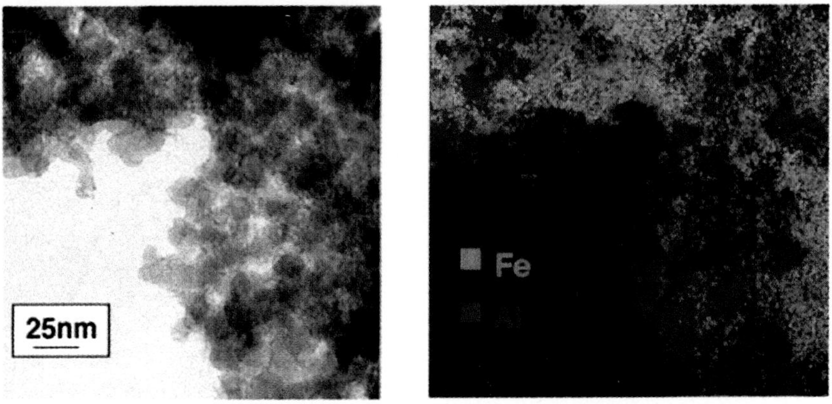

Plate 14.1. Transmission electron micrograph (TEM; left image) and energy filtered TEM (EFTEM; right) of a sol-gel derived Fe_2O_3/UFG Al aerogel. The EFTEM image is a color-coded map of the elemental distribution of iron oxide (green) and aluminum (red) in the energetic nanocomposite.

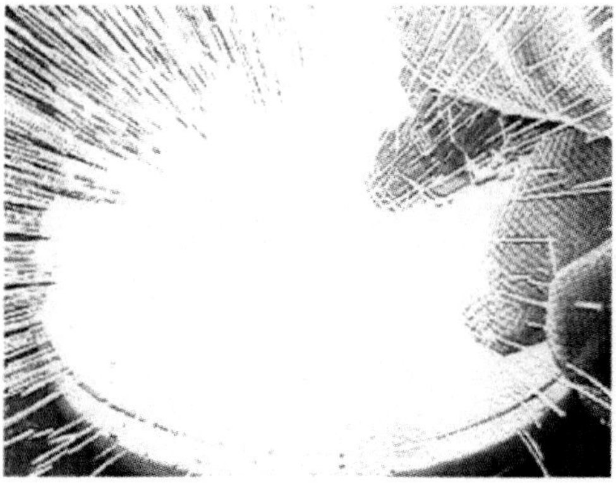

Plate 14.5. Photo of the thermal ignition of a Fe_2O_3/UFG Al nanocomposite.

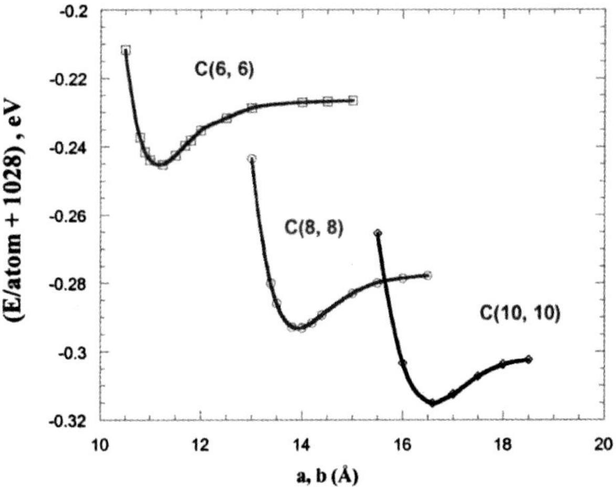

Plate 19.1. Total energy/atom as a function of intertube distance (Å). The energies are scaled by a value of 1028 (eV).

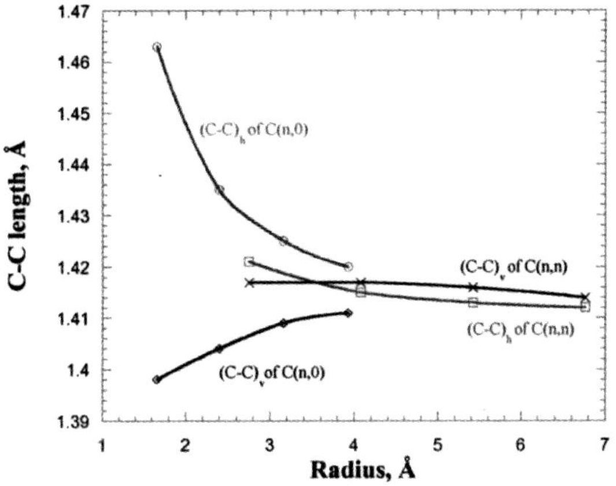

Plate 19.2. Bond lengths as a function of the tube radius (Å).

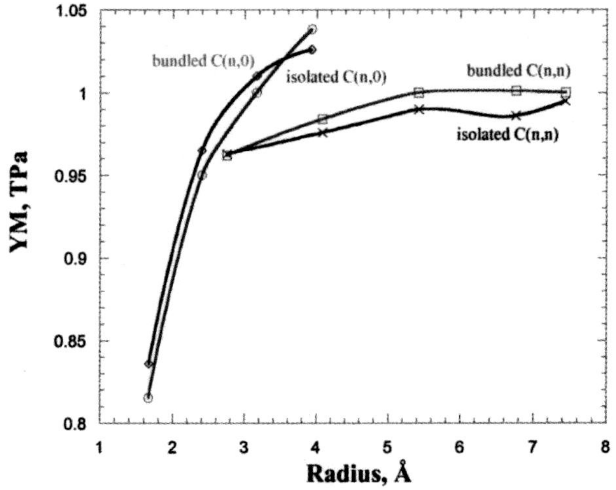

Plate 19.3. Young's modulus (TPa) vs. tube radius; 3.4 (Å) was used for the intertube separation in calculating V_0.

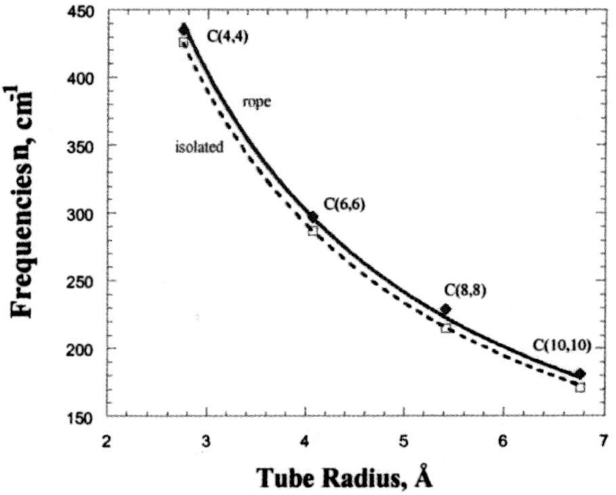

Plate 19.4. RBMs as a function of tube radius. The symbols are the calculated values and the lines are the fitted to the A/R curve.

Nanoenergetics

Chapter 13

Nanoenergetics Weaponization and Characterization Technologies

Alba L. Ramaswamy[1], Pamela Kaste[2], Andrzej W. Miziolek[2], Barrie Homan[2], Sam Trevino[2,3], and Michael A. O'Keefe[4]

[1]ECE Department, University of Maryland at College Park, College Park, MD 20742
[2]Army Research Laboratory, Aberdeen Proving Grounds, MD 21005–5001
[3]National Institute of Standards and Technology, 100 Bureau Drive, Stop 3460, Gaithersburg, MD 20899–3460
[4]NCEM, Materials Science Division, Lawrence Berkeley National Laboratory, 1 Cyclotron Road, MS 72, Berkeley, CA 94720

Nanoenergetic ingredients from nano-metal particles and carbon nanotube energetic material encapsulators to nano-particles of explosives and oxidizers are being investigated for incorporation in novel formulations for weaponization applications. Advanced characterization technologies such as LIBS (Laser Induced Breakdown Spectroscopy) & high resolution electron microscopy provide powerful tools for the advancement in the science of nanoenergetics.

The central component of modern weapon systems is the energetic material either in the form of an explosive or a propellant. The research in novel energetic materials and their weaponization is at the heart of defense. Nano-science and technology will not only provide new materials, propellants and explosives but also greatly enhance the understanding of these "powerful" materials.

Explosives and propellants have been in existence for many years prior to the discovery of dynamite by Nobel in 1867. However the initiation, sensitivity and safety aspects of explosives and some of the fundamental notions utilized by chemists in the synthesis of new energetic molecules are still to date primarily empirical. With the advent of new technologies for nanoscale characterization, major advances in our understanding of the atomistic/ molecular behavior of energetic materials are under way. This fundamental knowledge-basis is of assistance in the development of improved novel energetic materials and nanoenergetics for application in defensive weapons with increased effectiveness.

Nanoenergetic materials refer to nano-particles of energetics such as explosives, oxidizers and fuels either in "free" form or encapsulated. Nanoenergetic materials may provide a more rapid and controlled release of energy and can be used to improve various components of weapons or munitions such as the ignition, propulsion, as well as the warhead part of the weapon (*1*). In the latter application, nanoenergetics hold promise as useful ingredients for the thermobaric (TBX) and TBX-like weapons, particularly due to their high degree of tailorability with regards to energy release and impulse management (*1*). Finally novel energetic materials offer the promise of much higher energy densities and explosive yields in comparison to conventional explosives (*2*).

This article describes recent developments in nanoscale research studies on energetic materials. It begins by discussing fundamental aspects of initiation where the energetic reactions start at the "nano-scale". Control of the energy release may be obtained by targeted atomic/molecular-scale tailoring. The release of the energy in a rapid and controlled manner may furthermore be improved by utilizing nano-particles of the energetic components in propellant mixes. This is described in the section that follows. High-resolution electron microscopy imaging down to the atomic level of nano-aluminum is illustrated as an example of the characterization of the nanoenergetic ingredients. In order to incorporate the nano-particles safely in the formulations, carbon nanotube encapsulation is being investigated for these purposes. In addition carbon nanotubes may provide additional beneficial properties to novel propellant formulations. Finally LIBS (Laser Induced Breakdown Spectroscopy) characterization is described for in-line sensor development in future nano-metal manufacturing process optimization. LIBS is a technique that can provide in-situ information on the formation, growth, and passivation of the nanoenergetic metal particles during production and ultimately during their combustion.

Atomic/ Molecular Fundamentals of Energetic Materials

Nanoscale research studies were performed and revealed that the start or nucleus of a chemical reaction in energetic materials is localized at the nanometer scale and forms inhomogeneously on the surface of the material through the formation of nano-scale "reaction sites" with crystallographic structure or form (*3*). The energetic material may thus in itself be categorized in the class of "nanomaterials". The sensitivity to initiation of different crystalline polymorphic forms of the same energetic material is known to differ. Similarly single crystals of energetic materials such as PETN (penta-erythrotritol tetranitrate) and nitromethane show a shock initiation sensitivity anisotropy (*4*). The crystal structure of the energetic materials is thus known to control the initiation sensitivity of the same. How the crystal structure affects the initiation has been studied experimentally and theoretically (*5*) where it was found that the orientation of the molecules surrounding a given molecule in the crystal lattice appears to influence the initial decomposition reaction of the molecule.

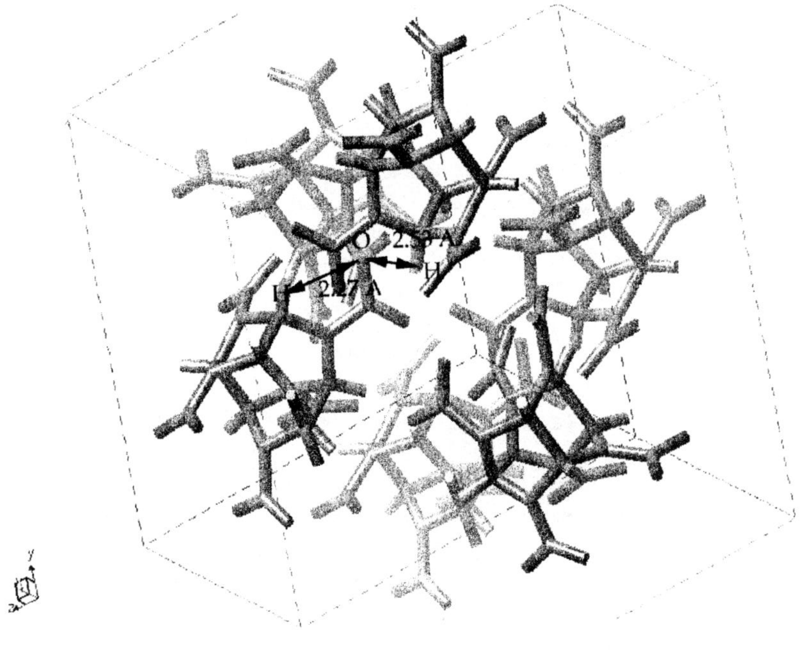

Figure 1 Metastable intermolecular trigger reaction mechanism in epsilon CL-20 by formation of a water molecule between two neighboring CL-20 molecules. The H-O-H angle is 109.56°.

This can be understood for example by comparing the crystal structure of the different polymorphs of HMX (cyclotetramethylene tetranitramine). The delta polymorph crystal structure is configured such that six out of eight of its oxygen atoms are positioned relative to the surrounding hydrogen atoms in an optimum position to form water molecules, one of the main reaction products of explosives and only two out of eight atoms in the alpha polymorph are in such favorable positions (*5*). Delta in fact is the most sensitive polymorph to initiation. Fig. 1 depicts the same phenomenon in the crystal lattice of ε-CL-20.

Further evidence comes from early work where the size of the smallest decomposition nucleus was being estimated (*6*). Experimentally (*7*) it was found that when rapidly moving atomic and nuclear projectiles are "fired" onto a lead azide surface, detonation occurs except when alpha-particles or electrons are used as projectiles. Alpha-particles and electrons have the smallest cross-section and would only activate a single molecule of lead azide. When ions of argon and mercury were used, which are large enough to activate neighboring molecules, explosion takes place. This supports the fact that two neighboring molecules are necessary for the formation of the first "reaction site" and this physio-chemical effect has been termed "metastable trigger intermolecular initiation mechanism"(*5*).

Standard Arrhenius kinetic analysis indicates that in the thermal decomposition of energetic materials the explosive reaction occurs as a result of the breaking of the first bond, say the C-N bond in the explosive molecule. However in the Arrhenius kinetic analysis a transitional excited state of the molecule, of which not much was known was assumed prior to bond breaking. Since the transitional excited state is short-lived, it can only be analyzed with the most advanced techniques of high-resolution temporal laser spectroscopy. Dr. A. Zewail, 1999 Nobel prize for chemistry for the development of femtosecond laser spectroscopy, studied the combustion reaction between hydrogen and carbon dioxide in the atmosphere. He found that the reaction, $H + CO_2 \rightarrow CO + OH$ crosses a transitional state of HOCO (1000 fs). The transitional excited state incorporates a bonding oxidation-reduction reaction, which is followed by bond rupture only 1000 fs later.

An oxidation-reduction reaction requires the re-arrangement of the electrons in the orbitals and is thus fundamentally electronic. This agrees with the band gap observations by Dr. B. Kunz and the detonation velocities of energetic materials. The beauty in the crystals of solid explosives is that the 3-D arrangement and repetition of atoms is "frozen" in space and reactions start to occur in the solid phase. Thus different arrangements provide for more or less favorable oxidation-reduction reactions and may be handled in the near-future by 3-D atomic/molecular predictive codes. Such tools should provide additional data for the chemical synthesizers to target and tailor novel energetic materials with improved properties.

Nanoenergetic Components of Propellants

The number and type of nano-energetic materials produced world-wide is increasing due to the potential impacts on future energetic formulations, micro-detonators, micro-initiators etc. Typical nano-energetic ingredients, which are being investigated for incorporation in formulation mixes can be sub-divided into inorganic nano-fuels, nano-oxidizers and/or nano-explosives (*8,9*). Their incorporation in formulations, has shown the potential to create or promote an extremely efficient and rapid combustion of propellants, leaving little or no un-reacted residue. The introduction of nano-energetic components may thus potentially improve the performance properties of propellant formulations by increasing the control and rate of the localized "energetic" chemical reactions as well as leading to more environmentally clean or benign formulations. Different architectures in which nano-scale energetic materials are embedded in energetic matrixes can be studied and designed with the added support of theoretical modeling at the molecular level.

Nanoaluminum particles manufactured by NSWC-IH have been examined at a resolution of 1.6Å using the NCEM (National Center for Electron Microscopy) Atomic Resolution Microscope (ARM). The high-voltage JEOL ARM-1000 microscope (*10,11*) was used at an electron energy of 800 keV and a magnification of 600kX to ensure adequate electron beam penetration of the whole (non-sectioned) nano-aluminum particles. The images obtained from the ARM were analyzed by computer imaging at the University of Maryland, and local areas of the micrographs enlarged. Traditional high-resolution transmission electron microscopy of samples at atomic resolution requires the sectioning of the material into fine slices. The ARM combined with digital imaging of the micrographs has the advantage over the traditional technique that no sectioning is required to obtain atomic-scale resolution imaging of the inside the particles.

Figure 2 shows two fused nanoaluminum particles. The crystalline lattice of the aluminum particles is evident. The interatomic spacing can be measured directly from the micrographs and is ~ 4 A, which agrees with the unit cell lattice spacing of 4.0496 A for the cubic close-packed unit cell of aluminum. A grain boundary is visible running vertically between the two fused nano-particles. The different orientations of the atomic planes in the two "grains" are observable. Finally the oxide coating (or partly hydroxide coating as evidenced from prompt gamma neutron activation analysis PGAA, see below) in the micrograph, appears as amorphous or lacking localized crystalline order except in an outermost crystalline "layer" which is composed of aluminum hydroxide due to the absorption of water molecules on the surface of the aluminum oxide. Analysis of other areas of the oxide/ hydroxide coating on different particles show locations of localized crystalline order, especially on the outermost surfaces and exfoliation of single molecular laminar sheets or foils of the hydroxide as they "peel" off. This is depicted in the micrograph of Figure 3 where the exfoliation of single molecular laminar sheets of surface aluminum hydroxide can be observed.

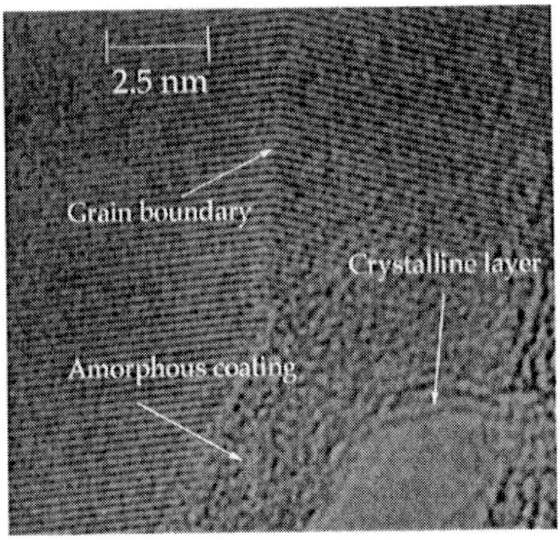

Figure 2. Two fused nanoaluminum particles revealing a grain boundary and oxide coating

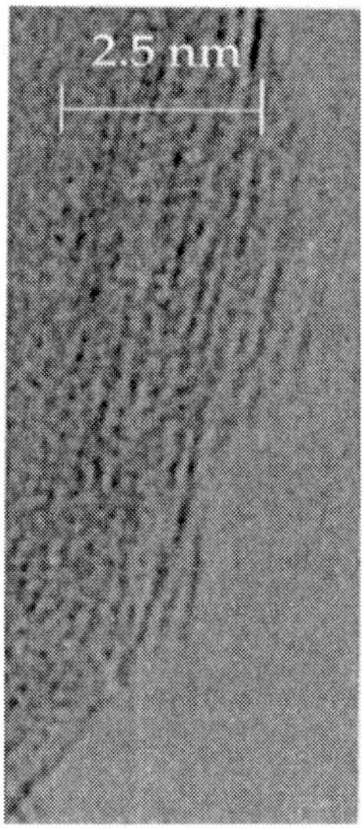

Figure 3.Oxide coating on nanoaluminum particle revealing exfoliation of aluminum hydroxide surface single molecular laminar sheets

The single "sheets or foils" can contain tens of aluminum hydroxide molecules along their length, as measurable from enlargements of the micrographs. The smaller is the length or size of the crystalline laminar sheets, the greater may be the exfoliation as evidenced by the surface layers, about 1to 3 nm in length, which appear to exfoliate first. This could be a clue to reducing the "barrier" to reaction of the nanoaluminum particles when incorporated in a formulation as a nano-fuel. Figure 4 depicts the unit cell of aluminum hydroxide with a possible exfoliation plane shown. X-ray diffraction measurements performed on the nanoaluminum indicate an overall amorphous coating. This is evidence for the fact that the areas of localized order, which are present in the laminar sheets, are very small or "nanoscopic".

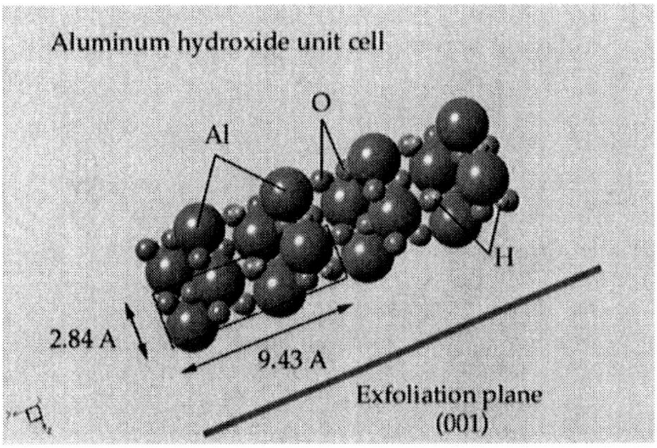

Figure 4. The unit cell of aluminum hydroxide

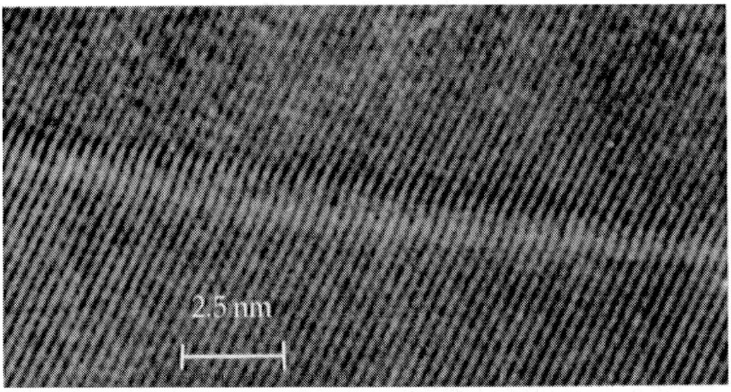

Figure 5. "Dislocation" inside aluminum nanoparticle

Finally Figure 5 shows the image of a dislocation found inside a nanoaluminum particle. The presence of dislocations appears rare.

In summary, the images at atomic level are evidencing nano-scale phenomena of coating exfoliation, which appears to depend on the amount of local crystallization, length or size of the crystalline layers/ "sheets" and perhaps chemistry i.e. hydroxide versus oxide. The form of the coating may thus be tailored to improve the aging characteristics as well as affecting the "barrier" to ignition and improving the combustion properties. In addition the nanoaluminum particles were found to be relatively defect free and single crystalline in nature.

Furthermore impurities such as Fe, Cu and K (and N for Russian ALEX) have been detected by x-ray photoelectron spectroscopy (12). Prompt gamma neutron activation analysis PGAA performed at NIST confirmed the presence of the same impurities as well as H in the form of hydroxide. These characterization techniques are providing evidence for the fact that the impurities appear to be concentrated in the coating. Tailoring of the impurities may be used to facilitate or trigger the oxidation reactions of the nanofuel as well as potentially triggering other "energetic" reactions. In fact a new class of "nano-fuels" which may be imparting improved burning rate characteristics to propellants is developing known as "mechanical alloys" (*13*). A "mechanical alloy" is defined as a metastable metal-based compound where a phase change is induced in the material, which affects the ignition delay and combustion rate of the formulation; the materials are rendered metastable by using solid solutions of a base metal as solvent and another component either a metal or gas such as hydrogen, as a solute forming nanoparticles of Al-Mg, Al-Mg-H, B-Mg etc. (*13*).

The technologies to produce "nano-oxidizers" and/ or "nano-explosives" are under development. Typical targeted compounds consist in PETN (penta-erythrotritol tetranitrate), RDX (cyclotrimethylene trinitramine), HMX (cyclotetramethylene tetranitramine) and HNS (hexanitrostilbene). The most common technique utilized is the RESS or Rapid Expansion of Supercritical Solutions. This technology is normally applied in the manufacture of nano-powders of inorganics and ceramics. It consists in the rapid expansion through a fine nozzle of a super-cooled solution of the energetic material in liquid super-critical carbon dioxide. The technique has been successfully applied by Dr. J. Kramer at LANL to manufacture nano-sized particles of PETN. The LANL laboratory plans to utilize the same set-up in the production of nano-sized HNS as well as up-scaling the production of PETN. Other Laboratories through-out the world are also developing techniques for the manufacture of nanosized explosives and oxidizers.

Functionalized Carbon Nanotubes for Nanoenergetic Applications

Recent groundbreaking advances in the development of novel nano-scale materials and technologies, have shown the potential to radically improve the safety and performance properties of propellant formulations. One of the major emerging technologies in the field of nano-materials is that of the carbon nano-tube. These long thin cylinders of carbon, consisting of graphitic single-atom layers "rolled" up into cylinders, were first discovered in 1991 by S. Iijima (*14*) using high-resolution electron microscopy. The cylinders are typically a few nanometers in diameter and several microns in length. Their longitudinal strength is reported (*15*) to be up to one hundred times that of steel! The carbon nano-tubes possess a very high electrical conductivity along their lengths reaching stable current densities (*16*) of up to 10^{13} A/cm^2 which corresponds to 120 billion to 3 trillion electrons per second! (a typical 18-gauge copper wire carries a current density of up to 10^2 A/cm^2) and the thermal conductivity is extremely high, approaching that of diamond (*17*), the material with one of the highest reported thermal conductivities. The properties of carbon nano-tubes clearly set them apart as a new candidate ingredient for energetic formulations.

Carbon nanotube (CNT) incorporation in propellant formulations offers the potential for burning rate and combustion tailoring due to the very high thermal conductivity of the CNTs. Furthermore the exceptionally high electrical conductivity of CNTs may be exploited for the improvement of propellant initiation for either an advanced plasma-based initiator (which is under development), or for use with more conventional electrical initiators. Finally the mechanical properties of the CNTs may be reflected in improved mechanical properties for the formulation as a whole. Research is being pursued to derivitize these compounds for incorporation in propellant mixes, ultimately with energetic functional groups that would improve performance and facilitate dispersion into a polymer matrix.

Figure 6 shows an electron micrograph of PEI (polyethyleneimine)-modified carbon nanotubes (magnification of 40,000x). By exploiting the naturally-occurring carboxylic acid impurities of the nanotubes, chemical titration was performed by Dr. M. Bratcher at ARL to attach the PEI polymeric chains at the defect sites. The tubes are distinct with a center core and an outer wall consisting of six to nine graphitic layers.

The lengths of the tubes are on the order of 20 µm, while the average diameter is approximately 20 nm. Clear surface features or defects occur at regular intervals of about 20-60 nm, typical of the "bamboo" CNT morphology and the PEI-functional groups have attached at the -COOH terminated carbon bonds in the "bamboo" defect sites (*18*) as evidenced by the micrograph of Figure 6 and depicted in Figure 7.

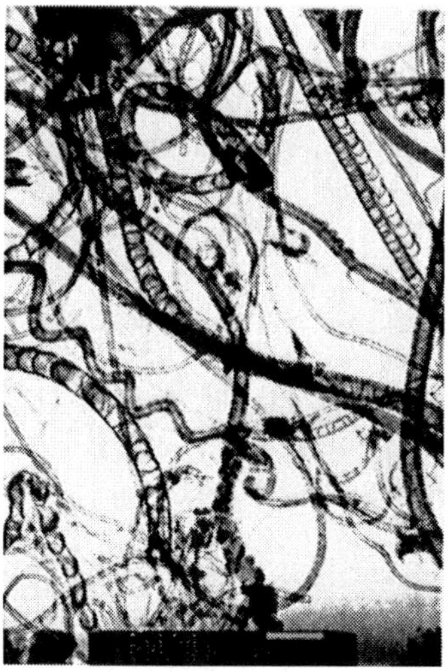

Figure 6. High-resolution transmission electron micrograph of polyethyleneimine-modified carbon nanotube (40,000 x)

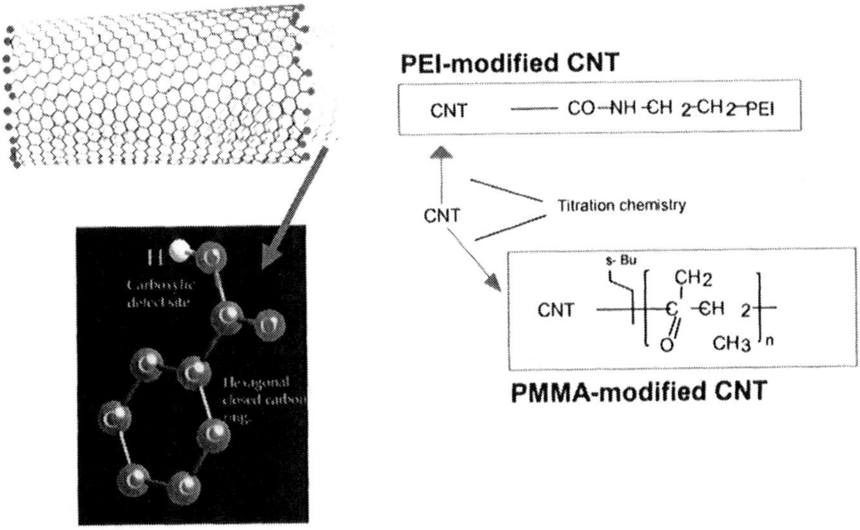

Figure 7.Titration "treatment" of carboxylic defect sites in CNTs to produce modified-CNTs

D-GC-MS (Desorption Gas Chromatography-Mass Spectroscopy) results for the virgin CNT samples showed no indication of organic solvents present, consistent with the fact that these materials are produced under conditions of high temperature and pressure with metal catalysts. The D-GC-MS analysis of the polyethylenimine-modified (PEI) and poly(methylmethacrylate)-modified (PMMA) CNT samples (*18*) was performed before and after vacuum exposure. The PEI vacuumed-samples desorbed with a 400 °C pulse yielding acetic acid. The non-vacuumed sample yielded species such as the THF solvent and 1,3-cyclohexadiene, and ethynlcyclobutane. Compared to the non-modified tubes, the PEI-modified tubes seem to be slightly resistant to thermal decomposition with the 400 °C pulse, particularly in the case of the vacuumed tubes. With the dual D-GC-MS 400/700 °C pulse treatment, the PEI-modified samples yielded larger fragments (multiple-ring compounds) than the unmodified nanotubes.

PMMA-treated samples had solvent levels that greatly exceeded either the PEI- or unmodified CNT samples, as exemplified by the very strong solvent peaks recorded, compared to no peaks for the unmodified CNTs. The levels diminished with vacuum treatment, but because the original level was so high, the solvent levels in the vacuumed PMMA samples still remained far higher than those discussed above. In summary the D-GC-MS experimental results show that the carbon nanotubes are quite thermally stable, and that decomposition can be very dependant on factors such as the occluded solvent level, used during modification and the nature and extent of the modification. This can be understood by comparing the different D-GC-MS behaviors for the pure CNTs and the modified-CNTs at the different pulse temperatures of 400°C and 700°C, both with and without vacuum, where the PEI-modified vacuum treated samples, as described above, were shown to have released most of the occluded solvents, whereas the PMMA-modified samples still retained some solvents after vacuum treatment. The affinity of carbon nanotubes for retaining solvents or species of interest will be further investigated. Such information may be important for assessing the chemical purity of the tubes and to assist in the design of the modified or derivatized carbon nanotubes.

Prompt gamma neutron activation analysis (PGAA) of the CNTs was performed at NIST and used to determine the C/H ratio, which was found to be 600:1. From the carbon-to-carbon bond lengths, assuming an hexagonal carbon closed-ring structure along the whole length of the CNTs and the fact that defect sites occur at regular intervals, the defect density was estimated from the PGAA data and was found to correspond with a defect repeat distance of 147 nm. The electron microscopy measurements gave a defect repeat distance of 20-60 nm. The mis-match in the estimate indicated that the CNTs did not consist of 100% hexagonal carbon rings but that additional defects may be present in the form of 5-membered or "squeezed" carbon rings amidst the 6-membered rings, reducing the repeat defect distance. In addition, the estimate did not take into account any tube chirality or "twist", which is common for CNTs and would further reduce the computed defect repeat distance along the length of the tube.

Novel techniques for incorporating "nano-energetic" particles into nanotubes are under investigation. The unique structure of carbon nano-tubes, may be used for confining the "nano-energetic" crystals in a nano-matrix with a positive effect on vulnerability properties. Thus, the possibility exists for simultaneously improving reactivity and stability of energetic materials, usually opposing characteristics. Figure 8 shows the concept being applied for the nanotube encapsulation of nanoenergetic particles.

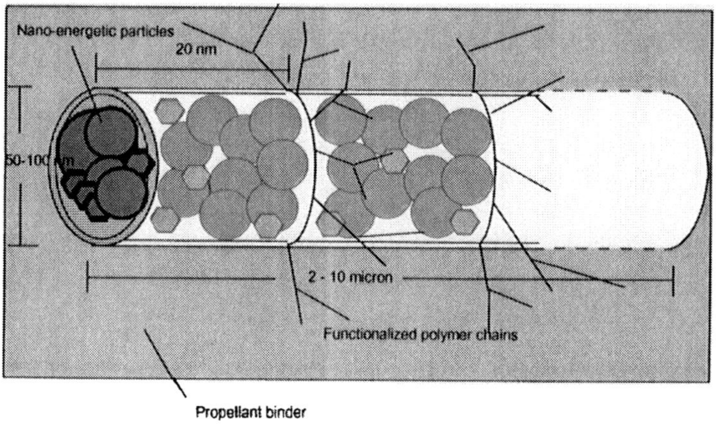

Figure 8. Concept for CNT encapsulation of nanoenergetic particles

Laser Induced Breakdown Spectroscopy

Understanding how nanoenergetic materials are both made and consumed requires the ability to monitor these processes with real time in-situ diagnostic techniques. Laser Induced Breakdown Spectroscopy (LIBS) is an optical technique that can detect all the elements simultaneously from very small samples of material. Only four elements are needed to implement this technique: an excitation source, delivery and collecting optics, a detector with wavelength dispersion capability, and a computer for control and analysis. Because of these relatively simple requirements, a complete LIBS system can be made compact, rugged, and fairly inexpensively. Spectrometers are now becoming commercially available that have the spectral range required to capture emissions from all the elements.

The LIBS technique measures the wavelength dependent emission from plasma generated from the interaction of a laser with the material. From this emission fingerprint, the chemical composition of the material can be determined in real-time and in-situ. For efficient plasma generation, the excitation laser needs to deposit a minimum 10 mJ of energy in less than 20 μs - well within the capabilities of current laser technology. Although different wavelengths may be more efficient for specific applications, the availability, low cost, and portability of lasers such as the Nd:YAG that emit in the infrared region are more than sufficient to reliably sample most materials.

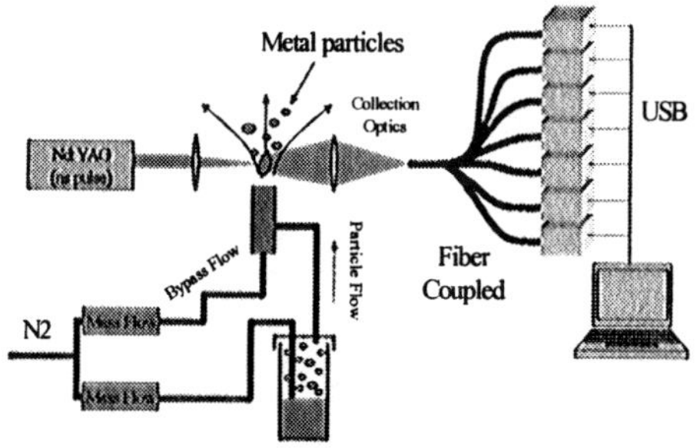

Figure 9. Metal Aerosol experimental

The excitation laser interacts with the bulk material by ablating a small amount of surface. This ejected material, stoichiometrically identical to the surface, is then ionized in the incident radiation producing ions and free electrons. Through a reverse Bremsstrahlung process, photon energy is coupled into the electrons thereby increasing their temperature. The hot electrons then transfer energy through collisions to the ions and neutrals raising their temperatures as well as creating more ion-electron pairs that can participate in further energy absorption. This process continues throughout the duration of the laser pulse and can raise the temperature to 20,000 K or more. The generated plasma has two components in its emission characteristics. One

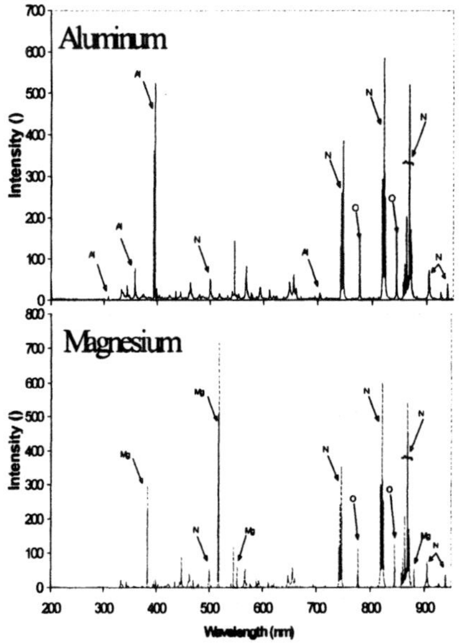

Figure 10.. LIBS spectra of metal aerosol

component is manifested as a broadband continuum radiation. Added to this background is a narrow recombination spectrum that contains the elemental information. The continuum radiation, again mainly due to the Bremsstrahlung process, is relatively short lived and can be removed from the spectrum using a temporal gate on the detection equipment. After a suitable delay, the emission from the plasma is mainly due to the recombination of the ions and electrons due to this process's longer lifetime. As each species has its own spectral signature, collection of the latter emission provides information about the identity of the original substance. Although the amount of sample needed is small, on the order of nano to pico-grams, the test does remove some of the sample. Therefore, it would be an advantage to be able to detect all the elements using a single laser shot. As all elements emit in the wavelength of 200-900 nm, full characterization of an unknown substance with a single laser pulse requires a spectroscopic technique that can simultaneously capture this full bandwidth. The application of LIBS for the detection and characterization of metal particles will provide another tool to probe the formation as well as the combustion processes. LIBS is a technique that can provide in-situ information on the formation, growth, and passivation of the energetic metals.

As a first cut, LIBS has been applied to a metal aerosol in open air. A diagram of the apparatus is shown in Figure 9. Flowing the gas through a container of material generates a particle-laden stream. This stream is then diluted, reducing the particle concentration to ensure only one particle in the measuring volume. Commercially supplied 20-40 μm aluminum or magnesium particles were studied. Both of these materials have a layer of oxide as they were exposed to air. Typical LIBS spectrums from these materials are shown in

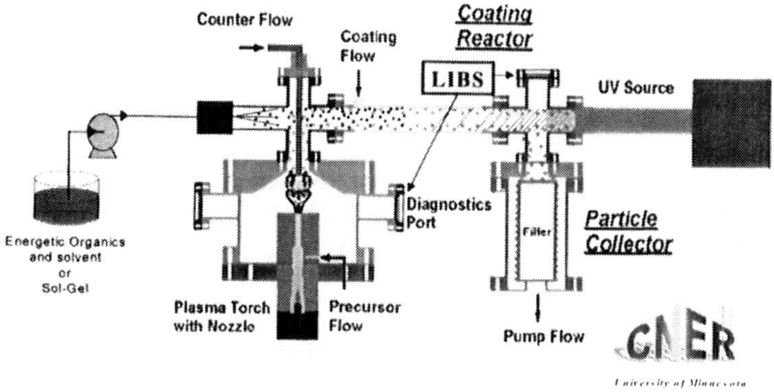

Figure 11. Nanometal generation and passivation facility at the University of Minnesota -Center for NanoEnergetic Research, directed by Prof. M. Zaccariah (1).

Figure 10. In both of the spectrums, peaks can be seen that are signatures for elemental metal along with the species nitrogen and oxygen. The oxygen peaks are due in part to the presence of the oxide layer surrounding the particles. Although the sample volume is small compared to the flow diameter, entrainment of atmospheric oxygen cannot be ruled out. To eliminate the signal due to the atmospheric oxygen, a chamber is being developed that will allow complete control of the environment.

To better understand the processes that produce nano-sized metals using a DC plasma torch, LIBS will be used to characterize the particles from the initial stages of growth to the passivation of the material by various coating technology. A schematic of the facility is shown in Figure 11. Sampling the articles using LIBS applied through the port nearest the plasma torch can monitor the growth processes. Because of the pyrophoric nature of elemental metals, a passivation coating will need to be applied. As both oxide and organic materials are under development, the broad spectrum capabilities of LIBS is well suited to provide diagnostic information as to the coating efficiency.

Conclusions

A summary of the nanoscale research studies on energetic materials described shows how fundamental aspects of the initiation of energetic materials are being unraveled. This is important in providing additional data for the chemical synthesizers to target and tailor novel energetic materials with improved properties and to assist them in the development of novel energetic materials with higher energy densities and explosive yields in comparison to conventional explosives. The "nanoscale" research studies revealed that the orientation of the molecules surrounding a given molecule in the crystal lattice of an energetic material appears to influence the initial decomposition reaction of the molecule. Two neighboring molecules appear necessary for the formation of the first "reaction site" through a physio-chemical phenomenon termed "metastable trigger intermolecular initiation mechanism". In the crystals of solid explosives the 3-D arrangement and repetition of atoms is "frozen" in space and reactions start to occur in the solid phase. Thus different arrangements may provide for more or less favorable oxidation-reduction reactions and thus better allow for the targeted atomic/ molecular scale tailoring of the energetic materials. This permits an improved control in the release rate of energy from the materials.

An improved control in the rate of energy release and combustion of propellant formulations may also be obtained by the incorporation of dispersed nano-particles of explosives, oxidizers and/or fuels. Thus for example an increase in the impulse achieved for a propellant may be obtained by the incorporation of nano-explosives, without the risk of a DDT (deflagration to detonation transition). For this purpose both the manufacture, characterization and practical testing of the nano-components is critical. Since the particles are "nanoscopic" in size, characterization techniques with atomic-scale resolution are highly valuable. An example of the characterization of nanoaluminum at atomic resolution was described together with the findings related to the exfoliation of the "crystalline" coating, an issue of relevance in the aging and combustion characteristics of the particles. Finally the safe incorporation of the nano-particles in formulations is being studied through the encapsulation of the nano-particles in carbon nano-tubes. For this purpose derivitization and dispersion of CNTs in propellants is under investigation. D-GC-MS findings have revealed the high affinity of carbon nanotubes for retaining solvents during the chemical titration process for derivitization. The data obtained may be important for assisting in the optimized design of the modified or derivatized carbon nanotubes.

Finally as described above the possibility to characterize the "nanoscopic" particles is of utmost importance for the development of the applications and science of the nanoenergetic materials. A novel technique known as LIBS (Laser Induced Breakdown Spectroscopy) is being developed into a compact, rugged and fairly inexpensive system or instrument for the in situ practical determination of the chemical transformations of nano-metals. This new characterization technology monitors the elemental composition of the nano-particles during the manufacturing processes, to characterize their initial stages of growth and passivation of the material. The system may also be used to characterize the combustion processes of the nano-metal particles. Such an instrument capability is extremely valuable for the in-situ real-time diagnostics, which will permit an improved control or optimization of the nanoenergetic manufacturing processes, determine coating efficiencies as well as increase the understanding of the combustion processes for the improved tailoring of these potential ingredients of the future.

References

1. Miziolek A.W. ; Nanoenergetics: An Emerging Technology Area of National Importance;AMPTIAC Quarterly: A Look Inside Nanotechnology, Vol. 6, No.1, Spring 2002.
2. Miller R.S., Research on New Energetic Materials, MRS Research Society Symposium Proceedings –Decomposition, Combustion and Detonation Chemistry of Energetic Materials, Nov. 27-30, 1995, page 3, Editors T.B. Brill, T.P. Russell, W.C. Tao and R. B. Wardle.

3. Ramaswamy, A.L., Microscopic Mechanisms Leading to the Ignition (Initiation), "Hot Spots" (Initiation Sites), Deflagration, Detonation in Energetic Materials, Proceedings of 11th International Detonation Symposium, Colorado, Aug., 1998.
4. Dick J.J., Mulford R.N., Spencer W.J., Pettit D.R., Garcia E., Shaw D.C., J. Appl. Phys. 70, 3572, 1991.
5. Ramaswamy, A.L., *Microscopic Initiation Mechanisms in Energetic Material Crystals*, Journal of Energetic Materials, Vol. 19, p.195-217, 2001.
6. Garner W.E., Trans. Faraday Soc., Vol. 34, 985-1008, 1938.
7. Kallman H., Shranker W, Naturwissenschaften, 21-23, Vol. 379, 1933.
8. Jianguo C., Zhanym Z., and Xiu D., *Microparticle Formation and Crystallization Rate of HMX with Supercritical CO_2 Antisolvent Recrystallization*, Chinese J. of Chem. Eng., 9, (3), p. 258-261, 2001.
9. Tillston T.M., Gash A.E., Simpson R.L., Hrubesh L.W., Satcher J.H., Poco J.F., *Nanostructured Energetic Materials Using Sol-Gel Methodologies*, Journal of non-crystalline solids 285, p.338-345, 2001.
10. O'Keefe M.A., Hetherington C.J.D., Wang Y.C., Nelson E.C., Turner J.H., Kisielowski C., Malm J.O., Mueller R., Ringnalda J., Pan M. and Thust A., *Sub-Ångstrom High-Resolution Transmission Electron Microscopy at 300keV*, Ultramicroscopy **89** (2001) 4, 215-241.
11. Kisielowski C., Hetherington C.J.D., Wang Y.C., Kilaas R., O'Keefe M.A. and Thust A., *Imaging Columns of the Light Elements Carbon, Nitrogen and Oxygen at Sub-Ångstrom Resolution*, Ultramicroscopy **89** (2001) 4, 243-263.
12. Hooton, I., *Nanopowders –Characterization Studies,* Proceedings from DIA New Materials III Symposium, 2-4 April, 2002.
13. Dreizen, E., *Metal-Based Metastable Solid Solutions as a New Type of High Energy Density Material*, Proceedings of the 8th International Workshop on Combustion and Propulsion- Rocket Propulsion: Present and Future, Pozzuoli, Naples 16-21 June 2002.
14. Ijima, S., Nature, 354 56, 1991.
15. Pan, Z.W., Sie, S.S., Chang, B.H., Wang, C.Y., Lu, L., Liu, W., Zhou, W.Y., Li, W.Z., Qian, L.X., Nature, 394 6694, p.631-632, 1998.
16. Yakobson, B. I., Smalley, R. E., *Fullerene Nanotubes: C(1,000,000) and Beyond,* American Scientist, 85 p324-337, 1997.
17. Berber, S., Kwon, Y., Tomanek, D., Unusually High Thermal Conductivity of Carbon Nanotubes, Phys. Rev. Lett. 84, 2000.
18. Bratcher, M.S., Gersten, B., Kosik, W., Ji, H., Mays, J., *A Study in the Dispersion of Carbon Nanotubes,* Proceedings of the Materials Research Society, Fall 2001.

Acknowledgements

The authors would like to thank Dr. Dave Carnahan of NanoLab for contributing the CNTs. Dr. Matthew Bratcher is thanked for supplying the modified nanotube samples. Dr. Michael Schroeder, ARL-retired (Voluntary Emeritus Corp), is thanked for providing numerous literature articles on nanotube research that helped to improve our understanding of these materials. Dr. Rose Pesce-Rodriguez is thanked for assistance in experimental and data processing techniques for the D-GC-MS characterization. Dr. Cheng Yu Song of NCEM is thanked for the ARM microscopy. Dr. John Kramer of LANL is thanked for the valuable scientific discussions. The CAD/PAD Dept. NSWC-IH is thanked for supplying the nanoaluminum. ARL is especially thanked for support through contract No. 020603-8046 and ONR is especially thanked for support of some of the fundamental work. The NCEM is supported by the Director, Office of Science, through the Office of Basic Energy Sciences, Material Sciences Division, of the U.S. Department of Energy, under contract No. DE-AC03-76SF00098.

Chapter 14

Direct Preparation of Nanostructured Energetic Materials Using Sol–Gel Methods

Alexander E. Gash, Randall L. Simpson, and Joe H. Satcher, Jr.

Energetic Materials Center, Lawrence Livermore National Laboratory, L–092 Livermore, CA 94551

The utilization of nanomaterials in the synthesis and processing of energetic materials is a new area of science and technology. At LLNL we are studying the application of sol-gel chemical methodology to the synthesis of energetic nanomaterials components and their formulation into energetic nanocomposites. Here sol-gel synthesis and formulation techniques are used to prepare sol-gel nano Fe_2O_3/ Al (nano) pyrotechnic nanocomposites. The resulting materials were found to have superb mixing of metal oxide and fuel phases at the nanometer level. Thermal analyses revealed that these nanocomposites underwent the thermite reaction at much lower temperatures than mixtures prepared with micron-sized Fe_2O_3 and Al. We have observed that the nanocomposites burn much more rapidly than their conventional analogs. We are currently investigating the role that the size of the oxidizer and fuel particle sizes play individually on the thermal properties of the material.

Since the invention of black powder, one thousand years ago, the technology for making solid energetic materials has remained either the physical mixing of solid oxidizers and fuels (e.g. black powder), or the incorporation of oxidizing and fuel moieties into one molecule (e.g., trinitrotoluene (TNT)). The basic distinctions between these energetic composites and energetic materials made from monomolecular approaches are as follows.

In composite systems, desired energy properties can be attained through readily varied ratios of oxidizer and fuels. A complete balance between the oxidizer and fuel may be reached to maximize energy density. Table I is a summary of the some of the energy densities of composite and monomolecular energetic materials (*1*). Current composite energetic materials can store energy as densely as 23 kJ/cm^3. However, due to the granular nature of composite energetic materials, reaction kinetics are typically controlled by the mass transport rates between reactants.

Table I. Energy densities for several composite and monomolecular energetic materials (SOURCE: reproduced with permission from reference 1).

Energetic Material	Energy Density (kJ/cm^3)
Ammonium dinitramide/ Aluminum	23
Compression moldable	19-22
Strategic propellants	14-16
CL-20(neat)	12.6
Tritonal	12.1
HMX(neat)	11.1
LX-14	10.0
TATB(neat)	8.5
Comp. C-4	8.0
LX-17	7.7
TNT(neat)	7.6

In monomolecular energetic materials, the rate of energy release is primarily controlled by chemical kinetics, and not by mass transport. Therefore, monomolecular materials can have much greater power than composite energetic materials. A major limitation with these materials is the total energy density achievable. Currently, the highest energy density for monomolecular materials is about half that achievable in composite systems.

In composite energetic materials, decreasing reactant sizes effectively increases the interfacial surface area contact between oxidizer and fuel phases. Both experimental and theoretical efforts by Brown *et al.* and Sukai *et al.* indicate that a decrease in particle size resulted in a qualitative increase in burn rates in solid-solid mixtures of oxidizers and fuels (*2,3*). In another example, Son and co-workers at Los Alamos National Laboratory (LANL) have shown that pyrotechnic nanocomposites of MoO_3 and Al burn at extremely rapid rates (>100m/s) and that the propagation mechanism is convective and not conductive, as is the mechanism of conventional pyrotechnics (*4,5*). Therefore, it is desirable to combine the excellent thermodynamics of composite energetic materials with the rapid kinetics of the monomolecular energetic materials. Thus, developing and or improving new methods for the synthesis and processing of nanometer-sized oxidizers and fuels are needed for tailorable energy and power.

The synthesis of nanomaterials is being pursued via a number of methods. Physical methods include sonication and ball milling of solids (*6*). Chemical techniques include vapor condensation methods (*7-9*), micellular synthesis, chemical reduction (*11*), sonochemical synthesis (*10*), and the sol-gel methodology (*12-20*). For nanomaterials to have a noteworthy impact in the area of energetic materials processes for their synthesis must satisfy significant production specifications such as cost, health and safety, and reproducibility. The sol-gel method provides another approach to nanomaterials synthesis for energetic nanocomposites and, in some respects, it is a more suitable candidate method than those described previously.

The sol-gel method

Sol-gel chemical methodology has been investigated for approximately 150 years and has been employed in the disciplines of chemistry, materials science, and physics. Sol-gel chemistry is a solution phase synthetic route to highly pure organic or inorganic materials that have homogeneous particle and pore sizes as well as densities. The method is commonly used to prepare metal-oxide based materials. However, sol-gel methods do exist for the preparation of organic fuel based materials so it can also be used to prepare nanomaterials of both oxidizers and fuels (*12-14*). Its benefits include the convenience of low-temperature preparation using general and inexpensive laboratory equipment. From a chemical point of view, the method affords easy control over the stoichiometry and homogeneity that conventional methods lack. In addition, one of the integral features of the method is its ability to produce materials with special shapes such as monoliths, fibers, films, and powders of uniform and very small particle sizes. There are several excellent references on the sol-gel method, where more complete and additional information can be obtained (*15-17*).

The process is summarized in the scheme shown in Figure 1. A sol can be formed through the hydrolysis and condensation of dissolved molecular

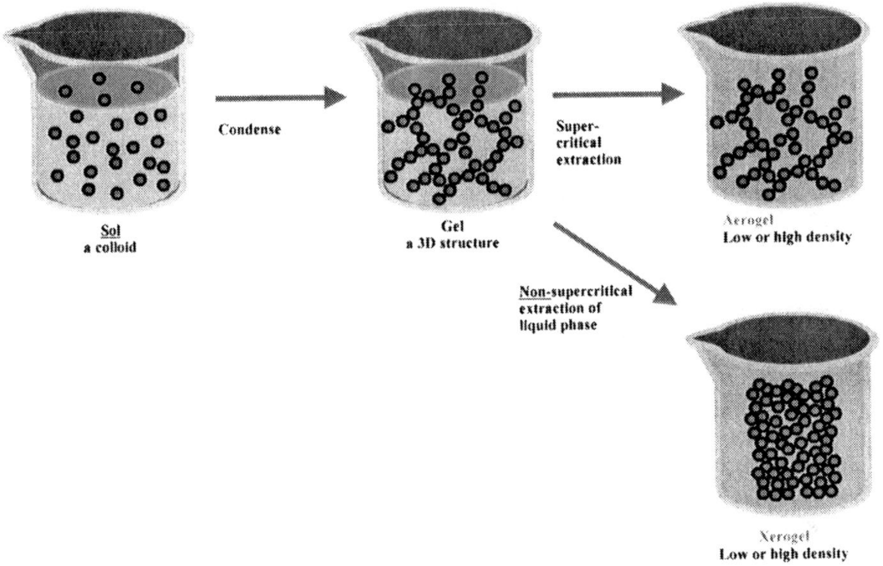

Figure 1. Steps in the sol-gel process.

precursors. This produces nanometer-sized particles, which aggregate to form clusters, with very uniform size, morphology, and composition. The pH of the solution, the solvent, the temperature, and the concentrations of reactants used dictate the size of the clusters, which can be from 1 nm to 1000 nm in diameter. By controlling the aforementioned solution conditions, the sol can be condensed into a robust gel. The linking together of the sol clusters into either aggregates or linear chains results in the formation of the stiff monolith. The gel can be dried by evaporation of the solvent to produce a monolithic xerogel or removed under the supercritical conditions of the pore liquid to produce an aerogel. Ambient drying results in the exertion of large capillary forces on the gel framework and causes in a significant amount of shrinkage of the material to produce a medium density material. Xerogel densities can be between 30-80% that of the bulk. With supercritical drying the capillary forces are effectively removed and aerogel densities are typically between 1-20% that of the bulk material.

Both aerogels and xerogels exhibit high surface areas and porosities. Typically the sol-gel materials are networks made up of nanometer-sized particles that are linked together in a network that contains pores of slightly larger dimensions (mesoporous ≈2-20 nm diameter). The homogeneous distribution and small diameter of the pores provide a convenient location for a second distinct chemical component to reside. We are interested in filling the pores with a phase (fuel or oxidizer) that will undergo rapid and highly energetic reactions with the skeletal gel component (oxidizer or fuel). In the example presented in Figure 2, the inorganic skeletal matrix is acting as the oxidizer, with the fuel particles embedded in the nanopores of this solid. Conversely, the skeletal component could be the fuel (e.g., organic sol-gel materials) with oxidizer in the pores. This figure illustrates the intimate mixing of oxidizers and fuels that is available via sol-gel methodology.

Scientists at LLNL have been actively investigating the application of sol-gel chemistry and nanoscience to the field of energetic materials for several years (*18-20*). The appeal of the sol-gel approach to energetic materials is that it offers the possibility to precisely control the composition, density, morphology, and particle size of the target material at the nanometer scale. These are important variables for both safety and performance considerations. The fine control of these parameters allows the chemist the convenience of making energetic materials with tailored properties. In addition, ambient temperature gelation and low temperature drying schemes prevent degradation of the energetic materials, and the water-like viscosity of the sol before gelation, allows casting to monolithic shapes.

Energetic Nanocomposites

Energetic nanocomposites are a class of material that have a fuel component and an oxidizer component intimately mixed on the nanometer scale. A portion of our work, in this area, has focused on the development sol-gel methods to synthesize porous monoliths and powders nano-sized transition metal oxides (i.e., Fe_2O_3, Cr_2O_3, WO_3, and NiO). When combined with metals such as aluminum, magnesium, or zirconium ensuing mixtures can undergo the thermite reaction (a scheme of the reaction is given below in **(1)**). In the thermite reaction the metal oxide ($M_{(1)}O(s)$) and

$$M_{(1)}O(s) + M_{(2)}(s) \rightarrow M_{(1)}(s) + M_{(2)}O(s) + \Delta H \quad \textbf{(1)}$$

($M_{(2)}(s)$) undergo a solid-state reduction/oxidation reaction, which is rapid and very exothermic (*21*). Some thermite reaction temperatures exceed 3000K. Such reactions are examples of oxide/metal reactions that provide their own oxygen supply and, as such, are difficult to stop once initiated. The energy densities for some thermite reactions are in the upper range of those in Table 1 (16-19 kJ/cc) (*22*). Thermitic compositions have found use in a variety of processes and products. They are used as hardware destruction devices, for welding of railroad track, as torches in underwater cutting, additives to propellants and high explosives, free standing heat sources, airbag ignition materials, and in many other applications.

Traditionally, thermites are prepared by mixing fine component powders, such as ferric oxide and aluminum. Mixing fine metal powders by conventional means can be an extreme fire hazard; sol-gel methods reduce that hazard while achieving ultrafine particle dispersions that are not possible with normal processing methods. In conventional mixing, domains rich in either fuel or oxidizer exist, which limit the mass transport and therefore decrease the efficiency of the burn. However, sol-gel derived nanocomposites should be more uniformly mixed, thus reducing the magnitude of this problem.

Preparation of nanosized metal oxide component by sol-gel methods

We have developed a sol-gel procedure for synthesizing monolithic and powdered aerogels and xerogels of nanostructured metal oxides from common inorganic salts (*23-25*). This is significant as historically, the sol-gel method has employed the use of metal alkoxide precursors. This synthetic route has proven to be an efficient, easy, and successful approach to the production of predominantly SiO_2, Al_2O_3, and ZrO_2-based porous materials. However, it is much less well developed for oxides of interest as pyrotechnics (i.e., Fe_2O_3, MoO_3, NiO, CuO, WO_3). The epoxide addition method has been successfully used at LLNL to prepare many different metal-oxide skeletons, of thermodynamically relevant oxides (e.g., Fe_2O_3, NiO, WO_3) for nanocomposite thermites. The addition of any one of several different epoxides to a solution containing the dissolved transition metal salt results in the formation of monolithic metal oxide gels. The as formed gel can be dried to either a xerogel or an aerogel whose characteristics

Iron oxide/aluminum nanocomposites

One transition metal oxide material of particular interest to this study is ferric oxide (Fe_2O_3). When ferric oxide powder is combined with powdered aluminum metal it forms the classic thermite reaction as described by Goldschmidt (*21*). The energy density for a stoichiometric mix of Fe_2O_3 and aluminum is 16.5 kJ/cc (*22*). We have reported previously on the formulation of Fe_2O_3/Al energetic nanocomposites via the *insitu* sol-gel synthesis of Fe_2O_3 in a suspension of ultra fine grain (UFG) Al nanoparticles, whose particle sizes range from 20-60 nm in diameter (*18,20*). The sol-gel Fe_2O_3 phase grows around and encapsulates the solid suspended Al particles to form an energetic nanocomposite like that shown in Figure 2. This process takes advantage of the viscosity increase of the sol-gel solution as it approaches its gel point. Once the gel forms the matrix is rigid and has effectively "frozen" the finely dispersed Al into place with in the gel network. The UFG Al was from LANL, and was prepared via dynamic vapor phase condensation. The gel nanocomposites have been dried to both aerogel and xerogel monoliths. Figure 3 is the photo of one such sol-gel Fe_2O_3/UFG Al aerogel monolith. Ambient and inert atmospheric drying of xerogels was done under ambient and elevated (~100°C) conditions. (**CAUTION**: In our hands, the wet pyrotechnic nanocomposites cannot be ignited until the drying process is complete. However, once dry, the materials will burn rapidly and vigorously if exposed to extreme thermal conditions. In addition, the autoignition of energetic nanocomposites has been observed upon rapid exposure of hot ~100°C material to ambient atmosphere.)

Qualitatively, the Fe_2O_3(*s*)/UFG Al(*s*) energetic nanocomposites appear to burn much more rapidly and are more sensitive to thermal ignition than conventional thermite powders. This is not unexpected as the ignition threshold of UFG aluminum powders depends upon its physical particle morphology (*7*). We are currently in the process of performing detailed burn rate measurements to more thoroughly quantify these observations.

The intimacy of mixing between oxidizer and fuel is very likely an important factor in the behavior of these materials. One problem that all nanomaterials suffer from is the tendency for agglomeration into larger aggregates. In such a case the properties of the resulting composite may be influenced by the size of the aggregate and not the size of the individual nanoparticles, thus defeating the potential benefit of the nanomaterial. Even though nanosized components are used, there is no guarantee that the sol-gel composite will have such mixing. To characterize this degree of mixing we have analyzed our composite material using energy filtered transmission electron microscopy (EFTEM) at LLNL.

EFTEM can be used to construct an elemental specific map of a given TEM image (*26*). Plate 1 contains a TEM image of a Fe_2O_3/UFG Al aerogel and the EFTEM maps for aluminum and iron respectively. The areas of the image representative of sol-gel Fe_2O_3 are colored green and those of Al nanoparticles are red. The EFTEM images in Plate 1 reveal the superb mixing that the sol-gel Fe_2O_3 and UFG Al in the composite. The two component phases are intimately mixed on this length scale with no evidence of agglomeration.

Some of the thermal properties of these materials have been investigated. Figure 4 contains the differential thermal analysis (DTA) traces for two energetic nanocomposites: sol-gel Fe_2O_3/UFG Al (top) and a dry powder mix of Fe_2O_3/Al (bottom), prepared from commercially available micron-sized powders. It is clear from these DTA traces that the thermal behavior of these materials is quite different. In the sol-gel Fe_2O_3/UFG Al nanocomposite there are thermal events at ~ 260, ~290, and ~590°C. We have determined that the two lower temperature events are related to a phase transition of hydrated amorphous iron oxide to nanocrystalline Fe_2O_3. We are currently evaluating the effect of this phase transition on the performance of the nanocomposite. The exotherm at ~590°C is the most interesting as it corresponds to the thermite reaction (confirmed by powder X-ray diffraction of reaction products). This

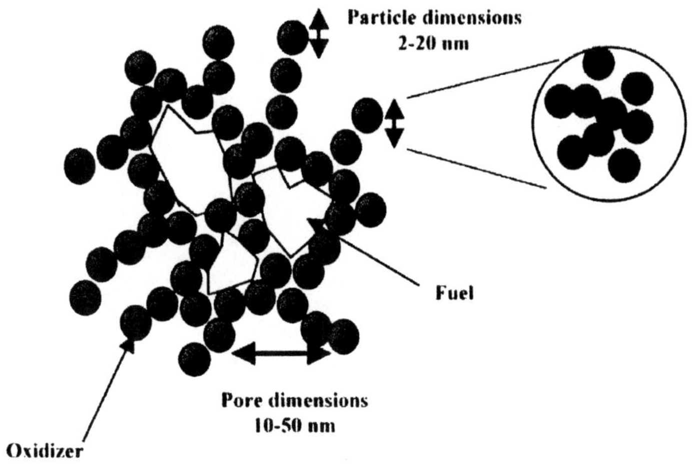

Figure 2. Schematic representation of sol-gel derived energetic nanocomposite, where the sol-gel skeleton is the oxidizer and the fuel particles are residing in the pores.

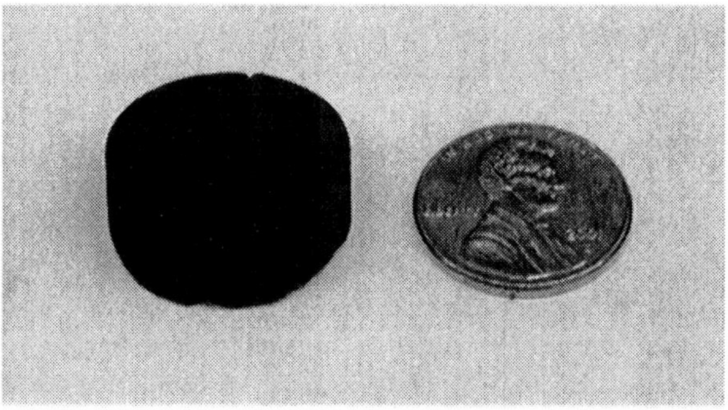

Figure 3. Sol-gel Fe_2O_3/UFG Al aerogel monolith.

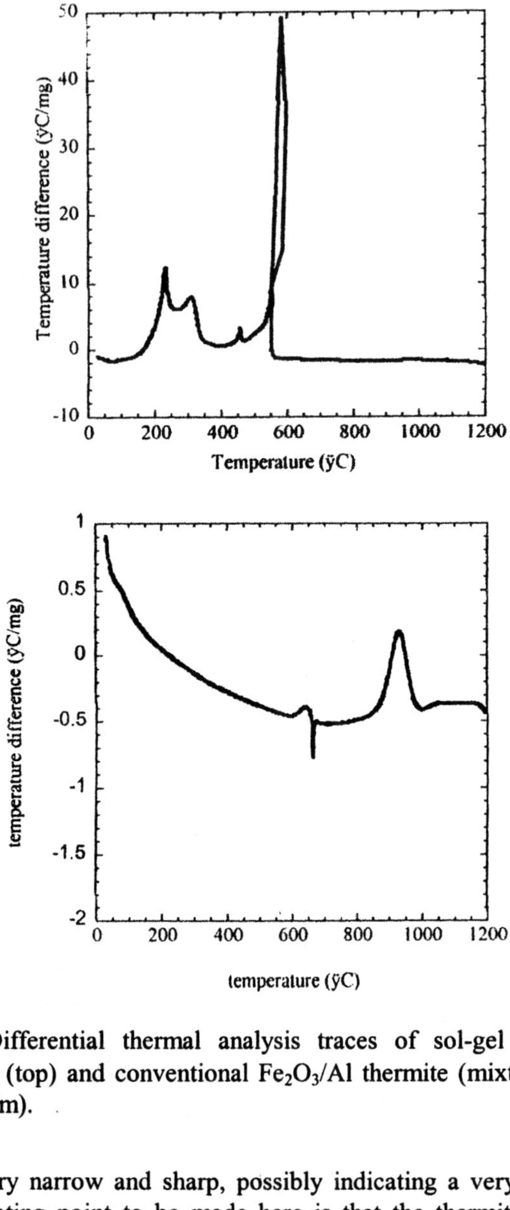

Figure 4. Differential thermal analysis traces of sol-gel Fe_2O_3/UFG Al nanocomposite (top) and conventional Fe_2O_3/Al thermite (mixture of μm-sized powders)(bottom).

exotherm is very narrow and sharp, possibly indicating a very rapid reaction. Another fascinating point to be made here is that the thermite reaction takes place at a temperature markedly below the melt phase of bulk Al (T_m = 660°C). This is very significant as in conventional thermite mixtures it is commonly thought that thermite reactions are initiated by the melting or decomposition of one of the constituent phases (*5*).

Alternatively, the DTA trace of the dry powder mix of Fe_2O_3/Al commercial micron-sized powders is significantly different than that of the nanocomposite. There are no low temperature events in this DTA, as the Fe_2O_3 phase is dry and crystalline. This is not surprising, as the oxidizer in this mixture is crystalline Fe_2O_3 and not amorphous hydrous sol-gel Fe_2O_3. The main

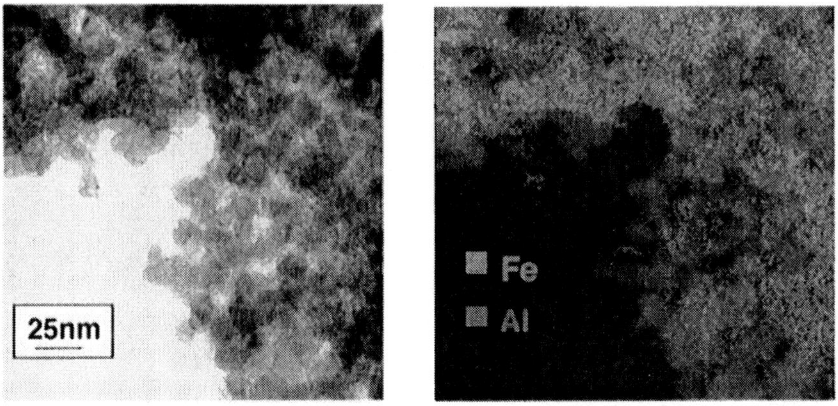

Plate 1. Transmission electron micrograph (TEM; left image) and energy filtered TEM (EFTEM; right) of a sol-gel derived Fe_2O_3/UFG Al aerogel. The EFTEM image is a color-coded map of the elemental distribution of iron oxide (green) and aluminum (red) in the energetic nanocomposite. *(See page 6 of color insert.)*

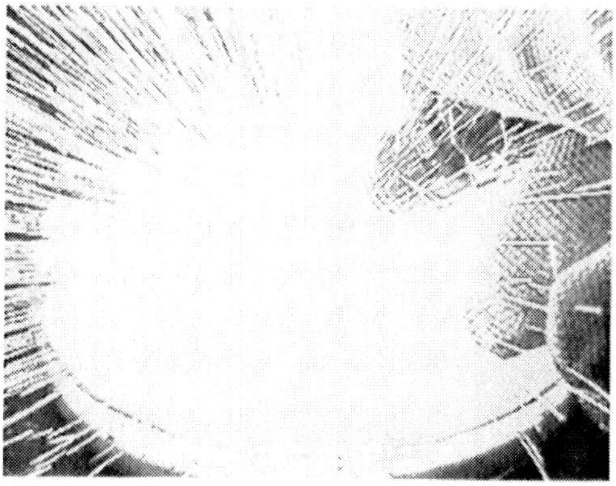

Figure 5. Photo of the thermal ignition of a Fe_2O_3/UFG Al nanocomposite.
(See page 3 of color insert.)

features of the trace occur at ~660°C and ~915°C. The 660°C endotherm corresponds to the melting of the micron-sized Al and the ~915°C exotherm is from the thermite reaction. In this case the thermite reaction takes place after the melting of Al. The ignition point of the traditional thermite material is ~325°C higher than that for the nanocomposite. This peak is also broader than that seen in the nanocomposite DTA and likely indicates a less rapid reaction. The DTA analyses indicate that the thermal behavior of the sol-gel Fe_2O_3/UFG Al and the conventional Fe_2O_3 composite are quite different. The grounds for these differences are still under active investigation in our laboratory.

Additional Materials:

As one can surmise, this sol-gel method allows for the addition of insoluble materials (e.g., metals or polymers) to the viscous sol, just before gelation, to produce a uniformly distributed energetic nanocomposite upon gelation. This process can be used for added materials with particle sizes from nanometers to millimeters and particle densities from low to high. In addition, to preparing Fe_2O_3(*s*)/UFG Al(*s*) nanocomposites we have prepared composites using micron-sized aluminum powders also. These materials also are readily ignited using thermal sources. They too seem to burn more rapidly and are more sensitive to ignition than conventional thermites. However, they burn more slowly and are less sensitive to ignition than the Fe_2O_3(*s*)/UFG Al(*s*) nanocomposites. The integration of polymers into the thermitic nanocomposites results in a new material with gas generating properties. This material may have potential use as a high temperature stable gas generator. It is important to note that under ambient conditions these materials are insensitive to standard impact, spark, and friction small-scale safety tests.

The sol-gel approach also allows for the relatively simple incorporation of other metal oxides into the metal-oxide matrix to make a mixed-metal-oxide material (*16,17*). Different metal-oxide precursors can be easily mixed into the solution, before the addition of the epoxide to dilute or enhance thermal or spectroscopic output.

Acknowledgements

This work was performed under the auspices of the U.S. Department of Energy by University of California Lawrence Livermore National Laboratory under contract No. W-7405-Eng-48.

References

1. Fried L.E., Howard W. M., Souers P. C. (1998) Cheetah 2.0 User's Manual Lawrence Livermore National Laboratory.

2. Brown, M.E.; Taylor, S.J.; Tribelhorn, M.J. *Pyrotechnics, Explosives, and Propellants* **1998**, *23*, 320-327.

3. Shimizu, A.; Saitou, J. *Solid State Ionics* **1990**, *38*, 261-269.

4. Son, S.F.; Asay, B. W.; Busse, J.R.; Jorgensen, B.S.; Bockmon, B.; Pantoya, M. *Proceedings of the 28th International Pyrotechnic Seminar*, **2001**, Adelaide, Australia, November 4-9, 2001.

5. Wang, L.L.; Munir, Z.A.; Maximov, Y.M. *J. Mater. Sci.* **1993**, *28*, 3693-3708.

6. *NANOMATERIALS: Synthesis, Properties, and Applications*; Edlestein, A.S.; Cammarata, R.C. Eds.; Institute of Physics, Bristol U.K., 1996.

7. Taylor T.N.; Martin J.A. *J. Vac. Sci. Technol. A* **1991,** *9*(3), 1840-1846.

8. Aumann C.E., Skofronick G.L., Martin J.A. *J. Vac. Sci. Technol.B* **1995,** *13*(2), 1178-1183.

9. Danen W. C. , Martin J. A. US Patent 5 266 132, 1993.

10. Suslick, K.S.; Price, G.J. *Annu. Rev. Mater. Sci.* **1999,** *29*, 295.

11. Klabunde, K.J.; Stark, J.V.; Koper, O.; Mohs, C.; Khaleel, A.; Glavee, G.; Zhang, D.; Sorensen, C.M.; Hadjipanayis, G.C. *Nanophase Materials,* Ed. Hadjipanayis, G.C.; Siegel, R.W.; Kluwer Academic Publishers, 1994, Netherlands, p.1-19.

12. Pekala, R.W. *J. Mater. Sci.*, **1989,** *24*, 3321-3227.

13. LeMay, J.D.; Hopper, R.W.; Hrubesh, L.W.; Pekala, R.W. *MRS Bull.*, **1990,** *XV*(12), 19-45.

14. Barral, K. *J. Non-Cryst. Solids* **1998,** *225*, 46-50.

15. Iler R. K. *The Chemistry of Silica*; Wiley: New York, **1979**.

16. Brinker C. J. , Scherer G. W. *Sol-Gel Science*; Academic Press, Inc.: San Diego, CA, **1990**.

17. Livage, J.; Henry, M.; Sanchez, C. *Prog. Solid St. Chem.* **1988,** *18*, 259

18. Simpson, R.L.; Tillotson, T.M.; Satcher, J.H., Jr.; Hrubesh, L.W.; Gash, A.E. *Int. Annu. Conf. ICT (31st Energetic Materials)*, Karlsruhe, Germany, June 27-30, **2000**.

19. Gash, A.E.; Simpson, R. L.; Tillotson, T.M.; Satcher, J.H., Jr.; Hrubesh, L.W. *Proc. 27th Int. Pyrotech. Semin.* Grand Junction, CO, July 15-21, **2000** p.41-53.

20. Tillotson, T.M.; Gash, A.E.; Simpson, R.L.; Hrubesh, L.W.; Thomas, I.M.; Poco, J.F. *J. Non-Cryst. Solids* **2001**, *285*, 338-345.

21. Goldschmidt H. *Iron Age* **1908**, *82*, 232.

22. Fisher S., Grubelich M. C. Proceedings of the 24th International Pyrotechnic Seminar **1998**, U.S.A., 231-286.

23. Tillotson, T.M.; Sunderland, W.E.; Thomas, I. M.; Hrubesh, L.W.H. *Sol-Gel Sci. Technol.* **1994**, *1*, 241-249.

24. Gash, A.E., Tillotson, T.M.; Satcher, J. H., Jr.; Hrubesh, L. W.; Simpson, R. L. *Chem. Mater.* **2001**, *13*, 999.

25. Gash, A.E., Tillotson, T.M.; Poco, J. F.; Satcher, J. H., Jr.; Hrubesh, L. W.; Simpson, R. L. *J. Non-Cryst. Solids* **2001**, *285*, 22-28.

26. Mayer, J. *European Microscopy and Analysis* **1993**, 21-23.

Chapter 15

Characterization of Nanoparticle Composition and Reactivity by Single Particle Mass Spectrometry

D. Lee[1], R. Mahadevan[1], H. Sakurai[1], and M. R. Zachariah[1,2,*]

[1]Departments of Mechanical Engineering and Chemistry, University of Minnesota, Minneapolis, MN 55347
[2]Current Address: Center for Nanoenergetics Research, University of Maryland, 2125 Glenn L. Martin Hall, College Park, MD 20742 (mrz@umd.edu)

In this paper we apply a recently developed single particle mass spectrometer (SPMS) to characterize the elemental composition of individual aerosol particles and study the kinetics of thermal decomposition of metal-nitrate aerosols. An aerodynamic lens inlet with high transport efficiency was implemented to deliver particles directly from the aerosol reactor to SPMS, where they are ablated and positively ionized by irradiation of a high energy pulsed laser. In this work we monitor the elemental composition of metal-nitrate particles and track changes during their thermal decomposition into corresponding metal-oxides. The results when modeled as first order reaction kinetics provide reaction kinetics parameters for the condensed-phase reaction. These results suggest that this technique may offer a new approach to study condensed-phase reactions including energetic material reactions with the benefits of minimized experimental artifacts linked with heat and mass-transfer effects due to the small sample size.

Introduction

Nanoparticles have been proved to have unusual and often desirable properties and are characterised by various methods, including traditional offline methods such as thermal analysis, electron microscopy, surface spectroscopies, elastic and inelastic light scattering, absorption, and diffraction. Online methods are primarily aimed at size and number concentration measurement (1,2). However, the quality of the resulting products made from nanoparticles is strongly coupled with both the particle physical and chemical properties.

Whereas online characterization of size is already a well developed method and morphology based on light scattering has received considerable attention in the last decade, chemical composition measurement is still an evolving technique (*3*). Typical offline measurement of the composition has disadvantages of a loss of temporal information and a distortion in particle properties during handling. On the other hand, recent developments of online aerosol mass spectrometer (OPMS) enable to measure the individual particle composition in real time. Aerosol mass spectrometry instrumentation can be classed into several groups based on particle inlet design, ionization techniques and types of mass spectrometers (*4*). A significant challenge in OPMS has been the efficient transmission of aerosol particles from atmospheric pressure to micro-torr pressures. The aerosol inlet to the mass spectrometer often constraints the particle size range which can be sampled and analyzed by the instrument. The interest in OPMS was triggered by a series of surface/thermal ionization methods (*5,6*) and followed by the first on-line laser desorption/ionization *time-of-flight* mass spectrometer (*7*) and an improved design with the help of a reflectron (*8*) which increased the resolution of the instrument.

Meanwhile, the measurements of condensed-phase reaction kinetics have been usually performed using conventional offline thermal techniques such as thermogravimetry (TG) and differential scanning calorimetry (DSC). Despite the fact that over 90 percent of condensed-phase reactions have been studied in this manner, there is still considerable controversy regarding the reliability of these methods (*9*). The literature is rife with examples indicating that traditional thermal analysis is plagued by experimental uncertainties associated with heat and mass transfer effects, which results in kinetic parameters that are dependent on experimental artifacts (*9,10*). Regarding the source of the artifacts, the aerosol mass spectrometer might be able to overcome the experimental artifacts because of minimal sample size in it.

We recently built a single particle mass spectrometer (SPMS) equipped with an aerodynamic lens inlet, a linear time-of-flight tube and a strong pulsed laser as an ionization source. Elemental compositions of a single particle are obtained with no molecular ions with the help of a very strong pulse laser. The most important benefit of the use of strong laser would be to achieve complete ablation and ionization of particles independent of relative sensitivity of constituent ions of particles, as was demonstrated by Reents et al. (*11,12*). The present study was aimed at two objectives: the first is to evaluate the ability of SPMS to quantatatively measure the elemental composition of individual particles and the second to apply SPMS for the first time to study reaction kinetics of nanoparticles. Although there is in principle no limitation on type of particle chose for study, we chose metal nitrates for our initial kinetic study, since they are used as precursors in production of metal oxides by spray-pyrolysis. We have also measured the elemental compositions of aluminum nanoparticles and organic coated aluminum particles as a preliminary step to study the reactivity of this class of energetic material.

Experimental

A. Single Particle Mass Spectrometer (SPMS)

A schematic of the SPMS is shown in Figure 1. The primary components of the system consist of an aerodynamic lens inlet, two differentially pumped vacuum chambers, a high power pulse laser, a linear Time-of-Flight (TOF) tube with a microchannel plate detector, and data acquisition system consisting of a fast digital sampling oscilloscope. The aerodynamic lens inlet is similar to the one developed by Liu et al. (*13, 14*) and is capable of focusing and transmitting aerosols with nearly 100% efficiency over a wide size range. The pressure inside the inlet after the pressure-reducing orifice was 1.6 Torr when the pressure upstream of the inlet was maintained at 1 atm, while the pressure in the first stage vacuum chamber, which was pumped by a 250 L/sec turbo molecular pump, was $\sim10^{-2}$ Torr. The second stage was pumped with a second 250 L/sec turbo pump to a pressure of $\sim10^{-6}$ Torr when the mass spectrometer inlet was open. Positioning of the aerodynamic lens inlet was adjusted with an X-Y translation stage mounted on a vacuum flange of the first-stage vacuum chamber. Alignment of the inlet tube with respect to the skimmer hole was accomplished with the X-Y translation stage by feeding charged particles into the vacuum system through the inlet, and maximizing the particle current detected with a Faraday cup detector connected to an electrometer while no voltage was applied to the extraction grids.

A frequency-doubled Nd:YAG laser operating at 10 hz was used with a pulse energy of 60 mJ/pulse. A spherical plano-convex lens with a focal length of 38 mm was mounted inside of the vacuum chamber to create a tight focus the at the center of the extraction field o the mass spectrometerand perpindicular to the particle beam. The laser power density at the focal point was approximately

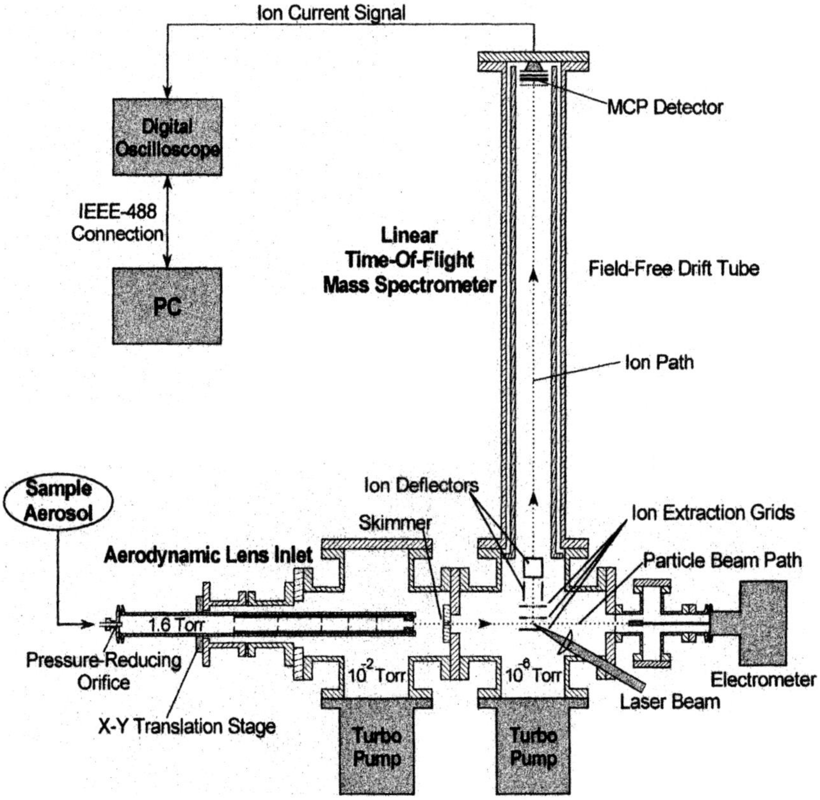

Figure 1. Schematic of the linear time-of-flight single particle mass spectrometer (SPMS)

1.7×10^{10} W/cm^2 assuming the beam diameter at the focal point be 0.3 mm and the laser pulse duration of 5 ns. Each laser shot was detected with a photodiode sensor to initiate time-of-flight measurement in the data acquisition system.

When the laser hit a particle, atomic ions within the particles were generated in the source region. Only positive ions, which had either one or multiple elemental charges, were accelerated by the high voltage supplied to extraction the plates, separated along the linear TOF tube, and detected by Z-stuck microchannel plates. The ion signal was stored in a 500-MHz digital storage oscilloscope which was capable of selectively recording spectra of only "successful" particle hits. A "successful" particle hit was defined as an event in which ion signal intensity exceeds a prescribed threshold.

B. Generation of Aerosols

Figure 2 shows the setup used to generate and process aerosols for both characterization of various aerosols and to study the kinetics of metal nitrates decomposition in the aerosol phase. A solution of metal nitrate or other samples, typically of 1 wt%, was prepared by mixing commercially available metal nitrates (> 99% in purity) or other commercial powdered samples with filtered deionized water, and atomized in air to generate aerosol droplets of the solution. The droplet-laden air was then passed through an aerosol dryer filled with dry silica, and delivered into a furnace reactor. The tube furnace had a heated length of 30 cm. The aerosol flow rate from the atomizer was about 2 L/min, and the flow rate in the furnace was monitored with a laminar flow meter in front of the furnace and was adjusted to 1.0 L/min at ambient pressure and temperature, by controlling the flow rate to the excess flow path split before the laminar flow meter. The output aerosol from the furnace was split into two, one which was delivered to the SPMS and the other for monitoring the aerosol concentration with an aerosol electrometer.

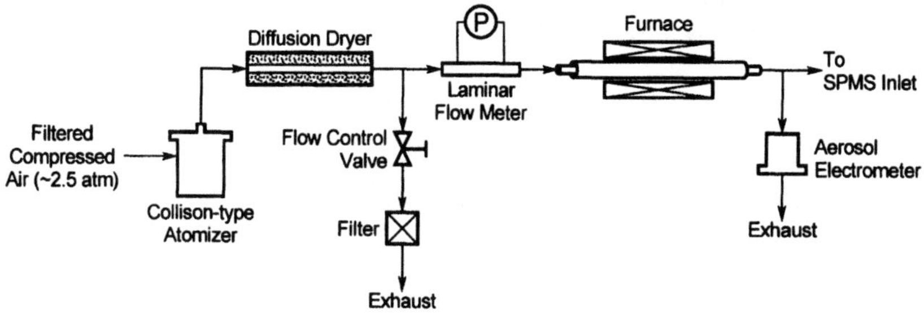

Figure 2. Schematic of the aerosol generator and themal reactor setup.

Characteristics of SPMS

A. Particle transmission efficiency and laser hit rate

The transmission efficiency of aerosol particles through the aerodynamic lens inlet and the skimmer in the SPMS was determined experimentally. Monodisperse dioctyl sebacate (DOS) particles of known sizes generated with a differential mobility analyzer were used for this characterization and assumed to carry a unit elemental charge. The aerosol current (number of particles per unit time) flowing into the pressure-reducing orifice was estimated from the particle concentration measured by the aerosol electrometer (see Figure 2) and the volume flow rate through the orifice. The aerosol current after the skimmer was measured by the Faraday cup as shown in Figure 1. The result is presented in Figure 3. Efficiencies greater than 60% were observed in the size range between 50 and 300 nm. Liu et al. (*13*) reported transmission efficiencies over 90 % for particles of 30-250 nm. The reduced efficiency for the SPMS inlet could be due to a broader beam diameter caused by slight misalignment of aerodynamic lens components and consequent beam blockage by the skimmer wall.

Kane et al. (*15*) defined the hit rate which was defined as the number of particles per unit time that were hit by the laser beam "successfully", as

$$Hit\ Rate = n_{aerosol} \cdot V \cdot T \cdot \frac{A_L}{A_P} \cdot L \cdot E_{ablation}$$

where $n_{aerosol}$ is the number concentration of sample aerosol before the mass spectrometer inlet, V is the volume flow rate through the inlet, T is the transmission efficiency through the inlet and differential pumping stages, A_L/A_P is the ratio of the cross sectional area of the laser beam to that of the particle beam, L is the laser duty factor which is defined as the particle transit time through the laser beam multiplied by the laser repetition rate, and $E_{ablation}$ is the ablation efficiency (the fraction of particles in the laser beam that produce a sufficient number of ions to be detected). The hit rate measured by using the sodium chloride aerosol without size selection was 0.33 hz when the particle number concentration and volume flow rate before the inlet were 10^6 #/cm^3 and 1 cm^3/s, respectively. The inlet transmission efficiency was 70% for 90-nm particles (see Figure 3) and the laser beam diameter was assumed to be 0.3 mm. The laser duty factor was calculated using the laser beam diameter of 0.3 mm, particle beam velocity of 170 m/sec, and the laser repetition rate of 10 Hz. In addition, the ablation efficiency is assumed to be 100%, which was believed to be reasonable since laser beam energy required for complete ablation and ionization was much less than the present beam energy. As a result, the particle beam diameter was estimated to be ~1.8mm and is larger than the Brownian-limit diameter (0.5 mm) calculated by Liu et al. (*13*). The discrepancy may reflect that our aerodynamic inlet was not perfectly aligned with respect to laser focus.

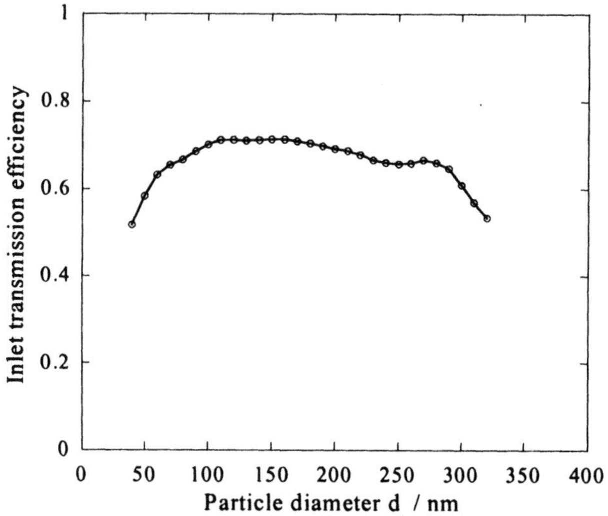

Figure 3. Particle transmission efficiency through the aerodynamic lens inlet and differentail pumping system.

B. Quantitative determination of the elemental stoichiometry of aerosol particles

Figure 4 shows examples of mass spectra obtained for an individual particle by a successful laser shot; (a) for sodium chloride and (b) for aluminum oxide. The aluminum oxide particles were generated by heating aluminum nitrate particles in the furnace at 800 °C. No molecular species were observed, and multiply charged atomic ion peaks were present in both spectra. Quantitation of the complete elemental composition of particles is possible if the particle can be completely ablated and ionized.

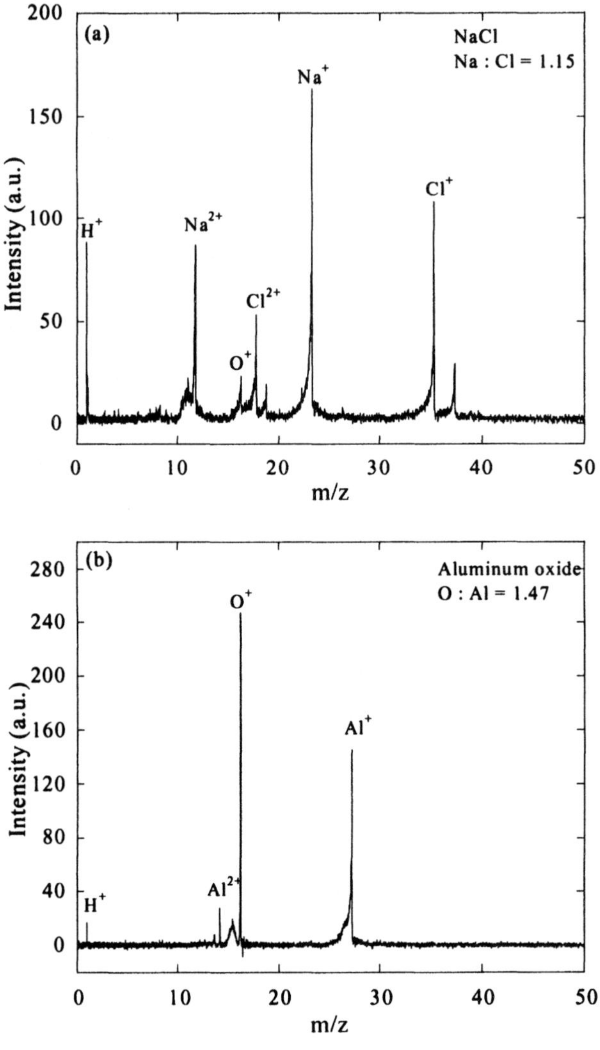

Figure 4. Single particle mass spectra for (a) sodium chloride and (b) aluminum oxide.

Figure 4(a) contains a considerable hydrogen peak which implies that the NaCl particles were not sufficiently dried. Metal nitrate had similar feature of hygroscopic. Therefore, the quantity of oxygen due to water present on the particle was subtracted from the total oxygen detected. Relative peak areas of each ion signals reflect directly their corresponding relative abundancesince the ion signals are proportional to the number of ions striking the detector. For the particles in Figure 5, the elemental stoichiometry of Na/Cl for sodium chloride, and O/Al for aluminum oxide were 1.15, 1.47, respectively. These values agree very well with the theoretical values, i.e. 1.00, and 1.50 for Na/Cl and O/Al, respectively. We summarize the measured elemental stochiometry of various aerosols (their spectra were not shown) in Table I. It should be noted that we excluded from analysis "unsuccessful" shots; those speactra that possess less than 10% doubly charged ions, according to the empirical criterion suggested by Reents and Schabel (*12*).

Table I. Stoichiometry ratios obtained by SPMS for various aerosols and comparison with theoretical values.

Particle	Ratio considered	Experimental value	Theoretical value	% Error
Aluminum nitrate	N : Al	3.08	3.00	2.7
Aluminum oxide	O : Al	1.47	1.50	-2.0
Sodium chloride	Na : Cl	1.14	1.00	14.0
	$^{35}Cl : ^{37}Cl$	3.00	3.13	-4.2
Strontium nitrate	N : Sr	1.95	2.00	-2.5
Silver	$^{107}Ag : ^{109}Ag$	1.12	1.08	3.7
Silver nitrate	N : Ag	0.98	1.00	-2.0
Iron nitrate	N : Fe	2.86	3.00	-4.7
Ammonium sulphate	N : S	2.27	2.00	13.5

From the uncertainty analysis, about 66% of the particles exhibited the sodium fraction which was within 20% of the theoretical value, while about 80% of particles had the aluminum fraction within 20% of the theoretical value. It should be noted that the experiment with sodium chloride is potentially one of the most challenging to determine the elemental stoichiometry, because of large difference in the ionization potential and electron affinity between the two elements. The ionization potential of sodium and chlorine are 5.1 and 13.0 eV, respectively, and the electron affinity are 0.55 and 3.6 eV, respectively. Therefore, the sodium fraction would always be expected to be at least equal to or greater than that of chlorine. Considering both the spatial distribution of laser power density and relatively wide aerosol beam, a slight shot-to-shot variation in the laser power would give a significant difference in this highly nonlinear, multiphoton processes, so the observation of such good agreement in stoichiometric is quite remarkable, and provided confidence that the laser power used in this study should be sufficient for any compoisition

Results and Discussion

Figure 5 shows the stoichiometry ratios of O/Al, N/Al, and N/O for decomposition of aluminum nitrate at varied temperatures in the spray pyrolysis setup shown in Figure 2. At 100 °C, the ratio of nitrogen to aluminum is about 2.7 (expected stoichiometry of 3:1) while oxygen to aluminum ratio was ~8.3, and reflect that reactant particles are just starting to decompose. Reduction of the nitrogen fraction, found in both N/Al and N/O, reveales decomposition of nitrate at higher temperatures. At 800 °C, the reaction seemed to be completed, and the O/Al ratio approached ~1.5, which corresponds to the theoretical stoichiometry of aluminum oxide, while the nitrogen fraction reached a negligible level. Note that, for a given temperature condition, some mass spectra which were classified "successful" according to the criterion described in the previous section were found to have very different stoichiometry compared to others at the same temperature. More specifically, some mass spectra showed particles had not reacted at all, while some others were found to have completely decomposed to oxide. This was probably due to difference in the heating histrory in the furnace reactor; since the flow in the heated reactor was laminar, particles which traveled near the center line stayed in the reactor for about a half of the average residence time, while those traveled near the tube wall spent significantly longer time in the heated tube. This was most significant when the temperature was near the onset of the decomposition reaction, and we excluded those extremes from the stoichiometry ratio determination. The error bars in Figure 5 represents the standard deviations, which are larger for lower temperatures. The extent of conversion, ($\alpha \equiv (w_0-w)/(w_0-w_f)$), where w is the sample mass and the subscripts 0 and f represent the initial and final states of the reaction, respectively, could be calculated from the atomic fraction of nitrogen in particles, $\langle N\rangle \equiv N/(Al+N+O) = 1/([Al/N]+1+[O/N])$, i.e., through estimation of a partially-reacted particle mass w from the nitrogen atomic fraction in the particle, with a known molecular fomulae for reactant and product. Overall, the relationship between α and <N> can be expressed for aluminum nitrate as $\alpha = ((13/3)\langle N\rangle-1)/((7/2)\langle N\rangle-1)$.

This allows us to estimated variations of the extents of conversion for four metal nitrates (aluminum, calcium, silver, and strontium) at different reaction temperatures. Silver and aluminum nitrates exhibited lower decomposition temperatures (~200 °C for $\alpha = 0.5$) whereas the 50 % conversion was observed at about 650 °C for calcium and strontium nitrates. The reaction rate constant, k(T), was calculated by using the relationship in traditional isothermal-heating thermal analysis among α, k(T) and the Arrhenius parameters, i.e. the activation energy (E) and pre-exponential factor (A) as:

$$\ln(g(\alpha)/t) = \ln k(T) = \ln A - \frac{E}{R}T^{-1}, \text{ where } g(\alpha) \equiv \int_0^{\alpha} \frac{d\alpha'}{f(\alpha')} = \int_0^t k(T)dt'$$

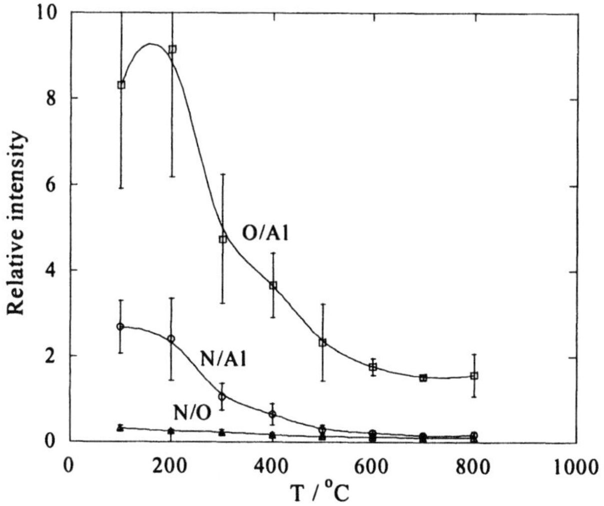

Figure 5. Change in relative intensities of nitrogen, oxygen, and aluminum ion peaks during thermal decomposition of aluminum nitrate.

The decomposition reaction was assumed to be single-step and first-order ($f(\alpha)=1-\alpha$), and the heating process was approximated to be isothermal. Results are shown in Figure 6. Each nitrate exhibited good linearity, which implies the single-step first-order assumption and use of the Arrhenius relationship were appropriate, although the number of points in the plots was limited. Table II lists the Arrhenius parameters obtained by this analysis.

Table II. Arrhenius parameters obtained by SPMS for thermal decomposition of four metal nitrates.

	E [kJ/mol]	*ln A [ln sec^{-1}]*
$Sr(NO_3)_2$	115.3 ± 23.1	14.1 ± 2.8
$Ca(NO_3)_2$	89.1 ± 10.5	11.5 ± 1.3
$Al(NO_3)_3$	15.6 ± 0.4	3.9 ± 0.1
$AgNO_3$	28.3 ± 2.3	7.1 ± 0.6

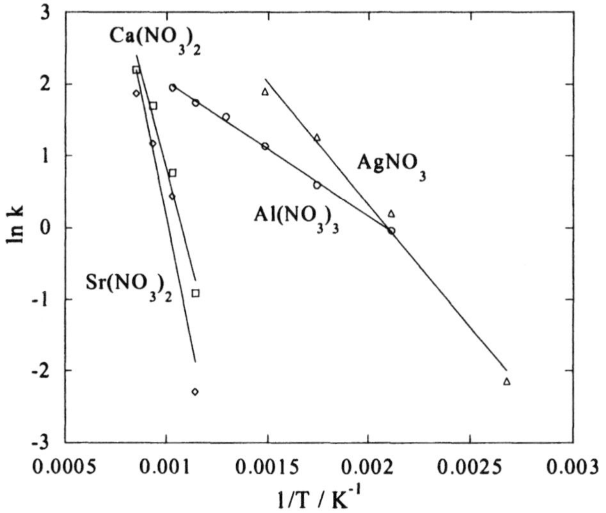

Figure 6. Arrhenius plot for thermal decomposition of metal nitrates.

Measurements of NanoAluminum

Defense applications of nano-metal, especially aluminum has received attention as an energetic fuel or an ingredients in solid propellants, with the prospect that the high surface to volume ratio will lead to enhanced reactivity and possibly even higher energy release. For example, addition of nano aluminum particles in place of micron sized particles in solid propellants increases the burning rate by a factor of approximately six with a resulting increase in the specific thrust. Of course this engnaced reactivity also requires some type of surface passivation and are usually produced with a a thin oxide or more recently with organic coatings. This in turn poses many questions about the thermal, oxidative, and humidity stability of these materials and is a major focus of our current activities using the SPMS. Example measurements of commercially available nanoaluminum are shown in Figure 7(a) and (b). In the first figure we see a pure aluminum particle with a small quantity if hydrogen and oxygen, indicating that the particle has a oxide/hydroxide coat. The second figure shows an SPMS corresponding to flowing the aerosol through a heated flow tube in oxygen at 900 °C. The residence time of aerosols in the reactor was roughly 1 second. The elemental ratio of oxygen to aluminum increased from approximately 2 % or less at room temp for the as-produced material, to 120 % when oxidized in air at 900°C. However, the shot-to-shot variabliity in oxygen measured content for the heated aluminum, was found to be quite large, indicating that the oxidation through these particle varied from particle to particle.

Figure 8 shows a new generation of aluminum nanoparticle with a fluorocarbon type of coating. The spectra clearly show the presence of carbon and fluorine. The intent of the coating is to passivate the coating with a more humid resistant coating. Our future investigations are aimed at understanding the stability of such passivation layers.

Conclusions

A time-of-flight single particle mass spectrometer has been built to characterize the elemental composition of aerosol particles. The results show that coupling the mass-spectrometer to an aerodynamic lens system provides a highly efficient transmission system for aerosols. The use of a high-energy laser ablation/ionization source enables us to eliminate any molecular fragments from the mass-spectra, making both interpretation of the spectra easy, but more importantly providing an opportunity to make a more quantitative characterization. Our results indicate that one should be able to determine the composition of an aerososl population with an uncertainty of less than 20%. The ability to determine the particle stoichiometry has been used to investigate the kinetics of chemical reactions for thermal decomposition of metal nitrates. The results are in qualitative agreement with the TGA results for the onset of decomposition. More importantly we have used the method to extract intra-particle reaction rates. These results indicate that the approach presented here should enable one to track the formation of new materials grown from the aerosol phase, which has found increasing importance in industry. Of a more general nature is that this approach may be a natural way in some cases to generically study condensed-phase reactions, where the use of an aerosol minimizes the usual heat and mass transfer effects that plague traditional methods.

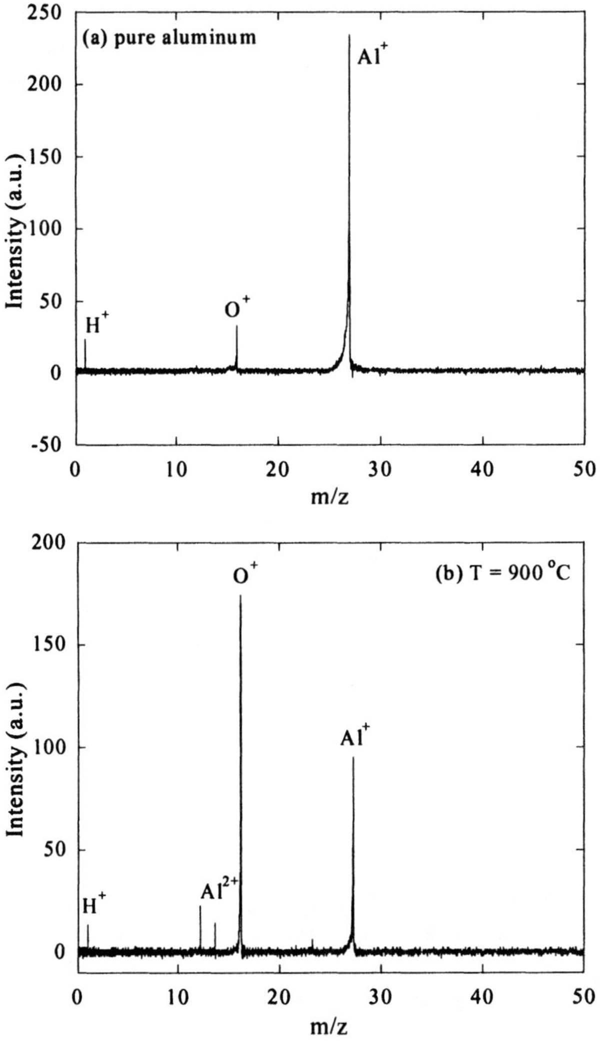

Figure 7 Single Particle Mass Spectra of as made (a) and oxidized (b) Aluminum nanoparticle.

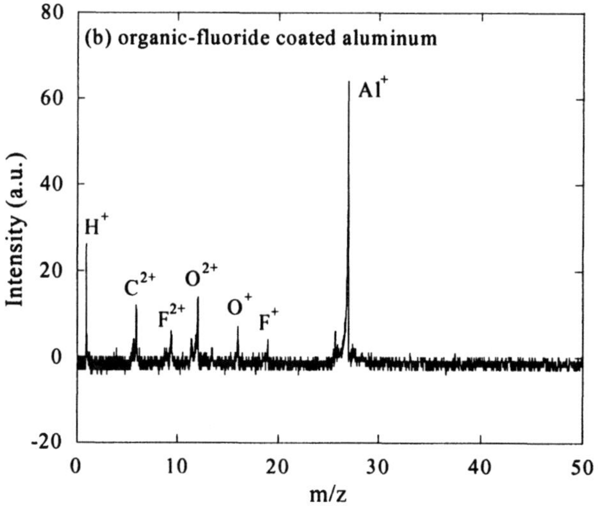

Figure 8 Single Particle Mass Spectra of an organic fluorode coated nanoparticle

References

1. *Aerosol Measurement: principles, techniques, and applications*; Willeke, K.; Baron, P. A., Eds.; Van Nostrand Reinhold: New York, NY, 1993.
2. Higgins, K., Jung, H., Kittelson, D.B., Roberts, J.T., and Zachariah , M.R., *J.* Phys. Chem. A. 106, 96 (2002)
3. Friedelander, S. K. *Smoke, Dust and Haze*; John Wiley and Sons: New York, NY, 2001.
4. Suess, D. T.; Prather, K. A. *Chem. Rev.* **1999**, 99, 3007-3035.
5. Davis, W. D. *J. Vac. Sci. Technol.* **1973**, 10, 278.
6. Davis, W. D. *Environ. Sci. Technol.* **1977**, 11, 587-592.
7. McKeown, P. J.; Johnston, M. V.; Murphy, D. M. *Anal. Chem.* **1991**, 63, 2069-2072.
8. Carson, P. G.; Johnston, M. V.; Wexler, A. S. *Rapid Commun. Mass Sp.* **1997**, 11, 993-996.
9. Ortega, A. *Int. J. Chem. Kinetics* **2001**, *343*, 1.
10. Anderson, H. *Thermochemical Acta* **1992**, *33*, 87.
11. Reents, W. D. Jr.; Ge, Z. *Aerosol Sci. Tech.* **2000**, 33, 122-134.
12. Reents, W. D. Jr.; Schabel, M. J. *unpublished* **2001**.

13. Liu, P.; Ziemann, P. J.; Kittleson, D. B.; McMurry, P. H. *Aerosol Sci. Tech.* **1995**, 22, 293-313.
14. Ziemann, P. J.; Liu, P.; Rao, N. P.; Kittleson, D. B.; McMurry, P. H. *J. Aerosol Sci.* **1995**, 26, 745-756.
15. Kane, D. B.; Oktem, B.; Johnston, M. V. *Aerosol Sci. Tech.* **2001**, 34, 520.

Chapter 16

Characterization of Metastable Intermolecular Composites

Michelle L. Pantoya[1], Steven F. Son[2], Wayne C. Danen[2], Betty S. Jorgensen[2], Blaine W. Asay[2], James R. Busse[2], and Joseph T. Mang[2]

[1]Mechanical Engineering Department, Texas Tech University Corner of 7th and Boston Avenue, Lubbock, TX 79409
[2]Los Alamos National Laboratory, Bikini Road, SM 30, Los Alamos, NM 87545

Nanocomposite energetic materials, such as Al/MoO_3, are highly exothermic and exhibit high energy densities. This chapter details microstructural features, such as morphology and size distribution, which affect ignition sensitivity and performance characteristics. Techniques such as small-angle x-ray or neutron scattering (SAXS or SANS) have been used to obtain size distributions of nano-scale materials. This chapter will also focus on the mechanisms and phenomena that govern thermal and chemical processes associated with nano-structured energetic material combustion. Specifically, the burning rates were measured experimentally for open combustion and confined burning. Reaction behaviors were characterized from these photographic data as a function of Al particle size, confinement and bulk density. These materials are shown to propagate at very high rates.

The term thermite reaction is used here to describe an exothermic reaction which involves a metal reacting with a metallic oxide to form a more stable oxide and the corresponding metal of the reactant oxide (*1*). Some thermite reactions offer energy release on a mass or volume basis that far exceeds typical explosives. However, their application has been limited because the energy release rate is generally slow. In recent years it has been demonstrated that considering reactants on the nano-scale reduces the diffusion lengths sufficiently to dramatically increase the energy release rate (*2*). These materials have been termed metastable intermolecular composites (MIC). Here we use this term to refer to any energetic material utilizing nanoscale energetic constituents. In a sense, the rules are changed at the nano-scale and the door is opened for the consideration of new applications for these materials. Reaction rates are orders of magnitude faster than conventional formulations (*2*). As a consequence these materials have application directly for such things as primers, or nano-materials can be formulated with other energetic materials, including gas generating materials, and used for a much wider variety of applications. In this chapter we present our recent efforts to better characterize the morphology and to understand the propagation physics of these materials.

Material Characterization

Minor differences in microstructure can lead to dramatic changes in the performance of energetic materials (*3*). In classical energetic materials microstructural differences may be induced in formulation and processing, or generated by physical insult, chemical insult, or aging. Detecting and understanding these differences requires the ability to probe microstructural features with sizes ranging from millimeters to angstroms. No single characterization method can cover this entire range of length scales. Figure 1 shows that incorporating a variety of techniques will allow material characterization over a broad dimensional range. In combination, these techniques allow quantitative identification of features and differences in materials over length scales from 10^{-10} to 10^{-2} meters.

Scanning Electron Microscopy (SEM)

SEM is used to probe microstructural features from 10 nm – 1 mm. A Field Emission Scanning Electron Microscope (FESEM) allows imaging of many non-conductive materials at relatively low voltages (200 eV – 30 keV) without applying a conductive coating. An energy dispersive x-ray spectrometer (EDS) and a wavelength dispersive x-ray spectrometer (WDS) allow additional analysis, quantification, and mapping of elements.

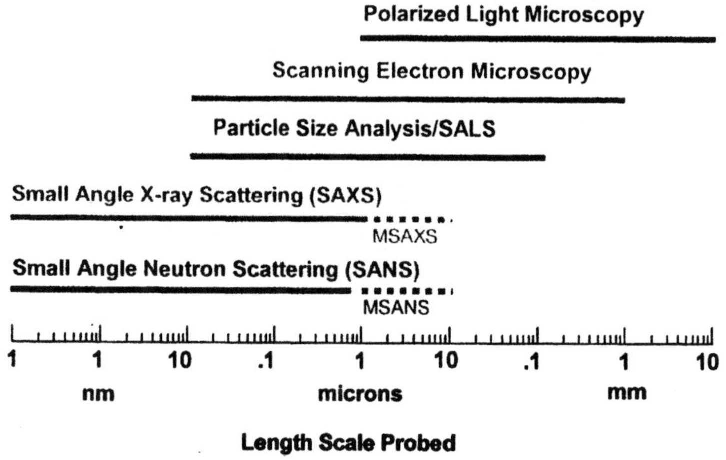

Figure 1. A comparison of particle characterization techniques in terms of length scale probed.

Aluminum and MoO_3 powders were characterized using these techniques to determine the particle size distribution and degree of particle agglomeration. Figure 2 shows an SEM image of a typical aluminum sample. Based on experience with these materials, as the average particle size increases, the particle size distribution broadens. Also, as observed in Fig. 2, aluminum particle agglomeration can occur in the form of chains or branches. Figure 3 is an SEM of Al/MoO_3 MIC. The MoO_3 images reveal sheet-like and rod-like structures while the aluminum is spherical.

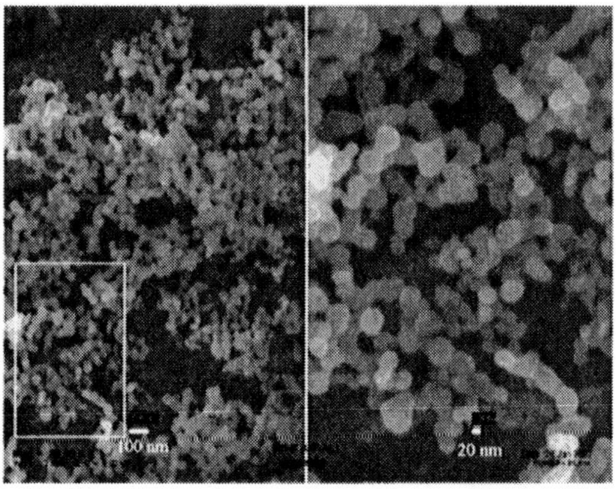

Figure 2. SEM image of nano-aluminum made at Los Alamos.

Figure 3. SEM image of Al/MoO3 (MIC).

Small Angle Scattering

In a SAS experiment, a beam of neutrons or x-rays impinges upon a sample characterized by a scattering length density, $\rho(\mathbf{r})$, which reflects the microscale structure. Fluctuations in $\rho(\mathbf{r})$ give rise to small-angle scattering. The intensity of the scattered radiation, I(Q), is measured as a function of the scattering vector, **Q**, of magnitude $Q = (4\pi/\lambda)\sin\theta$, where λ is the wavelength of the incident radiation and θ is half of the scattering angle. I(Q), for a monodisperse system of non-interacting particles, dispersed in a uniform media, can be expressed as (*4*),

$$I(Q) = N\Delta\rho^2 V^2 \langle P(\mathbf{Q}) \rangle$$

where P(Q) is the normalized, single particle form (shape) factor and is related to Fourier transform of $\rho(r)$, V is the particle volume, and N is the number of scatterers per unit volume. Also, $\Delta\rho$ is the scattering length density contrast between the average scattering length density of the particle, and that of the surrounding media, which is ambient air. The complete interpretation of I(Q) in terms of the sample structure, $\rho(r)$, ultimately involves careful comparison with calculations of the scattering expected from model structures.

An advantage of SANS and SAXS is that a statistically significant number of particles is sampled in a single measurement, which allows accurate assessment of particle size distributions. SANS measurements were performed on the Low-Q Diffractometer (LQD) at the Manuel Lujan Jr., Neutron Scattering Center at Los Alamos National Laboratory (*5*). SAXS measurements have been made at the University of New Mexico (*6*). In both cases, data were reduced by conventional methods and corrected for empty cell and background scattering. Absolute intensities were obtained by comparison to a known standard and normalization to sample thickness.

SAXS can provide a quantitative measure of sheet thickness for the MoO_3. Figure 4 shows data obtained from SAXS measurements of a sample of MoO_3. The solid line in the figure is a result of a fit to a model (developed based upon SEM images) of polydisperse sheet-like structures with large aggregates. The model is consistent with the measured data. An average of 15.5 nm was measured for the sheet-like structures seen in the SEM image.

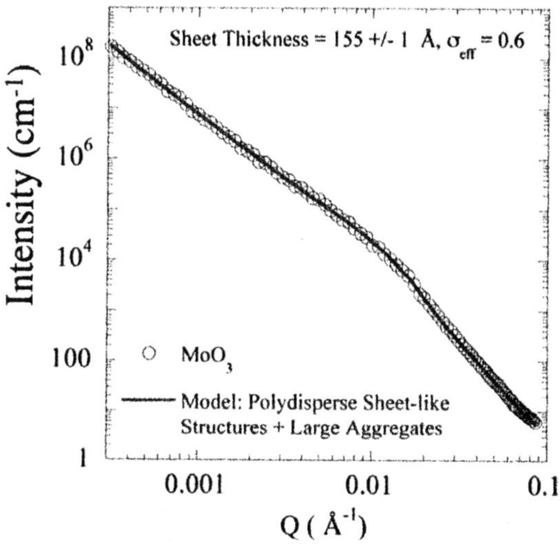

Figure 4. SAXS results obtained from MoO_3.

Figure 5 shows the distribution obtained from SAXS measurements of three Los Alamos produced nano-aluminums. The dispersal media for the SAXS measurements is air. Based upon the SEM images, a model of mass fractal aggregates was used to interpret the scattering data (see Fig. 2 for example). This model was found to be consistent with the SAXS data. The volume-weighted size distribution obtained from the analysis is shown.

Figure 6 is a comparison of size distributions for an aluminum powder obtained from a SAXS measurement and particle size counting from SEM images. The comparison is reasonable considering the limitations of the sample size considered in the SEM measurements (only a few hundred particles), possible halo effect (electron diffraction) in the SEM images, and the difficulty determining the diameter of agglomerates. In contrast, the SAXS measurement includes ~ 10^{18} particles, and therefore may be more representative of the bulk material. The SEM distribution shows an average particle diameter of 39.2 nm while the SAXS measurement indicates 26 nm.

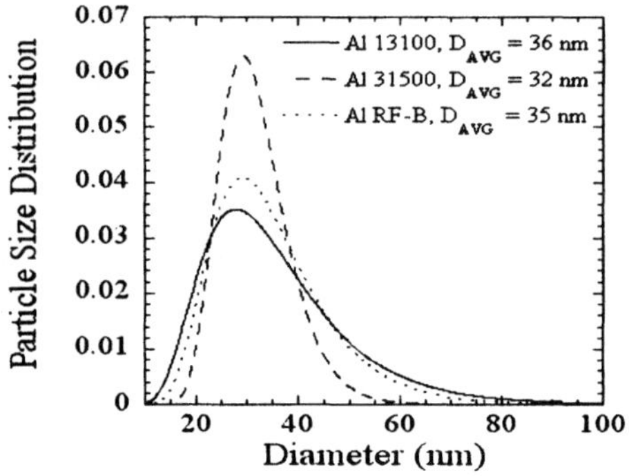

Figure 5 Volume weighted distributions of three Los AlamosAl samples.

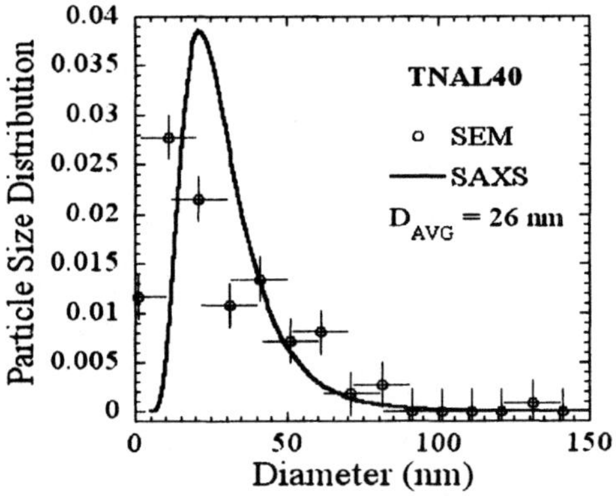

Figure 6 Comparison of size distributions obtained from SAXS and SEM.

Combustion Behaviors

These materials are found to be spark and friction sensitive. Consequently, only small samples (i.e., less than one gram) are handled at a time and appropriate personal protection is used. In all cases, the Al/MoO_3 composite was prepared by immersing the powders in a solvent and using an ultrasonic mixer. The particles became suspended in the solvent and once thoroughly mixed, the mixture was poured into a pan and heated to a few degrees above ambient, allowing the solvent to evaporate. Open combustion and confined burning were considered. Al/MoO_3 powder or pellet samples were ignited at one end and a high-speed imaging system recorded the flame propagation. Reaction behaviors were characterized from these photographic data. The goal was to understand flame propagation mechanisms.

Pressure Cell Experiments

Figure 7 is a schematic of a constant volume (13 cc), cylindrical chamber in which the reaction pressure and light intensity were measured. A fixed volume of powder was placed in the chamber under ambient conditions and laser ignited. The powder sample bulk density ranged from 7 to 10 % of the theoretical maximum density (TMD). Ignition occurred via a fiber-optic cable by a 30 ns, 20 mJ pulse from a Nd:YAG laser. Pressure was measured from Two PCB Piezotronics$^{INC.}$ high-frequency pressure transducers that were installed in the wall of the chamber; one with a 50 psi range and the other with a 250 psi range. A fiber-optic cable mounted in the wall of the chamber monitored light emission from each test.

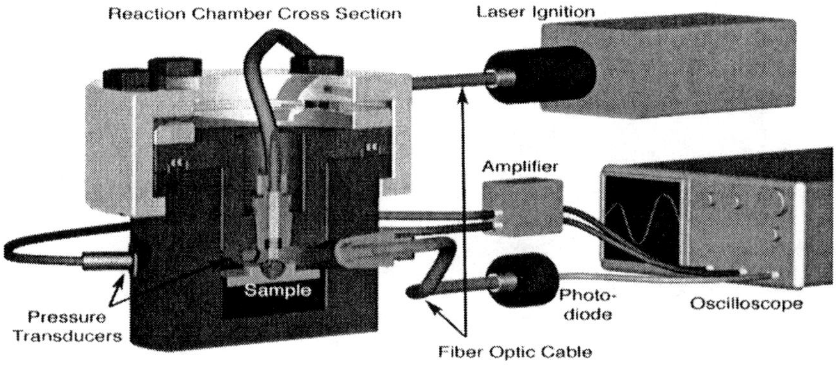

Figure 7. Schematic of the pressure cell apparatus.

The pressure cell experiment was used to compare the performance of powder samples of MIC materials. Samples were placed in the same volume for each experiment. It must be noted that the bulk density can vary between experiments but efforts were made to ensure consistancy. The bulk density is determined for each test by measuring the sample mass while keeping the sample volume fixed. Ignition time is defined as the time required for the reaction to produce 5% of the maximum pressure. This time is measured from the initial laser pulse. This parameter is indicative of the reactivity of the material.

Nano-composite Al/MoO_3 mixtures exhibit high spark-sensitivity, especially in dry environments. This phenomenon led to questions about the possible effect of electrostatic charge on material ignition behavior. Consequently, a series of tests was conducted in the pressure cell to quantify possible performance variations caused solely by electrostatic charge. The composition of the mixture and the origins of the Al and MoO_3 were held constant. Seven mixtures with different charges were examined. The first two mixtures were not altered, the third and fourth were neutralized with a destatic gun. An attempt was made to positively charge the fifth and sixth samples, while the seventh sample received a negative charge, both using the destatic gun in either the positive or negative mode.

Figure 8 shows that no significant variations of any of the major performance characteristics (i.e., ignition delay, pressurization rate, and peak pressure) were evident as a function of the electrostatic charge on the powder. The pressure histories of all seven tests overlaid consistently. These results imply that the electrostatic charge on the sample does not appear to play a significant role in the overall performance of the material. However, there may be an effect on the spark sensitivity.

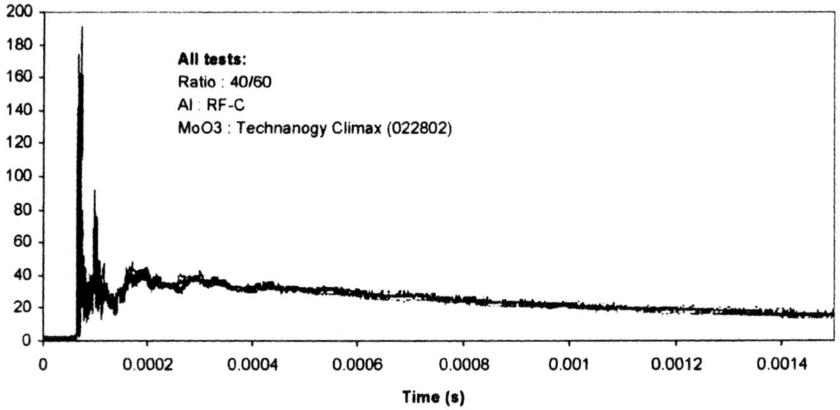

Figure 8. Pressure cell experimental results from several tests of the same material with different static charge induced.

When the aluminum particle size is changed, an effect on the reaction performance is expected. In Fig. 9 the aluminum particle size was changed from approximately 40 nm particle diameter (LA-Al-RF-B, see Fig. 2 & 5 for size distribution) to 120 nm particle diameter. There are measurable differences in the ignition times and maximum pressurization rate. These differences are likely the result of varying Al particle size. As seen in Fig. 9, MIC 1 and MIC 2 correspond to 30 nm and 121 nm Al particle size and therefore reducing the particle size decreases the ignition delay time. Also, the TEFMC curve corresponds to an Al/MoO_3/Teflon mixture which generates more gas than MIC without Teflon. Using Teflon as a gasifying agent causes the pressure history to naturally exhibit higher pressure sustained over a longer duration, as observed in the pressure traces in Fig. 9.

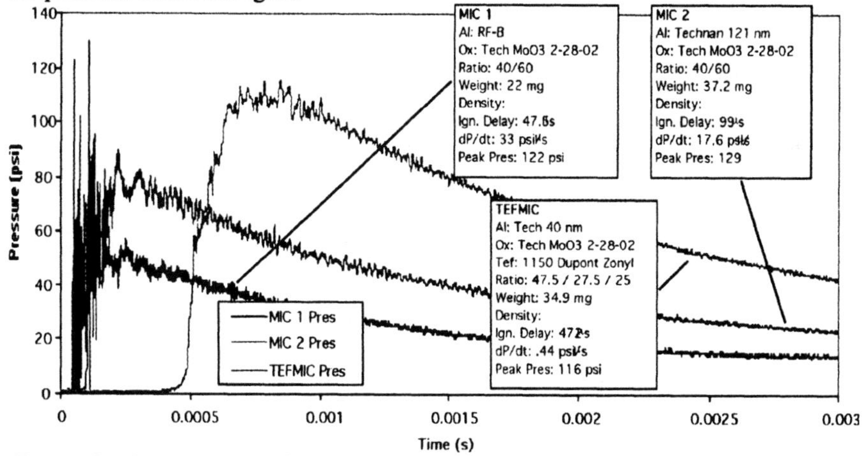

Figure 9. Comparisons of three types of MIC compositions in the pressure cell.

In Figure 10 the pressure response of two materials that are more similar than those in the previous example are compared. This figure illustrates how the performance differences are nearly indistinguishable between the two Al/MoO_3 mixtures. In these two mixtures, the Al particles were slightly different (see Fig. 8 for size distributions), but the MoO_3 and composition ratios were held constant. The density differences alone may account for the differences observed.

Changing the stoichiometry affects the propagation rate significantly. This is confirmed in Fig. 11 where the percentage of aluminum was varied and both open loose powder burn propagation rate and maximum pressurization rate is presented. Relatively small changes in the percentage of aluminum have large effects on the propagation rate and pressurization rate. In fact, the percentage of aluminum was varied by only 14% (Fig. 11). Higher pressurization rates are an indication of faster propagation rates. Pressure build up in a confined space accelerates the reaction front and increases the flame propagation rate.

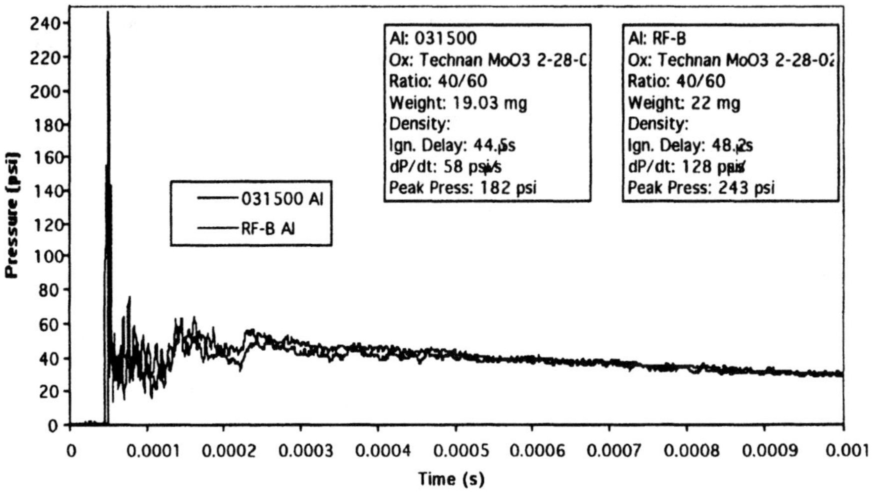

Figure 10. Pressure traces from two similar nano-aluminum samples.

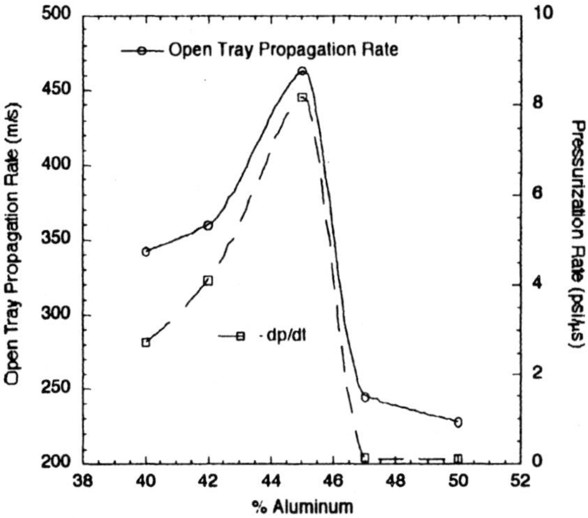

Figure 11. Shows the dependence of the percent of Al on the maximum pressurization rate and open tray burn.

Pellet MIC Experiments

Pellets of Al/MoO_3 MIC where produced in an axial press, 4.4 mm diameter and 5 mm in length. Various bulk density pellets can be obtained. The lower density pellets (<2.4 g/cc) generally exhibit nonplanar burning. Presumably, small unobserved flaws are accessed and flames spread rapidly through the pellet. When flaws exist in the pellet, flames can spread in and spread convectively through the pellet. This is very similar to the combustion of energetic materials where nonplanar convective burning can be observed in damaged material at pressures sufficient for flame incursion. Convective burning in powders and lower density pellets may be the dominant propagation mechanism and work is proceeding to eliminate other possibilities (*7*).

Flame spreading down the side of pellets is also sometimes observed. Initially, applying a thin coat of epoxy to the sides of the pellet, which is the usual method of inhibiting side burning of pellets or strands of energetic materials, was attempted. This inhibited flame spread on the sides of the pellet, however the epoxy apparently mixed with a small amount of MIC material to create an extremely opaque layer so the intense burn front could not be observed. Both of the above effects are illustrated in Fig. 12 where the burning front cannot be seen through the epoxy layer and the flame has spread through to the bottom the pellet and is jetting hot products.

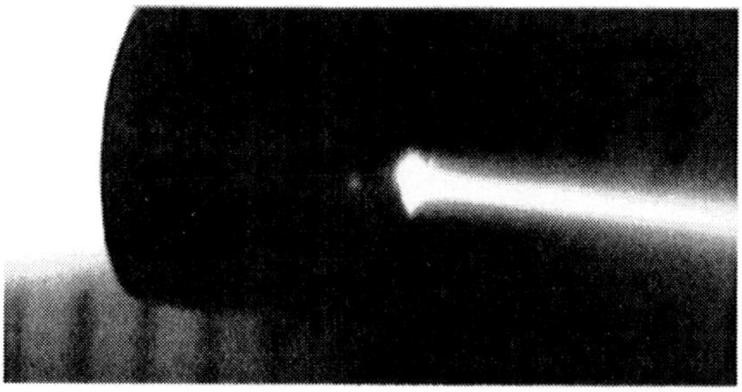

Figure 12. Image of burning lower density Al/MoO_3 pellet that has been inhibited with epoxy.

Spraying a very thin layer of polyurethane adequately inhibited side flame spread, yet allowed observation of the flame front. Igniting the entire pellet surface initially is also important and was accomplished using a Nd:YAG laser. A planar burning MIC pellet is presented in Fig. 13.

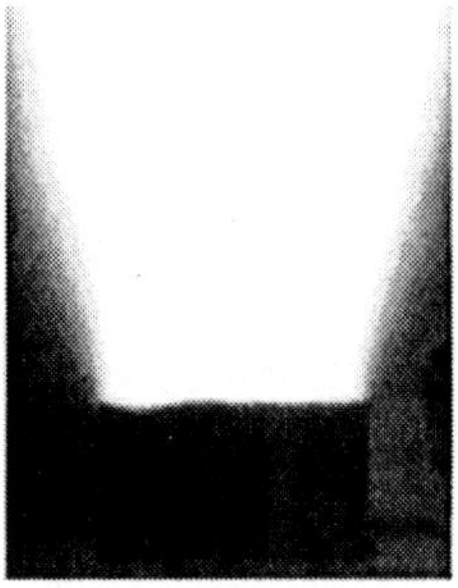

Figure 13. Normal planar burn of a high density pellet (2.9 g/cc). Note that the experiment was actually performed horizontally.

At higher densities, planar normal burns were achieved. The density of this pellet is 2.93 g/cc and the burn rate was measured at 20.6 cm/s. This rate is three orders of magnitude slower than the propagation rate observed in the loose powder. This is consistent with the transition from convectively dominated burning to conductive (or normal deflagration) burning. This propagation rate is about two orders of magnitude faster than HMX burning at one atmosphere and an order faster than the fastest burning organic energetic material under the same conditions. We anticipate further studies at different pressures and particle sizes. We have observed that ignition is more difficult when the MIC is pressed compared to loose powders.

Confined MIC Burning Experiments

One objective of this study was to determine the physical mechanism controlling reactive wave propagation. Possible mechanisms that control propagation rate include conductive (classical deflagration), convective (advective energy transport), acoustic (involving shock waves, as occurs in detonations) and radiative or compressive (mechanical dissipation). A wide variety of experiments were performed to reveal the mechanism(s) that dominate reactive wave propagation in these materials. Propagation experiments were performed using powder confined in 3.8 mm inside diameter glass tubes. This arrangement allowed for observation of the flame propagation in a confined system. If convective processes are important, a pressure gradient is needed to drive the advective transport of energy forward. Although the reactants and products of pure Al/MoO_3 materials are both solids, intermediates or products may be gaseous before they condense as they cool. Also, initial gas between particles is heated and could behave as a working fluid to transport heat forward. Observation of particulates ejected from the ends of the tube would indicate a possible convective mechanism controlling the reaction propagation. An accelerating wave front is also typical of convective flame spread.

Conduction is the mechanism of reaction propagation in traditional, slower burning, thermite mixtures. These thermite mixtures typically burn at relatively slow velocities, i.e. between 2-20 mm/s. Reactions that are dominated by conduction are also characterized by relatively steady propagation rates when burned at constant pressure and usually exhibit a planar reaction front. In a convective dominant reaction, the reaction front will propagate much faster with noticeable acceleration. When confined, convectively dominant reactions will demonstrate pressure build up and may explode. To test convective influences, glass tubes 6 cm in length and 3.8 mm inside diameter were filled with a 40/60 mixture of Al/MoO_3 powder. The confined material was ignited at one end of the tube and the reaction propagated through the powder.

Figure 15 shows a sequence of images of from this reaction. The first image was captured immediately following material ignition and each consecutive image was captured after a time lapse of approximately 0.03 ms. All images show highly luminescent plumes ejected from the tube. The plumes are likely composed of gas and high temperature particulates. The radiant emission of these particles, possibly still burning, illuminates the plume. The expansion of the exit plume indicates pressurization of the tube. However, the pressure was low enough that the tube was not destroyed. Although Al/MoO_3 nano-particles undergo the solid/solid reaction

$$2Al + MoO_3 \rightarrow Al_2O_3 + Mo$$

gaseous transport is clearly present and illustrated in the plumes on the tube ends. These plumes may result from intermediate gaseous products of the reaction and from initial air pockets within the powder that when heated, expand and thrust particulates out the tube. In either case, plumes of particulates are suggestive of significant pressurization generated by the reaction such that convection may be dominant. If the reaction produced gas, as in high explosive burning, gases would be expelled. Although these materials are not explosives they do appear to expel gas and illustrate the possible significant convective mechanisms driving the reaction forward. Conduction does not involve the bulk motion of a fluid but rather heat transfer by random atomic or molecular activity. The images in Fig. 14 indicate that the bulk motion of a fluid may be integral in the reaction, suggesting convection as a dominant mechanism controlling the reaction.

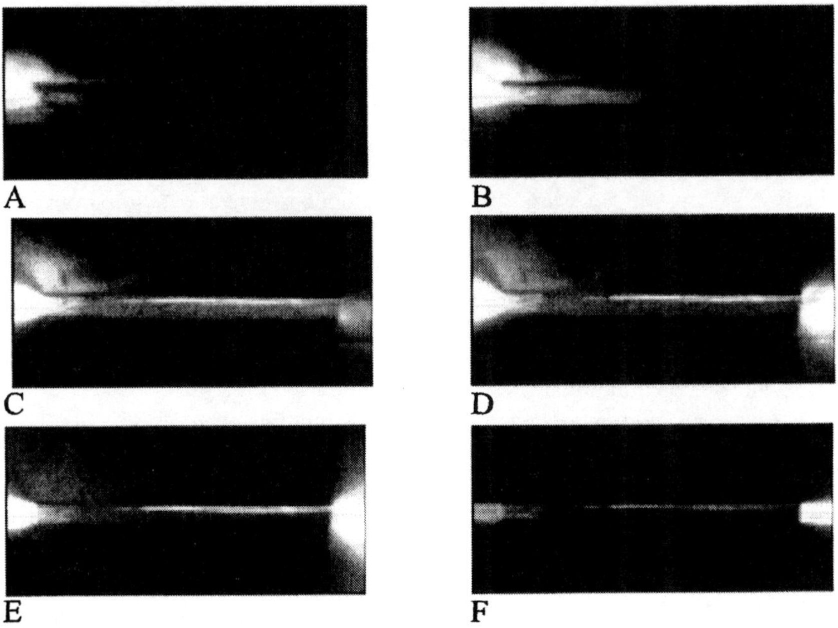

Figure 14. Al/MoO₃ powder ignited at one end in a glass tube.

References

1. L. L. Wang, Z. A. Munir, Y. M. Maximov, *Journal of Material Science*, 28:3693-3708 (1993).
2. Aumann, C.E., Skofronick, G.L., and Martin, J.A., *J. of Vac. Sc. & Tech.* B **13**(2): 1178-1183 (1995).
3. H. Moulard, Kury, J.W. "The effect of RDX particle size on the shock sensitivity of cast PBX formulations" *Eighth Symposium (International) on Detonation*, p 902, Albuquerque, NM 15-19 July (1985).
4. Glatter, O. and Kratky, O., *Small Angle X-ray Scattering*, Academic Press, London (1982).
5. Seeger, P. and Hjelm, R., *J. Appl. Cryst.*, **24**, 467 (1991).
6. Rieker, T. P., Hubbard, P. F., *Rev. of Sci. Inst.*, **69**, 3504 (1998).
7. S. F. Son, B. W. Asay, J. R. Busse, B. S. Jorgensen, B. Bockmon , and M. Pantoya, "Reaction Propagation Physics of Al/MoO_3 Nanocomposite Thermites," The 28th International Pyrotechnics Seminar (2001).

All authors would like to acknowledge support from Los Alamos National Laboratory Advanced Energetics Initiative; M. Pantoya additionally acknowledges the U. S. Army Research Office (DAAD19-02-1-0214).

Nanoelectronics and Photonics

Chapter 17

Chemical Synthesis and Directed Growth of Photonic Bandgap Crystals

M. L. Breen[1–3], S. B. Qadri[2], and B. R. Ratna[2]

[1]Spectrolab, Inc., 12550 Gladstone Avenue, Sylmar, CA 91342–5373
[2]Naval Research Laboratory, 4555 Overlook Avenue, SW, Washington, DC 20375–5320
[3]During the period that this work was conducted the author was a National Research Council Postdoctoral Fellow

Research is underway to make microscopically ordered, three-dimensional structures with large refractive index contrast at sub-micron length scales. These materials, known as photonic crystals, can localize electromagnetic waves within a narrow range of wavelengths, then direct them to propagate through a crystal along discrete channels. Such materials could have important applications in the field of optoelectronics. Primarily, this chapter focuses on aqueous solution methods for preparing colloidal particles of a diameter near that of infra–red or visible light and directing the self-assembly of those particles into 3-D matricies in a variety of packing arrangements.

Introduction

Many great minds have been stimulated by the possibilities of photonic bandgap materials since their utility was first demonstrated by Eli Yablonovitch in 1991. (1,2) Defects are selectively creating in these remarkable, crystalline materials which allow them to localize electromagnetic waves and direct those waves to propagate along a predefined path. Within photonic materials, it has been shown that "light" can be bent at 90° angles (3,4) and due to the microscopic dimensions involved, they could enable standard semiconductor electronic devices to be replaced with new optical and optoelectronic analogs.

Three-dimensional, periodic modulation of the refractive index (n) in a material can produce a band structure in waves. As a result, light propagation and emission is restricted to specific directions within these materials and the density of photon states (hence, the rate of spontaneous emission of light by fluorophores) can be controlled. Materials with carefully tailored structures and sufficient refractive index contrast can have a complete photonic band gap (PBG) – a range of electromagnetic frequencies prohibited from propagating in *any* direction within the material. A complete bandgap for optical wavelengths is much more than Bragg scattering (which only occurs in specific directions). It requires a large refractive index contrast (at least 2.8:1) with a structure that is periodic at sub-micron length scales. PBG materials prevent spontaneous emission of light with frequencies in the gap, potentially a very valuable tool for making efficient, threshold-free lasers. (4) Furthermore, defects within a PBG material can allow gap frequencies to propagate only in the region of the defect, resulting in microscopic wave-guides or resonant cavities.

While photonic bandgaps have been demonstrated in one- and two-dimensions at the optical and infrared wavelengths, (2) there are many challenges in making a microscopically ordered, three-dimensional structure with significantly large refractive index contrast. Some success has been achieved through lithographic methods to produce materials active in the infrared region. Using standard microelectronic fabrication techniques, structures with periodic features of 1-2 μm were prepared in an arrangement, which resulted in a bandgap for infrared light of 10-14.5 μm wavelengths. (5) Samples made through this layer-by-layer method exhibit adequate spatial resolution, but are difficult to prepare in extended 3-D structures. More importantly, to produce crystals with PBG's in the visible or near-infrared it will be necessary to pattern features at the sub-micron level. This degree of structural resolution stretches the limits of lithographic techniques and would be prohibitively expensive to achieve.

Colloidal self-assembly is a simple and elegant path to the formation of 3-D periodic structures. (5) A variety of methods have been explored to make photonic crystals out of porous, ceramic materials templated to sub-micron balls of SiO_2, (6) polystyrene latex, (7,8) and oil emulsions, (9) arranged into 3-D ordered structures. Solutions of metal alkoxides (7,8) or phenolic resins (6) were coated over the templates, filling the interstitial spaces. The metal alkoxides were hydrolyzed to metal oxides and the phenolic resin was reduced to graphitic carbon. The materials were annealed to increase the density of their ceramic/carbon matrices. Unfortunately, the resulting strain from shrinkage distorted the lattices and substantially degraded the optical properties of the colloidal crystals. A silicon photonic crystal with a complete bandgap near 1.5 μm was prepared using a close-packed opal crystal of silica spheres. (10) The voids in this crystal were backfilled with silicon using CVD and the silica template was subsequently acid-etched to produce a 3-D periodic, inverse fcc crystal of silicon.

In most of the methods discussed above, an assembly of low refractive index material is backfilled with high refractive material. Alternatively, photonic materials could be made directly by organizing colloidal particles of high-n into 3-D arrays. These structures would not need to be annealed. Therefore, they would not suffer from the problems associated with lattice shrinkage and distortion. Unfortunately, most high refractive materials are far too dense to form well-behaved self-assemblies. The techniques described in this chapter refer to a hybrid approach where refractive material is uniformly coated on to polystyrene microspheres, prior to self-assembly.

1. Sonochemically Produced ZnS-Coated Polystyrene Core-Shell Particles

Following work conducted at the United States Naval Research Laboratory (Washington, DC), a process was previously reported (11) to prepare 3-dimensional photonic crystals using zinc sulfide/polystyrene core-shell (c-s) particles. Zinc sulfide has a high refractive index ($n_{488} = 2.43$) and relatively low absorption in the visible spectrum, making it an excellent material for use in photonic crystals. Furtuitously, ZnS can be prepared in an aqueous bath. Therefore, this method of preparation has the added benefits of being both low in cost and environmentally-benign. A unique feature of the core-shell approach is its ability to provide a structure in which the high-n material is continuous (within each sphere) and present in a low overall volume fraction. This allows for an optimum fraction of 0.2 – 0.3 of high refractive material as prescribed by theoretical calculations. (12,13) We further showed that it is possible to remove polystyrene (PS) before self-assembling the photonic crystal by solvent extraction, thus eliminating issues associated with lattice shrinkage and post-acid etching. This simple and inexpensive approach enabled the growth of large, photonic crystals. Lessons learned from ZnS were later adapted to employ even higher refractive index materials, such as PbS, discussed in Section 2.

2. Preparation of Lead Sulfide-Coated Polystyrene Core-Shell Particles

Lead sulfide materials are actively studied in the field of optoelectronics because of their relatively high refractive index and photoconductivity in the infrared region, utilized in fiber optic telecommunications. The formation of metal sulfide compounds (e.g. CdS, ZnS, PbS) is easily achieved by combining metal salts with sodium sulfide in an aqueous bath. Resulting complexes are highly insoluble and rapidly precipitate out of solution. To successfully prepare uniform films, the rate of precipitation must be carefully controlled to direct nucleation on the surface of a substrate rather than in solution. Highly uniform CdS films have been prepared on a variety of substrates (e.g. glass, silicon, graphite, copper, steel, polyethylene, and Teflon) with minimum formation of aggregate CdS powder. (14) The rate of film deposition is controlled by reaction temperature and through the slow hydrolysis of thiourea or thioacetamide, a base-catalyzed reaction. Chelating agents, such as ethylenediamine and triethanolamine, are added to sequester metal cations, further reducing the concentration of free ions in solution. By utilizing these measures, the ion-by-ion growth of CdS films on a mica surface has been studied and characterized by scanning electron and atomic force microscopy. (14) This method of growth is not as easily controlled for ZnS and PbS which are less soluble than CdS in water. (15) A considerable amount of ZnS precipitate is formed when zinc salts are combined with thioacetamide and ethylenediamine under reaction conditions similar to those used to produce good CdS films. Fortuitously, highly uniform films of ZnS can be deposited on silica (16) and polystyrene microspheres (11) through sonochemical deposition, as discussed in Section 1. Comparable reactions with lead can be used to deposit PbS on glass, but are not suitable for generating uniform core-shell particles.

Numerous papers have been published describing films of PbS grown on glass substrates from heated aqueous baths. (17–19) Grain sizes reported for these films are in the hundreds of nanometers to micrometer size range. Some oxidants were shown to affect film morphologies and reduce grain size. Ammonium halides were especially successful at reducing grain sizes, following the trend: diameter = $NH_4Cl > NH_4Br > NH_4I$. Poorly soluble lead halide complexes (e.g. PbX_2, PbOHX, PbX^+, $PbHX_3^-$, and PbX_4^{2-}; X = halogen) form which compete with sulfide ions to further slow the reaction. (20) Unfortunately, the smallest grains were still 150 - 200 nm in size. Such crystals are far too large to deposit on polystyrene or silica spheres less than a micrometer in diameter.

Herein is described a novel method for preparing core-shell (c-s) particles of polystyrene coated with lead sulfide nanocrystals. The reaction is performed at room temperature, in an aqueous bath, from lead acetate and thioacetamide in the presence of ethylenediamine.

2.1 Experimental Section

Materials. Lead acetate [99.99%] and thioacetamide [99+%] were purchased from Aldrich and used without further purification. Carboxylate-modifided polystyrene (PS-CO_2) microspheres were obtained from Seradyn (Indianapolis, IN). Microspheres were used, as is, without removing the proprietary surfactant. The surfactant was found not to interfere with deposition and removing the surfactant lead to increased aggregation of the particles. Anhydrous ethanol was obtained from Burdick & Jackson (Muskegon, MI).

Preparation of PbS/PS-CO_2. Lead acetate (130 mg, 3.2 mmol) was added to 50 mL of Millipore Milli-Q water and agitated in a sonic bath for 30 s. The solution was then combined with (1.0 g, 10% wt./wt. solids) 0.827 μm PS-CO_2 and ethylenediamine (510 mg, 8.49 mmol). Thioacetamide (25.6 mg, 3.2 mmol) was diluted to 20 mL, then added dropwise to the reaction solution with stirring at room temperature. The crude product was washed with Milli-Q water and centrifuged several times to remove unreacted starting material.

Preparation of PbS on Glass and ZnS/PS-CO_2. In order to directly compare PbS films with PbS c-s particles, the same reaction bath was used to coat a glass slide suspending in the solution. The glass was cleaned with No-Chromix and H_2SO_4, then rinsed with Milli-Q water. PbS/PS-CO_2 particles were also compared with sonochemically grown ZnS/PS-CO_2 particles, prepared as previously described, (11) and uncoated microspheres.

Annealing of PbS/PS-CO_2 Particles. Suspensions of the core-shell particles were dried in a vacuum desiccator (80°C). The dry powder was heated in a Thermal Analysis Hi-Res TGA 2950 thermal gravimetric analyzer under a stream of nitrogen gas. Temperature was increased at 2°C/min, using a dynamic heating algorithm which held temperature isothermally during weight losses > 0.5 wt.%/min.

Characterization. Particle suspensions were studied with a Nikon Optihot optical microscope. Transmission electron micrographs of the PbS/PS-CO_2 core-shell particles were taken with a Hitachi H-8100 TEM (accelerating voltage = 200 kV). Shell thickness was estimated from TEM micrographs. X-ray diffraction patterns were collected on a Rigaku two-circle diffractometer using CuK_α radiation from a Rigaku 12 kW rotating anode x-ray generator. The peaks were analyzed with PeakFit, spectroscopic analysis software used to obtain peak positions, peak widths, and other structural information. Particles were further studied by LEO 1550 scanning electron microscope (SEM). Optical data was collected on a Cary 2400 UV/VIS/NIR spectrometer.

2.2 Results and Discussion

Initially, several attempts were made to slow the reaction by adjusting pH, adding chelating agents, changing the reaction temperature, forming lead complexes with alkyl halides, and using different lead and sulfur reagents. Sonochemical deposition, previously utilized successfully for making ZnS c-s particles, was also employed. (11,16) Ions in solution have a considerable amount of kinetic energy. Partly due to this fact, it is energetically favorable for crystals to nucleate on a surface or seed crystal rather than in suspension. Traditional approaches in the chemical deposition of thin-film semiconductors typically involve the manipulation of reaction conditions like those mentioned above to favor the more thermodynamically-stable formation of films vs. precipitation, which potentially is more kinetically favorable. (11,14) Surprisingly, the method found to produce the best core-shell particles occurred rapidly. In all previous attempts, a variety of lead precipitates (e.g. needles, cubes, amorphous agglomerations) were seen but little to no film deposited on the polystyrene beads. Rapid precipitation produced nanocrystals of lead sulfide small enough to form an aggregate film on the surface of the polystyrene beads. Also, ethylenediamine is a bidentate chelator that may bridge between lead atoms to improve adhesion of the lead sulfide crystals to the surface of the beads. By slowly adding thioacetamide from a dropping funnel, just enough sulfide was introduced to form the nanocrystals which adhered to the polystyrene beads. This allowed for limited control over the rate of film formation. If the thioacetamide is added all at once, a catastrophic precipitation of large aggregates of PbS results.

X-ray Powder Diffraction Pattern. Analysis of the PbS/O.827 μm PS-CO_2 core-shell particles confirmed that deposited films assumed the Galena mineral structure (Figure 1). Information on the crystallite size was obtained from the diffraction peaks (111), (200), (220), (311), and (222). After correcting for instrumental broadening, the average crystallite size, calculated from full widths at half maximum (FWHM) of the diffraction peaks and Scherrer's formula, (21) was found to be 240 Å or 24 nm. A least square refinement of the data revealed a lattice parameter of a = 5.937 Å. The powder pattern was in good agreement with the published values for bulk PbS [PDS 05-0592]. No presence of lattice strain was indicated, as was previously observed for ZnS/polystyrene core-shell particles. (11)

Film Characterization. The thickness of lead sulfide shells, estimated from TEM micrographs, was 40 - 60 nm (Figure 2). Since the electron density of PbS is much greater than that of polystyrene, the metal sulfide shell appeared as a dark ring around a core of lighter material. SEM micrographs provide a topographical perspective showing the morphology of the PbS film (Figure 3). The shell is actually a monolayer of 40 - 60 nm grains composed of smaller 24 nm crystallites.

UV/VIS/NIR Spectroscopy. In previous work with ZnS c-s particles, we observed oscillations in the UV/VIS spectrum caused by interference patterns between the shell and core materials. (11) The phenomenon is a result of the difference in refractive index between the two layers. Such behavior is predicted by Mie light-scattering theory which can be used to calculate shell thickness and refractive index from UV/VIS data. (22) Oscillations increase in amplitude and the transmission window undergoes a red shift as these properties increase in magnitude.

A series of samples were washed, centrifuged, and resuspended in anhydrous ethanol, rather than water which has several absorption bands in the NIR. Data was collected for uncoated, ZnS-coated, PbS-coated 0.827 μm PS-CO_2, and PbS films on glass (Figure 4). Similar absorption bands (actually, the light is scattered or reflected) were seen for both the ZnS and PbS-coated beads, which were "red shifted" from the absorption band of the plain, uncoated beads by ~80 nm. As previously observed, oscillations or interference patterns were seen for the ZnS c-s particles but they were not seen for the PbS c-s particles in suspension. However, the oscillations in the ZnS spectrum were only distinguishable in the visible region where the PbS is strongly absorbing. Discontinuities in the PbS shells may also mask the interference patterns

Thermal Gravimetric Analysis. In our previous work with ZnS c-s particles, it was possible to remove the polystyrene core, leaving intact shells of ZnS. (11) Hollow shells of PbS would also be desirable, so the experiment was repeated with dried PbS c-s powder. At 270°C, the sample lost approximately 40% of its mass as the critical temperature of polystyrene was approached (Figure 5). Upon reaching this temperature and in the absence of oxygen, the material depolymerizes to release the monomer, styrene. A total mass loss of 48.82 wt. % was recorded as the sample was heated out to 350°C. This weight percent approximates that of the polystyrene core based on the relative thickness and density of the polystyrene and lead sulfide layers. Unfortunately, SEM analysis showed that the shells do not survive removal of the polystyrene support. The ZnS shells were thicker and consisted of smaller crystals which packed together more densely than the larger PbS crystals.

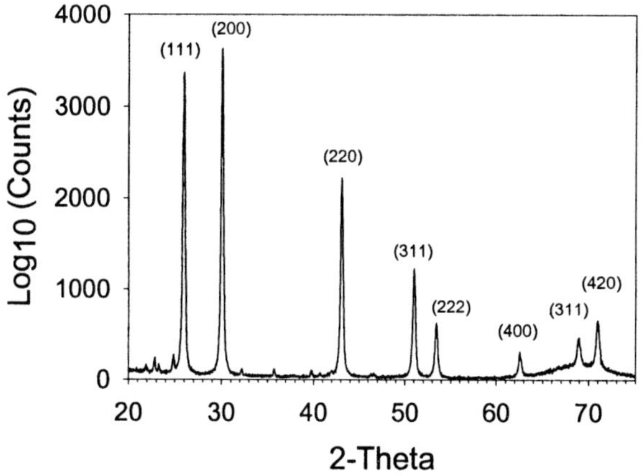

Figure 1. X-ray powder pattern of PbS/0.827 μm PS-CO_2 core-shell particles.

Figure 2. TEM image showing cross-section of PbS film coating 0.827 μm polystyrene core.

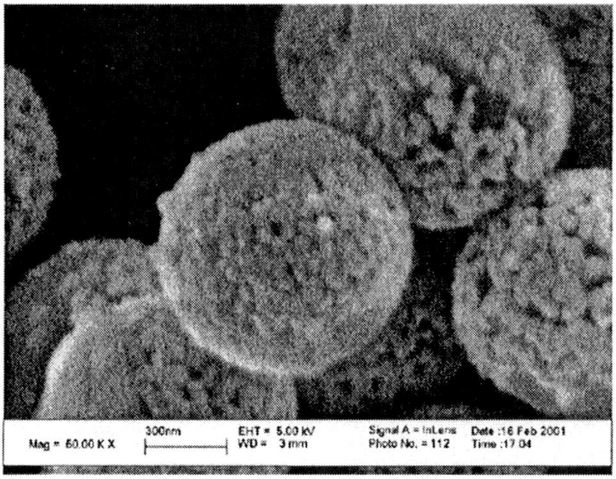

Figure 3. SEM image showing morphology of PbS coating on core-shell particles.

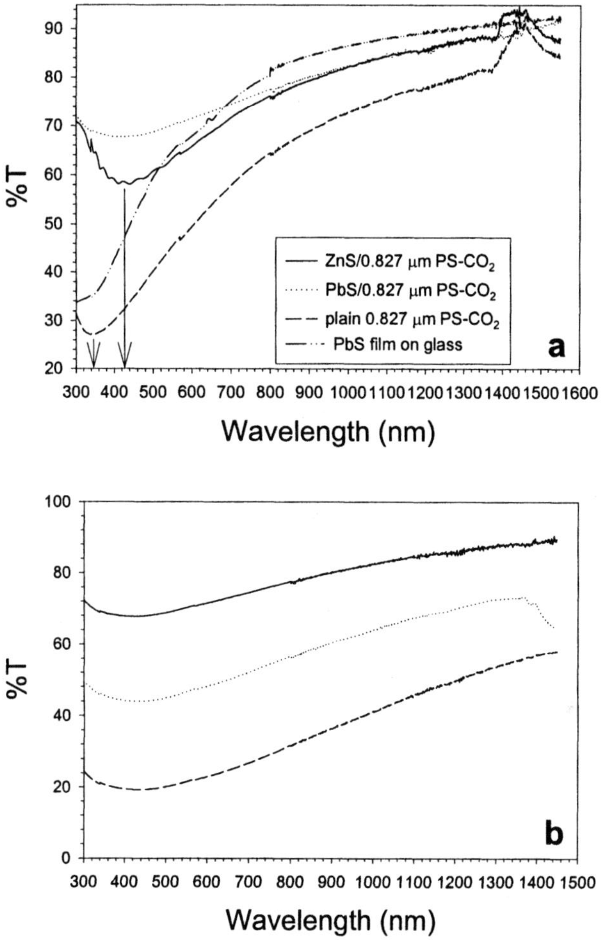

Figure 4. UV/VIS/NIR spectrum of (a) uncoated, ZnS-coated, or PbS-coated 0.827 μm PS-CO_2 microspheres in EtOH and PbS film on glass showing ~80 nm "red shift" and (b) PbS/0.827 μm PS-CO_2 in EtOH at higher concentrations. Oscillations seen in the ZnS spectrum, used to calculate particle size, were not visible for the PbS core–shell particles.

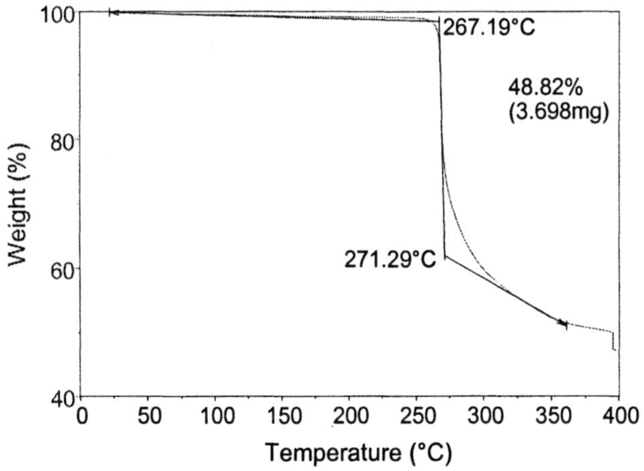

Figure 5. Thermal gravimetric analysis showing loss of polystyrene from PbS/0.827 μm PS-CO_2 core-shell particles.

2.3 Conclusions

A novel method was demonstrated for preparing PbS/polystyrene core-shell materials. Numerous experimental parameters were varied to carefully control film deposition. In the end, a rapid precipitation of PbS in the presence of ethylenediamine controlled by the dropwise addition of thioacetamide proved most effective at generating thick, uniform films of PbS over core polystyrene micorspheres.

Acknowledgement. This work was supported by the Office of Naval Research and the National Research Council through their Research Associateships Program. Special thanks to Joseph Perrin (Naval Research Laboratory) for the TGA analysis, also to Anthony Dinsmore (Naval Research Laboratory/Harvard University) and Roger Pink (George Mason University) who participated in the early development of this project. Of course, appreciation goes to the American Chemical Society, Andrzej Miziolek, Matt Mauro, all the editors, authors and others who helped make the 2001 Symposium on Defense Applications of Nanomaterials a resounding success.

References

1. Yablonovitch, E. *Phys. Rev. Lett.* **1987**, *58*, 2059.
2. Foresi, J.S.; Villeneuve, P.R.; Ferrera, J.; Thoen, E.R.; Steinmeyer, G.; Fan, S.; Joannopoulos, J.D.; Kimerling, L.C.; Smith, H.I.;Ippen, E.P. *Nature* **1997**, *390*, 143.
3. Noda, S.; Yamamoto, N.; Kobayashi, H.; Okano, M.;Tomoda, K. *Ap. Phys. Lett.* **1999**, *75*, 905-907.
4. Lin, S.Y.; Fleming, J.G.; Hetherington, D.L.; Smith, B.K.; Biswas, R.; Ho, K.M.; Sigalas, M.M.; Zubrzycki, W.; Kurtz, S.R.;Bur, J. *Nature* **1998**, *394*, 251-253.
5. Xia, Y.; Gates, B.; Yin, Y.;Lu, Y. *Adv. Mater.* **2000**, *12*, 693.
6. Zakhidov, A.A.; Baughman, R.H.; Iqbal, Z.; Cui, C.; Khayrullin, I.; Dantas, S.O.; Marti, J.;Ralchenko, V.G. *Science* **1998**, *282*, 897-901.
7. Wijnhoven, J.E.G.J.;Vos, W.L. *Science* **1998**, *281*, 802.
8. Holland, B.T.; Blanford, C.F.;Stein, A. *Science* **1998**, *281*, 538-540.
9. Imhof, A.;Pine, D.J. *Nature* **1997**, *389*, 948.
10. Blanco, A.; Chomski, E.; Grabtchak, S.; Ibisate, M.; John, S.; Leonard, S.W.; Lopez, C.; Meseguer, F.; Miguez, H.; Mondia, J.P.; Ozin, G.A.; Toader , O.;Driel, H.M.v. *Nature* **2000**, 437-440.
11. Breen, M.L.; Dinsmore, A.D.; Pink, R.H.; Qadri, S.B.;Ratna, B.R. *Langmuir* **2001**, *17*, 903-907.
12. Biswas, R.; Sigalas, M.M.; Subramania, G.;Ho, K.-M. *Phys. Rev. B* **1998**, *57*, 3701.
13. Lin, S.-Y.; Chow, E.; Hietala, V.; Villeneuve, P.R.;Joannopolous, J.D. *Science* **1998**, *282*, 274.
14. Breen, M.L.; Woodward IV, J.T.; Schwartz, D.K.;Apblett, A.W. *Chem. Mater.* **1998**, *10*, 710-717.
15. Weast, R.C., *CRC Handbook of Chem. and Physics*. 54 ed. 1974, Cleveland, OH: CRC Press.
16. Dhas, N.A.; Zaban, A.;Gedanken, A. *Chem. Mater.* **1999**, *11*, 806.
17. Blount, G.H.; Schreiber, P.J.; Smith, D.K.;Yamada, R.T. *J. Appl. Phys.* **1973**, *44*, 978-981.
18. Kothiyal, G.P.; Chosh, B.;Deshpande, R.V. *J. Phys. D: Appl. Phys.* **1980**, *13*, 889-872.
19. Torriant, I. *Thin Solid Films* **1981**, *77*, 347-356.
20. Markov, V.F.; Maskaeva, L.N.;Kitaev, G.A. *Inorg. Mater.* **2000**, *36*, 657-659.
21. Cullity, D., *Elements of X-ray Diffraction*. 1978, Reading, MA: Addison-Wesley Pub. Inc.
22. Bohren, C.F.;Huffman, D.R., *Absorp. and Scattering of Light by Small Particles*. 1983, New York: John Wiley and Sons, Inc.

Chapter 18

Nanostructured Polymeric Nonlinear Photonic Materials for Optical Limiting

James S. Shirk[1], Richard G. S. Pong[1], Steven R. Flom[1], Anne Hiltner[2], and Eric Baer[2]

[1]Optical Sciences Division, Naval Research Laboratory, 4555 Overlook Avenue, SW, Washington, DC 20375
[2]Department of Macromolecular Science, Case Western Reserve University, 2100 Adelbert Road, Cleveland, OH 44106

A new type of nonlinear nano-layered polymeric optical material is described. The material comprises alternating polymer layers with thousands of layers and a layer thickness down to 30 nm or less. A nonlinear response is introduced by including an appropriate dye in alternate polymer layers. Such materials have several uses, including optical limiting. Several nano-layered materials using optical limiter dyes in alternate layers were fabricated. Some of the linear optical properties and a verification of nonlinear transmission and reflectivity at high intensity for example materials are reported.

Introduction and Background

Optical limiters are devices with a transmission that decreases as the light intensity increases. Ideally, the transmission is high at normal light intensities, and it instantaneously decreases to minimal levels when intense beams are present. The intensity dependent transmission limits the output intensity so that it is always below some maximum value. All-optical, passive optical limiters

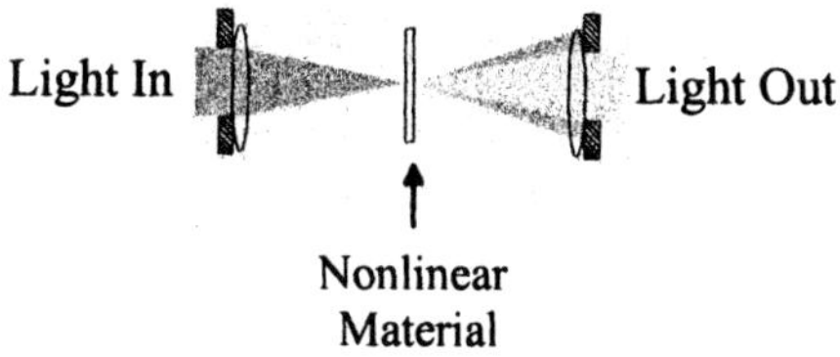

Figure 1. A typical optical limiter.

normally rely on materials with nonlinear absorption and refraction. (1) A typical optical limiter configuration is shown in Figure 1. In this figure, the nonlinear element is shown placed between a lens pair, near the focus of the first lens. Positioning the nonlinear element in the focal region is required to provide the higher intensities necessary to drive the nonlinear response.

The nonlinear material is critical to the operation of a limiter. Several different nonlinear effects can be used in the nonlinear element to provide a limiting function. Nonlinear absorption, where the material absorbance increases with the intensity of the incident light, is the most useful mechanism to remove light from the transmitted beam. A large number of organic and organometallic materials have been reported to exhibit an intensity dependent increase in absorption, sometimes called a reverse saturable absorption. The most effective are phthalocyanines (2, 3, 4, 5) and porphyrins. (6,7)

Nonlinear refraction can also be used to deflect or scatter light out of the transmitted beam thereby decreasing the device transmission. An example was illustrated in Figure 1. When the nonlinear material has a refractive index that depends on the light intensity or fluence, the intensity distribution across the beam generates a refractive index variation in the nonlinear material that behaves as a lens with strong aberrations. This lens will deflect some of the light into the wings of the transmitted beam where an exit aperture blocks it. This serves to limit the energy in the transmitted beam at high input fluences

A potentially more effective way to use refractive index changes for optical limiting is by using a nonlinear photonic crystal. (8,9) A photonic crystal is a material having a spatially periodic modulation of the refractive index. Photonic crystals possess a set of band gaps or wavelength ranges where light cannot propagate but is diffracted or reflected instead. The transmission spectrum of a photonic crystal depends on the magnitude of the refractive index modulation.

The reflectivity and width of the band gaps increase as the refractive index modulation increases. The basic principle for optical limiting by a 1D nonlinear photonic crystal material can be understood from Figure 2. Initially, (A), the refractive indices of the different layers are matched, and the sample

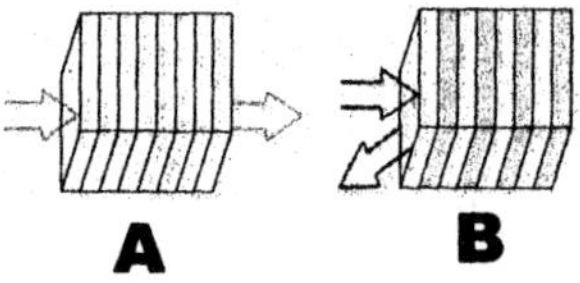

Figure 2. The concept of a nonlinear photonic crystal optical limiter.

transmission is high. As the light intensity increases, the refractive index of the nonlinear layer changes. The resulting index modulation makes the material an effective dielectric mirror (B) and the transmission falls.

A difficulty arises because the photonic crystal structures for the visible region are not easy to fabricate. However, in this chapter we describe a facile extrusion method for fabricating polymers with a 1D structure like that shown in Figure 2. The nanolayered polymeric structures can consist of many thousands of layers and have a modulation in the *nonlinear* refractive index in the direction normal to the surface of the layers. Such materials are the nonlinear analogue of polymeric multilayer interference mirrors. (10,11,12) They are also the 1D analogue of the 2D photonic crystals studied by Lin et. al. (13) The latter workers demonstrated that photonic crystals do indeed provide an effective method for converting an intensity dependent refractive index into an intensity dependent transmission.

Nonlinear absorption provides the most useful mechanism for optical limiting, so it is desirable to use the induced reflectivity in conjunction with nonlinear absorption. The target material is then one in which alternate layers contain a dye that possesses both nonlinear absorption and nonlinear refraction. This material has a modulation in both the real and imaginary (absorptive) parts of the index. This paper describes the progress towards such a material.

Experimental Apparatus

For the optical characterization, the samples were cut directly from the extruded material. Absorption spectra were recorded on a Perkin-Elmer Lambda 9 spectrophotometer and on an Ocean Optics fiber optic spectrometer. The reflectance spectra were recorded using the fiber optics spectrometer with a reflectance probe.

The laser source for the nonlinear experiments was an optical parametric oscillator (OPO) pumped by the third harmonic of a Nd/YAG laser.

The input beam was a spatially filtered beam from one of the laser sources described below. The spatial profile of the input beam usually showed a greater than 96% correlation with a gaussian profile. The pulse width was 2.5 nanoseconds. The experiments at low energies were conducted at 10 Hz. For incident fluences above about 10 mJ/cm^2, the repetition rate was reduced to 0.5 Hz. This removed any effect of persistent thermal effects. The laser intensity was controlled by wave plate/polarizer combination. The input beam was focused onto the sample using approximately f/45 optics. The focal spot size (beam radius) was measured to be 2.6 ±0.1 μm at 550 nm.

The sample was mounted on a translation stage and translated through the focus to find the maximum change in reflectivity or transmission. For some nonlinear reflection measurements the reflectivity was measured as a sample was translated through the focus of the laser. A schematic of the experiment is shown in Figure 3. It can be thought of as the reflection analog of the well-

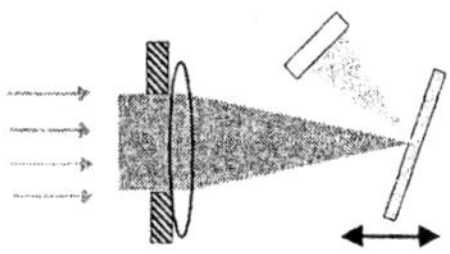

Figure 3. *Reflection Z-Scan.*

known transmission Z-scan experiment. A reversible nonlinear reflectivity is observed as a change in reflectivity as the intensity increases when the sample passes through the focus. The reflectivity was measured at about a 45-degree angle of incidence.

In some experiments, the samples were annealed to, or held at, 120 °C. We found this reduced some microscopic inhomogeneities in the extruded films.

Results and Discussion

Fabrication of the Nanolayered Polymers

The preparation of layered structures of polymers by coextrusion is well known. It has been used to fabricate materials with combinations of physical properties. Materials with three to seven layers are produced commercially. More recently, the development of layer multiplying dies has allowed the preparation of nanolayer polymers with hundreds or even thousands of layers. (14, 15, 16, 17)

The technique is illustrated in Figure 4. Two polymer melt streams are

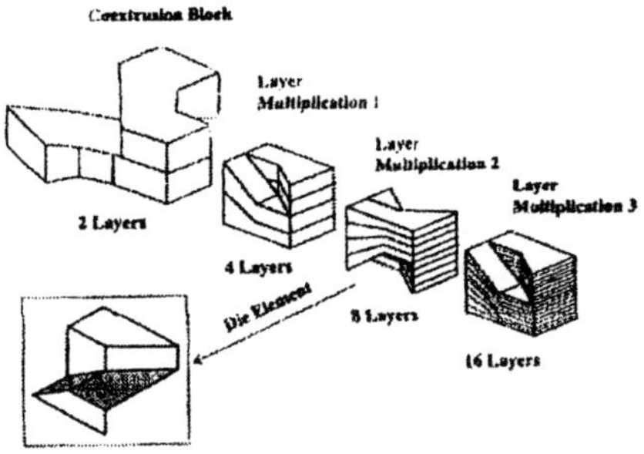

Figure 4. Layer multiplication of extruded polymers.

combined in the coextrusion block to form a two-layered polymer. The first multiplying element slices this bilayer polymer vertically, spreads the melt horizontally, and then recombines the two flowing melt streams. The illustration shows how the flowing streams are recombined to double the number of layers. Subsequent multiplying elements multiply the number of layers further. An assembly of n multiplier elements produces an extrudate with the layer sequence $(AB)_x$ where x is equal to 2^n. Layered polymeric structures with up to 4096 layers have been successfully processed and more are possible. By altering the relative flow rates or the number of layers, while keeping the film or sheet thickness constant, the individual layer thickness can be controlled.

When two materials of differing refractive indices are extruded into a nanolayer structure that has an effective layer thickness of approximately 1/4 the wavelength of visible light, the material is effectively a 1D visible photonic crystal. The nano-layer extrusion process is a continuous process that yields photonic crystals that are flexible sheets of a polymer. This is a great advantage since it can easily and rapidly produce photonic materials with a much larger size than typical photonic crystal fabrication techniques.

The details of the layered structure control the optical properties of the nanolayer films. The wavelength dependence of the photonic band gaps formed depends on the number of layers, the layer spacing and the refractive index variation that is induced. For an optical limiting application, a broadband reflectivity is required. This can be achieved in the extrusion process by chirping the layer thickness or by using a distribution of layer spacing in the nanolayered material. (18)

Nonlinear Polymeric 1D Photonic Materials

The nonlinear nanolayered materials are fabricated by introducing nonlinear dyes into alternate layers of the nanolayered polymer. Nonlinear dyes with large refractive and absorptive nonlinearities have designed, synthesized, and characterized at NRL over the past several years. The nonlinear optical chromophore is typically a phthalocyanine. Polymer miscibility was achieved by peripheral substitution of the phthalocyanine. Nonlinear dyes compatible with styrenes, epoxies, (19,20) and urethane polymers (21) are available.

For this work, nonlinear nanolayered polymeric materials were fabricated using SAN, a styrene-acrylonitrile copolymer The nonlinearity is introduced into the alternate layers of SAN by dissolving either lead tetrakis(cumylphenoxy)-phthalocyanine ($PbPc(CP)_4$) or nigrosine in one of the initial polymer streams first. $PbPc(CP)_4$ is known to have a strong nonlinear absorption and refraction coefficient. In addition, there is a thermal contribution to the nonlinear response in $PbPc(CP)_4$. (2) It is one of the best visible optical limiter materials available. The nigrosine dye was of interest because it has a broad absorption in the visible and an excited state that was reported to relax rapidly with the conversion of the absorbed energy into heat. (22) The resulting rise in temperature causes a change in the refractive index of the host. Nigrosine does not exhibit a substantial reverse saturable absorption. It was of interest because it had been used as the nonlinear material in a previous photonic crystal optical limiter.

Nanolayer films consisting of layers of SAN25 (styrene-acrylonitrile with 25% acrylonitrile) doped with a dye alternating with an undoped SAN20 polymer (styrene-acrylonitrile with 20% acrylonitrile) layer were fabricated by the extrusion technique described. A slightly different copolymer, SAN20, was chosen for the second layer to improve the index matching of the layers at low intensity. Adding the dye to the SAN25 polymer will raise the index of refraction of the dyed polymer. We compensate for this by using a polymer with a slightly higher index for the undyed layer.

Characterization

A SAN25/SAN20 nanolayered sample was cleaved perpendicular to the plane of the film. The thickness of the individual layers was then measured by atomic force microscopy (AFM). The AFM image is shown in Figure 5. This

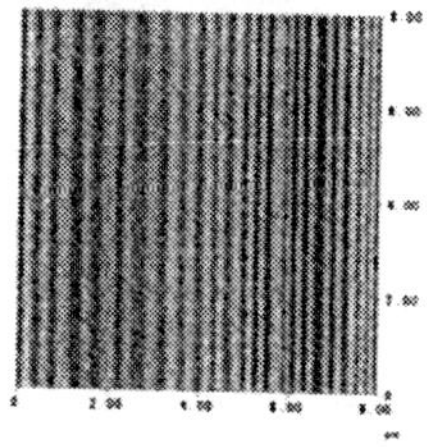

Figure 5. AFM phase image of nano-layered SAN25+dye/SAN20 film.

figure shows the layer structure clearly. It also shows that there is a distribution of layer thickness across the sample. The layer thickness distribution gives rise to a broad band gap, i.e. a broad band reflectivity, when there is an index contrast between the layers.

Figure 6 shows the measured reflectivity of two different nanolayered samples with different distributions of layer thickness. The upper trace in this

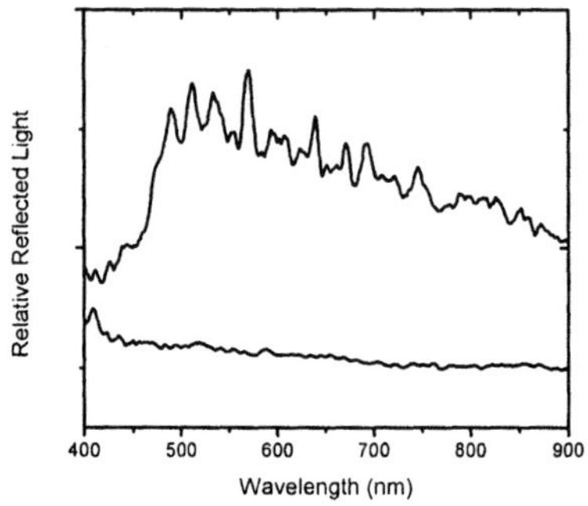

Figure 6. *The spectrum reflected from a nano-layered SAN polymer with an* average *layer thickness of (A) 87 nm and (B) 31 nm.*

figure shows the spectrum of the reflected light from a 4096 layer samples, one with an average layer thickness of 87 nm. The lower trace is a reflection spectrum of a sample with an average layer thickness of 31 nm. In each sample, the index difference between the alternate layers is approximately 0.0045. The 87 nm layered film has a first order band gap in the visible and shows a band of reflectivity from about 450 nm to beyond 900 nm. The band gap for the 31 nm layered film is at 197 nm, well outside the visible spectrum. The sample with an average layer thickness of 31 nm shows only the normal Fresnel reflection near 500 nm. The layer thickness distribution derived from the AFM measurements is qualitatively consistent with the width of the observed reflection spectrum in Figure 5.

There is a noticeable fine structure on the reflection spectrum of the sample with an 87 nm spacing. This fine structure is reproducible. Such structured reflectivity is expected to arise in a layered sample with a random distribution of layer thickness. (23) The fine structure will be discussed in more detail in a later publication. Figure 6 demonstrates that the broad-band reflectivity that is essential for optical limiter applications can be achieved. In fact, the reflectivity of the 87 nm layer sample is somewhat broader than is optimum for a practical limiter. It is possible to adjust the fabrication process to produce a narrower distribution of layer thickness.

Nonlinear Optical Properties

The nonlinear reflectivity of a nanolayered sample was demonstrated by measuring the reflectivity as a sample was translated through the focus of a nanosecond pulsed laser. A schematic of the experiment was shown in Figure 3.

The reflectivity as a function of sample position for two samples comprised of 4096 layers alternating SAN25 and SAN20 polymer is shown in Figure 7. In

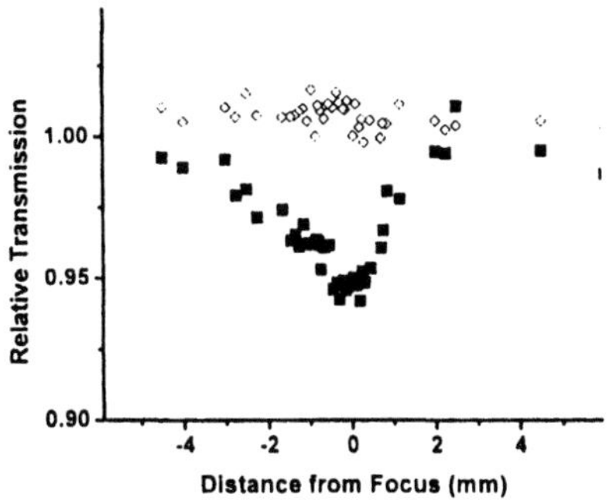

***Figure** 7. Reflectvity as a function of distance to the focus. A (filled squares) only alternate layers contain dye; B(open circles) all layers are dyed the same.*

the first sample (curve A) the SAN25 layers are dyed with nigrosine while the SAN20 layers are left undyed. In Figure 7, curve A, the reflectivity of this material was observed to decrease as the sample approaches focus. In this sample, the refractive index, n_0, of the dyed layers is initially slightly larger than that of the undyed layer. The index of the dyed layers will decrease with fluence so that the index modulation and the resulting reflectivity should decrease with increasing incident energy. This is consistent with the decrease in reflectivity when the sample is near the focus in Figure 7. The second sample (curve B) was identical to the first, except that every layer was dyed with the same concentration of nigrosine. In this sample there is no modulation in the nonlinear index. The reflectivity of this sample does not vary outside of the experimental uncertainty as the sample is translated through the focus. This confirms that observed nonlinear reflectivity is observed in the layered polymer only when there is a modulation in the nonlinear index.

Nanolayered samples with a modulation in both the nonlinear absorption and refraction were also fabricated and studied. Figure 8 shows the transmission as a function of incident energy for a sample comprised of 4096 layers alternating SAN25 and SAN20 polymer. In this case, the SAN25 layers were dyed with $PbPc(CP)_4$, a strong nonlinear absorber. The SAN20 layers are left undyed. Figure 8 shows that this sample exhibits a strong nonlinear absorption, it behaves as an optical limiter. The relative reflectivity of this sample as a

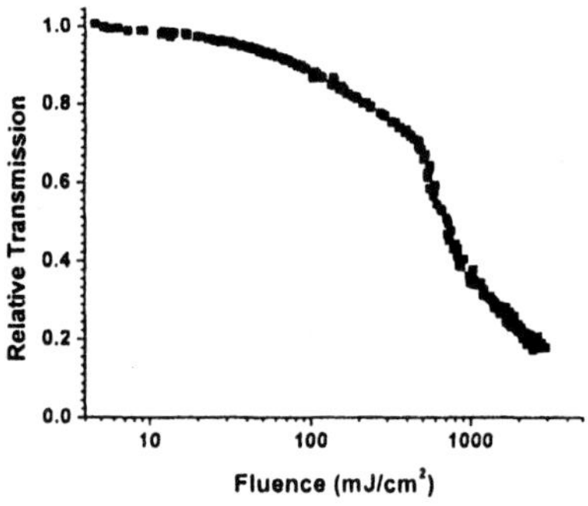

Figure 8. Relative transmission as a function of incident fluence at 532 nm for a 4096 layer sample of SAN with alternate layers dyed with PbPc(CP)4. Average layer thickness approx. 87 nm.

function of incident fluence also measured. The reflectivity decreases with incident intensity.

Summary

A new type of nano-layered polymeric nonlinear optical material for optical limiting applications has been fabricated. It comprises alternating polymer layers with thousands of layers and a layer thickness down to 30 nm or less. By introducing appropriate dyes into alternate polymer layers, a material with a modulation in the real and/or imaginary part of the nonlinear refractive index in the direction normal to the surface is possible.

We described a facile multilayer extrusion technique for fabricating these novel polymer film materials. The presence of a discrete nanolayered structure was verified by an AFM study. The AFM study also demonstrated the presence a distribution of layer thickness within the polymer film. The distribution is essential to controlling the bandwidth of the induced reflectivity. A broad band reflectivity was demonstrated in a layered material with a permanent index modulation of ~ 0.0045. This wavelength dependence of the reflectivity showed a fine structure characteristic of dielectric stack with a distribution of layer thicknesses.

We demonstrated the fabrication of nonlinear versions of these nanolayered materials. One such material used nigrosine as the nonlinear dye. This introduces a thermal nonlinear refraction into the dyed layers. We demonstrated that an input fluence of a few mJ/cm^2 was sufficient to produce a nonlinear reflectivity from this material when alternate layers were dyed. In a control sample, all layers were dyed and no nonlinear reflectivity was observed. This confirmed that a modulation in the nonlinear index was necessary to produce the observed reflectivity.

Another nonlinear nanolayered material was fabricated with the phthalocyanine dye, $PbPc(CP)_4$, in alternate layers. This dye possesses a strong nonlinear absorption and a nonlinear refraction. In the layered material, both are modulated at a spatial frequency corresponding to the layer thickness, which was an average of 87 nm. The material showed both a nonlinear absorption and a nonlinear reflection and provided optical limiting. The modeling of this nonlinear response of this material was not attempted here. To our knowledge, there is little available theory on the response of such nonlinear structures to pulsed lasers.

This work demonstrates that these new materials have the potential to provide enhanced optical limiting by the more efficient use of nonlinear refractive processes. Clearly, it is necessary to understand in more detail how the magnitude of the different components, real and imaginary (absorptive), of the nonlinearities in the layers and the distribution of layer thickness influence the nonlinear response. Experiments to this end are in progress, however, a better theoretical model for the response of multilayer structures with both nonlinear absorption and refraction is desirable. Such a model would aid in the design of more efficient layer structures.

Acknowledgements

We thank the Office of Naval Research and the Army Research Office for support of this work.

References

1. Nalwa, H.S., Shirk, J.S., in *Phthalocyanines: Properties and Applications* (Eds.: C.C. Leznoff, A.B.P Lever, VCH Publishers, Inc., New York 1996, vol. 4, p. 79ff.
2. J.S. Shirk, R.G.S. Pong, F.J. Bartoli, A.W. Snow, Appl. Phys. Lett., 63 (1993), 1880.
3. J.W. Perry, K. Mansour, I.Y.S. Lee, X.L. Wu, P.V. Bedworth, C.T. Chen, D. Ng, S.R. Marder, P. Miles, T. Wada, M. Tian, H. Sasabe, Science, 273 (1996), 1533.
4. G. de la Torre, P. Vazquez, F. Agullo-Lopez, T. Torres, J. Mat. Chem., 8 (1998), 1671.
5. T.H. Wie, D.J. Hagan, M.J. Sence, E.W. Van Stryland, J.W. Perry, D.R. Coulter, Appl. Phys. B, 54 (1992), 46.

6. W. Su, T.M. Cooper, Chem. Mater., 10 (1998), 1212.
7. D.N. Rao, S.V. Rao, F.J. Aranda, D.V.G.L.N. Rao, M. Nakashima, J.A. Akkara, J. Opt. Soc. Am. B, 14 (1997), 2710.
8. H.-B. Lin, R.J. Tonucci and A.J. Campillo, Opt. Lett., **23**, 94-96, 1998.
9. Shirk, J.S. and A. Rosenberg,. Laser Focus World, 2000. 36(4): p. 121.
10. T. Alfrey, E.F. Gurnee, W.J. Schrenk; Polym. Eng. Sci. 9, 400 (1969)
11. J.A. Radford, T. Alfrey, W.J. Schrenk; Polym. Eng. Sci. 13, 216 (1973)
12. Weber, M.F. et. al. Science, 287, 2451, 2000
13. H.-B. Lin, R.J. Tonucci and A.J. Campillo, Opt. Lett., **23**, 94-96, 1998.
14. Ebeling, T.; Hiltner, A., and Baer, E.; J. Appl. Polymer Sci. 68, 693 (1998)
15. Kerns, J.; Hiltner, A., and Baer, E.; "Processing and Properties of Polymer Microlayered Systems" in "*Structure Development During Polymer Processing*" , NATO-ASI Conference on Polymers, Universidade do Minho-Azurem, Guimaraes, Portugal May 17-28, 1999
16. Nazarenko, S., Dennison, M., Schuman, T., Stepanov, E.V., Hiltner, A., and Baer, E.; J. Appl. Polymer Sci. 73, 2877 (1999)
17. Schuman, T., Nazarenko, S., Stepanov E.V., Maganov, S.N., Hiltner, A., and Baer, E.; Polymer 40, 7373 (1999)
18. Brzozowski, L. and E.H. Sargent,. IEEE Journal of Quantum Electronics, 2000. **36**, 1237.
19. George R.D., Snow, A.W., Shirk, J.S., Flom, S.R., Pong, R.G.S.; Polymer Preprints, 35, 236 (1994)
20. Flom, S.R., Pong, R.G.S., Shirk, J.S., Bartoli, F.J., Cozzens, R.F., Boyle, M.E., and Snow, A.W.; Proc. Mat. Res. Soc. 479 (1997)
21. S. R. Flom, J. S. Shirk, R. G. S. Pong, J. R. Lindle, F. J. Bartoli, M. E. Boyle, J. D. Adkins, and A. W. Snow, SPIE Proc., 2143 (1994) 229; and M. E. Boyle, J. D. Adkins, A. W. Snow, R. F. Cozzens, and R. F. Brady, J. Appl. Polym. Sci., 57 (1995) 77
22. Justus, B.L., Huston, A.L., Campillo A.J., Appl. Phys. Lett. 63: 1483 (1993)
23. Yoo, K.M. and R.R. Alfano,. Physical Review B, 1989. **39**, 5806

Chapter 19

Properties of Single Wall Carbon Nanotubes: A Density Functional Theory Study

B. Akdim[1], X. Duan[2], Z. Wang[1], W. W. Adams[3], and R. Pachter[1,*]

[1]Air Force Research Laboratory, Materials and Manufacturing Directorate, Wright-Patterson Air Force Base, OH 45433
[2]Aeronautical Systems Center, Major Shared Resource Center for High Performance Computing, 3005 P Street, Wright-Patterson Air Force Base, OH 45433
[3]Center for Nanoscale Science and Technology, Rice University, Houston, TX 77005
[*]Corresponding author: ruth.pachter@wpafb.af.mil

We present a study of single-wall carbon nanotubes (SWCNTs) for C(n,n) and C(n,0), n=(4,6,8,10), modeled as isolated tubes or crystalline-ropes, using a full-potential linear combination of atomic orbitals (FP-LCAO) density functional theory (DFT) approach. Structural and mechanical properties agree well with previous theoretical and experimental work. The study provides further insight, for example, into structural aspects of SWCNTs, where we note a different trend for bond lengths in armchair and zigzag tubes. An up-shift of the Raman radial breathing mode (RBM) due to intertube interactions was calculated and compared to experiment and previous theoretical studies. Our calculations provide an example for a computational approach to be used in studying SWCNT materials for defense related applications.

Studies of carbon nanotubes, recently reviewed (*1*), suggest a wide range of applications (*2*). Moreover, with the advent of the synthesis, controllable growth, manipulation and characterization of these materials (*3,4,5,6*), it has become possible to gain significant insight from the calculation of properties, due to the possibility of comparison with experiment. Indeed, resonance Raman spectroscopy is well known for measuring the RBMs of SWCNTs and characterizing tube radii *(7,8)*, in comparison with theory (*9,10*).

The rolled structure of a graphene sheet gives rise to unprecedented electronic properties of carbon nanotubes, where the semiconducting or metallic characteristics depend on chirality and radii. These properties were first predicted by theory (*11,12)*, and then confirmed by experiment (*13,14*). Indeed, electronic nanodevices such as field-effect transistors were proposed (*15,16,17*), holding promise also for defense related applications. In addition, carbon nanotubes also attracted considerable attention for field emission, and could possibly be utilized for defense applications such as field emission cathodes in high-power microwave tubes (*18)*, also studied theoretically *(19*).

Furthermore, the exceptional mechanical properties of nanotubes, as demonstrated by their tensile strength and elastic moduli (*20*), render them appropriate for composites applications (*21)*. A Young's modulus (YM) measurement for multi-wall carbon nanotubes (MWCNTs) was carried out by the thermal vibrational amplitudes technique (*22*), reporting a value of 1.8TPa. Krishnam et al. (*23*) performed a similar experiment for SWCNTs, and obtained a value of 1.3 ± 0.5TPa. An experiment by Selvetat and co-workers (*24)* for single-wall crystalline-ropes using an AFM with a special substrate to allow for a direct measurement of Young's modulus, resulted in an average value of 1.28 ± 0.59TPa. The recently reported experiment by Demczyk et al. (*25*), estimated the tensile strength and YM to be 0.15TPa, and 0.9TPa, respectively.

In this study, we report an evaluation of structural, mechanical, and vibrational properties of SWCNTs [C(n,n), and C(n,0), n=4,6,8,10], applying a FP-LCAO method *(26,27)*, as it has previously been pointed out that all-electron full-potential local density approximation (LDA)-based calculations are appropriate for treating van der Waals interactions in systems such as fullerenes, graphene and graphite (*28,29*). Although the full-potential all-electron scheme is computationally intensive, our work contributes to the further validation of previous theoretical studies. Large-scale pseudopotential DFT calculations were also recently performed by Yoon et al. *(30)*, in order to study the effects of intertube interactions on the transport properties of SWCNT junctions.

We also studied crystalline-ropes, optimizing the intertube distance, in comparison with other theoretical studies using the pseudopotential LCAO approach (*31*), tight-binding methods (*32,33*), or an empirical force-constant model (*34)*, as well as with experimental data whenever available.

Computational Details

In all SWCNTs calculations we applied DMOL3 (*26)*, using a double-numeric polarized basis set, with an atomic cutoff radius of 10.4 (a.u.). The exchange-correlation term follows the Perdew and Wang parameterization (*35*). Calculations were carried out with full geometry optimization, assuming a hexagonal symmetry to reduce computation time. Different k-point samplings of the irreducible Brillouin zone were tested, adopting 15 k-point for crystalline-ropes (bundles) and 5 k-point for an isolated nanotube. Periodic boundary conditions were used in 3-D to simulate a crystalline-rope in the hexagonal structure, with a triangular unit cell in the plane perpendicular to the tube axis. A wide lateral separation, larger than 45Å, was used to model isolated tubes. In addition, for the C(6,6) SWCNT, we performed calculations of a super-cell of five unit-cells and found insignificant changes in the structural properties. Thus, all reported results were carried out using only one unit-cell with no defects.

Results and Discussion

Structural and Mechanical Properties

Some of the early theoretical work on carbon nanotubes includes that of Lu (*34*), using an empirical force-constant model to calculate elastic moduli for SWCNTs and MWCNTs, assuming that the interwall distances are those of the graphite interlayer distance (3.4Å) with no further optimization. In addition, Hernandez et al. (*32*) performed a comparative tight-binding study of isolated carbon nanotubes and other compositions ($B_xN_yC_z$), using 3.4Å to predict the properties of a crystalline-rope with no further optimization, whereas Sanchez-Portal and co-workers (*31*) used a pseudopotential LCAO method to calculate properties of isolated carbon nanotubes. Notably, interwall carbon nanotubes interactions (*36*), were also most recently studied (*37,38,39,40*).

Intertube distances for the SWCNTs considered were optimized by calculating the energy as a function of intertube separation, allowing for a full relaxation of atom coordinates at each intertube distance. The results were taken as the intertube distance evaluated from the center of one tube to the center of an adjacent one and listed for selected tubes (Figure 1, Table I). With the exception of C(4,0), where we obtained a value of 12% lower than $2R_{average}$+3.4Å, our results differ by about 2-4.5% when compared to other theoretical work. Interestingly, a previous DFT study of carbon nanotube packings has shown that hexagonal packing does result in intertube distances smaller than 3.4Å (*41).*

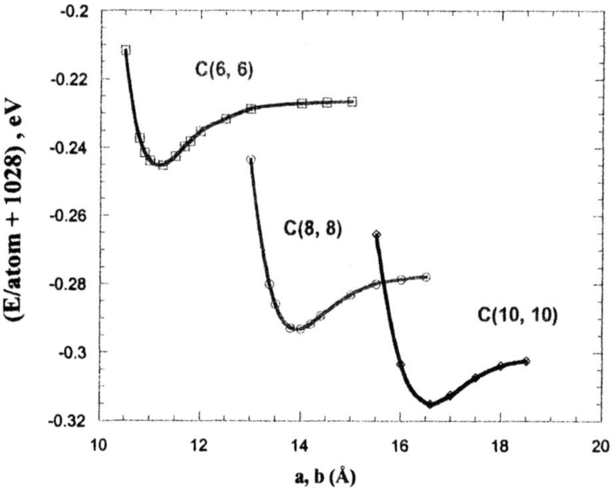

***Figure 1**: Total energy/atom as a function of intertube distance (Å). The energies are scaled by a value of 1028 (eV). (See page 7 in color insert.)*

The nanotube crystalline-rope interaction energies were found to be, on average, 0.2 (eV/Å) and 0.17 (eV/Å), for armchair and zigzag tubes, respectively. These results agree well with a previous theoretical study (*42*). The average radii ($R_{average}$), and the $(C\text{-}C)_h$ and $(C\text{-}C)_v$ bond lengths, derived from the optimized geometry of the equilibrium structure, are summarized in Table I. The subscripts h and v designate horizontal and vertical bonds, respectively; the vertical bonds in zigzag tubes correspond to the double bonds, and the horizontal bonds to the single ones, whereas in the armchair tubes the correspondence of the bonds is reversed. Radii were shown to be within 1.5% of tight-binding (*32*) and pseudopotential LCAO (*31*) calculations.

In Figure 2, we plot the carbon-carbon bond lengths of various nanotubes. For zigzag tubes, the horizontal bond lengths, in the direction of the circumference of the tube, were shown to be larger for small-diameter tubes, whereas the vertical bonds, in the direction of the tube axis, have a smaller bond length, increasing with tube radius. Both the horizontal and vertical bond lengths converge to the graphite carbon-carbon bond length experimental value (*43*) of 1.419Å. In the case of armchair tubes, the difference is insignificant, and their values are shown to be within the DFT graphite bond length value.

Young's modulus was calculated by carrying out a full self-consistent field calculation for each strain imposed along the tube axis, allowing for a full relaxation of atomic positions. The tubes were then compressed and stretched to about –4% and +4%, respectively. A smaller range of strains [-1%, +1%], as previously suggested (*31*), resulted in incorrect predictions of Poisson's ratio.

The total energy was then fitted to a third order polynomial as a function of the strain (ε), from which Young's modulus was derived (*31*) (Table II).

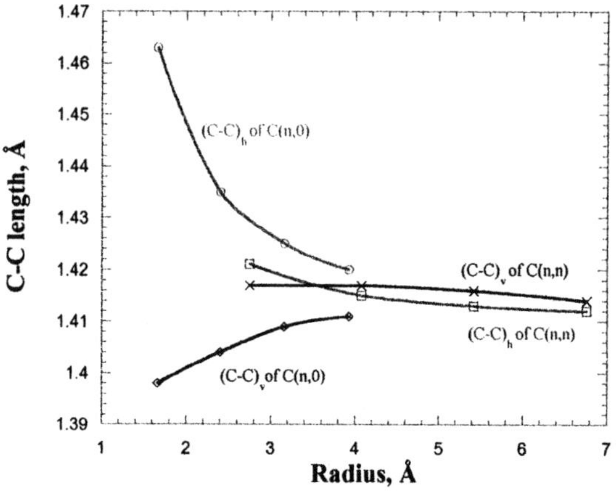

***Figure 2**: Bond lengths as a function of the tube radius (Å).*
(See page 7 of color insert.)

We note that, in general, calculated values are within the experimental estimation, and in particular, in good agreement with recent data (*25*). In addition, we examined the effects of intertube interactions on tube radii and Young's moduli, using 3.4Å (Figure 3a) and optimized distances (Figure 3b). Our results show that intertube interactions have only a small effect on Young's moduli. Young's modulus increases with the tube radius for zigzag nanotubes, while only small variations are shown for armchair tubes. In both cases, the values converge to the in-plane C_{11} elastic modulus of a graphite sheet, measured as 1.02TPa. Young's moduli using the FP-LCAO optimized intertube separations are shown in Figure 3b and summarized in Table II. We obtained larger values when compared to those evaluated with the interlayer graphite distance of 3.4Å. Clearly, these results are due to the lower values obtained for the intertube distances (Table I). It is interesting to note that for C(4,0), with a radius smaller than 2.4Å, we predict a different trend, compared to the results using the interlayer graphite distance (Figure 3a), thus suggesting a different strength. This is in agreement with Stojkovic et al. (*44*), reporting a high stiffness for small radii tubes.

We further investigated the effect of strain (ε) on the tubes calculating Poisson's ratio (σ), as listed in Table III. Intertube interactions appear to have overall a small effect on Poisson's ratio, although a higher response to an external force for small radii tubes was calculated. Our results were found to be lower than the tight-binding values (*32,33*).

Table I. A summary of calculated structural parameters. ITD denotes the intertube distance (Å).

SWCNT		*Crystalline-rope*	*Isolated*
C(4,0)	$(C\text{-}C)_h$	1.460	1.463
	$(C\text{-}C)_v$	1.400	1.398
	$R_{average}$	1.660	1.665
	ITD	6.03	--
	$(2R_{average}+3.4)$	6.72	--
(6,0)	$(C\text{-}C)_h$	1.435	1.435
	$(C\text{-}C)_v$	1.405	1.404
	$R_{average}$	2.402	2.399
	ITD	7.980	--
	$(2R_{average}+3.4)$	8.200	--
(4,4)	$(C\text{-}C)_h$	1.421	1.421
	$(C\text{-}C)_v$	1.418	1.417
	$R_{average}$	2.746 [2.794(*31*)]	2.746
	ITD	8.600	--
	$(2R_{average}+3.4)$	8.900 [8.99(*31*)]	--
(8,0)	$(C\text{-}C)_h$	1.422	1.425
	$(C\text{-}C)_v$	1.411	1.409
	$R_{average}$	3.158	3.160
	ITD	9.497	--
	$(2R_{average}+3.4)$	9.720	--
(10,0)	$(C\text{-}C)_h$	1.421	1.420
	$(C\text{-}C)_v$	1.411	1.411
	$R_{average}$	3.933 [3.979(*31*), 3.955(*32*)]	3.929
	ITD	11.04	--
	$(2R_{average}+3.4)$	11.27 [11.36(*31*), 11.31(*32*)]	--
(6,6)	$(C\text{-}C)_h$	1.415	1.415
	$(C\text{-}C)_v$	1.417	1.417
	$R_{average}$	4.070 [4.14(*31*), 4.1(*32*)]	4.077
	ITD	11.20	--
	$(2R_{average}+3.4)$	11.54 [11.68(*31*), 11.60(*32*)]	--
(8,8)	$(C\text{-}C)_h$	1.413	1.413
	$(C\text{-}C)_v$	1.416	1.416
	$R_{average}$	5.420 [5.498(*31*)]	5.416
	ITD	13.90	--
	$(2R_{average}+3.4)$	14.24 [14.40(*32*)]	--
(10,10)	$(C\text{-}C)_h$	1.412	1.412
	$(C\text{-}C)_v$	1.414	1.414
	$R_{average}$	6.760 [6.864(*31*), 6.8(*32*)]	6.761
	ITD	16.60	--
	$(2R_{average}+3.4)$	16.92 [17.13(*31*), 17.00 (*32*)]	--

Table II. Young's moduli (TPa)[a] of SWCNTs compared to other theoretical studies.

SWCNT	*Cystalline-rope*			*Isolated*			
	YM *δR[b] (TPaÅ)	YM[c]	YM[d]	YM *δR[b] (TPaÅ)	YM[c]	YM[d]	
(4,0)	2.77	0.82	1.06	2.84	0.84	1.04	
(6,0)	3.23	0.95	1.03	3.28	0.97	1.01	
(4,4)	3.27	0.96	1.03	3.27	0.96	1.03	
(8,0)	3.54	1.01	1.09	3.43	1.01	1.08	
(10,0)	3.53	1.04	1.12	3.49	1.03	1.14	1.22 (*32*) 0.975 (*34*)
(6,6)	3.34	0.98	1.08	3.32	0.98	1.07	1.22, 1.09 (*31*)
(8,8)	3.40	1.00	1.11	3.37	0.99	1.10	
(10,10)	3.40	1.00	1.09	3.35	0.99	1.10	1.24(*32*) 0.972(*34*)

[a] $Y = \frac{1}{V_o}\frac{\partial^2 E}{\partial \varepsilon^2}$, where $V_o = 2\pi L R \delta R$; δR edge-to-edge intertube separation; *L* unit cell length, *R* radius of the tube; [b]Young's modulus*δR (TPa*Å) values are provided due to the controversy regarding the definition of δR (*20*); [c]YM values using δR =3.4Å; [d]YM values using δR =ITDÅ, ITD values are listed in Table I.

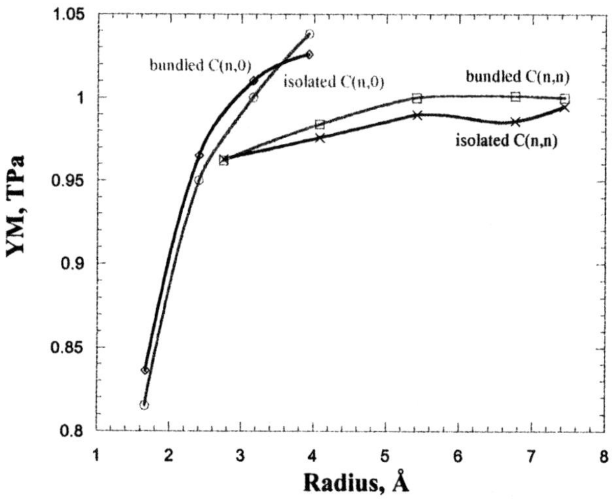

Figure 3: *Young's modulus (TPa) vs. tube radius; 3.4 (Å) was used for the intertube separation in calculating V_0. (See page 8 of color insert.)*

Table III. Calculated Poisson's ratio[e] values.

SWCNT	*Crystalline-rope*	*Isolated*
(4,0)	0.28	0.30
(6,0)	0.23	0.22
(8,0)	0.22	0.25
(10,0)	0.19	0.19 [0.275(*32*), 0.28(*34*)]
(4,4)	0.17	0.18
(6,6)	0.17	0.18 [0.247(*32*)]
(8,8)	0.15	0.18
(10,10)	0.16	0.18 [0.256(*32*), 0.278(*34*)]

[e] $\sigma = -\frac{1}{\varepsilon}\frac{R - R_{eq}}{R_{eq}}$

Radial Breathing Modes

Since structural changes in carbon nanotubes can be important for an accurate prediction of RBMs, it was interesting to note that Henrard et al. (*45*) obtained a 10% increase in the RBM due to intertube interactions, which are higher than the observed experimental values, as reported by Kuzmany et al. (*46*). In contrast, our results (see Table IV and Figure 4) show a lower up-shift, varying from 2% to 6% for the armchair nanotubes.

An A/R (R, tube radius) dependence (Figure 4) was calculated, where A (Å/cm) is the fitted constant, found to equal 1168 (Å/cm) and 1207 (Å/cm), for isolated and crystalline-ropes SWCNTs, respectively. This difference exemplifies the sensitivity of the RBM to the intertube distance and the importance of its optimization. Moreover, with an optimized crystalline-rope structure we calculated a RBM value of 287 (cm^{-1}) at an intertube distance of (2R+3.4)(Å) for C(6,6), which is the value obtained for an isolated tube (see Table IV). Therefore, adopting the value of the interlayer graphite distance as the intertube separation should be applied with care. It is interesting to note that our RBMs for isolated tubes are in good agreement with recent experiments (47). We also compared our results to a recent study by Rao et al. (*48*), where the Raman spectra of isolated SWCNT in a CS_2 solution, and bundled nanotubes in powder form, were measured. The measurements show a 10 (cm^{-1}) decrease in the RBMs due to the intertube coupling, contrary to our and other calculations. This contradiction is due (*48*) to excitation of different diameter tubes in the sample. Electronic structure and RBM calculations to model the expansion of the van Hove singularities were published elsewhere (*49*).

Conclusions

In this study we report SWCNT properties calculated by using a full-potential all-electron DFT approach, showing that structural and mechanical properties agree well with other theoretical results and experiment, with only small effects of intertube coupling. The RBMs are shown to be sensitive to the intertube separation, and the (2R+3.4Å) estimate is to be used with care when simulating SWCNT bundles.

Table-IV: RBMs (cm^{-1}) of selected SWCNTs.

SWCNT	*Crystalline-rope*	*Isolated*	*Exp.*
C(4,4)	435	426	
C(6,6)	297	287	
C(8,8)	229	215	206 (*7*)
C(10,10)	181	171	165 (*7*) 177(*9*)

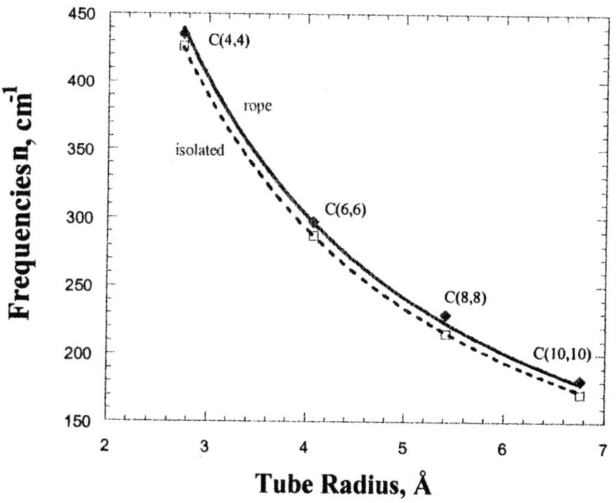

***Figure 4**: RBMs as a function of tube radius. The symbols are the calculated values and the lines are the fitted to the A/R curve. (See page 8 of color insert.)*

References

1 Dresselhaus, M. S., Dresselhaus, G., Avouris, Ph., "Carbon Nanotubes: Synthesis, Structure, Properties, Applications", (M. S. Dresselhaus, G. Dresselhaus, Ph. Avouris, eds.), Springer-Verlag Berlin, (2000).
2 Baughman R. H., Zakhidov, de Heer W. A., Science (2002) 297, 787.
3 Chhowalla, M., Teo, K. B. K., Ducati, C., Rupesinghe, N. L., Amaratunga, G. A. J., Ferrari, A. C., Roy, D., Robertson, J., Milne, W. I., J. Appl. Phys. (2001) 90, 5308.
4 Ci, L., Xie, S., Tang, D., Yan, Xi., Li, Y., Liu, Z., Zou, X., Zhou, W., Wang, G., Chem. Phys. Lett. (2001) 349,191.
5 Saito, R., Dresselhaus, G., Dresselhaus, M. S., "Physical Properties of Carbon Nanotubes", Imperial College Press, 1998.
6 Dai, H., Acc. Chem. Res. (2002) 35, 1035.
7 Kuzmany, H., Burger, B., Thess, A., Smalley, R. E., "Fullerenes Carbon Based Materials" (1997) 68.
8 Dresselhaus, M. S., Dresselhaus, G., Jorio, A., Souza Filho, A. G., Saito R., Carbon (2002) 40, 2043.
9 Kurti, J., Kesse, G., Kuzmany, H., Phys. Rev. B, (1998), 58, R8869.
10 Kurti, J., Kuzmany, H., Burger, B., Hulman, M., Winter, J., Kresse, G., Phys. Rev. B (1999) 103, 2508.
11 Mintmire, J. W., Dunlap, B. I., White, C. T., Phys. Rev. Lett. (1992) 68, 631.
12 Hamada, N., Sawada, S.-I., Oshiyama A., Phys. Rev. Lett. (1992) 68, 1579.

13 Wildoer, J. W. G., Venema, L. C., Rinzler, A. G., Smalley, R. E., Dekker, C, Nature (1998) 391, 59.
14 Odom, T. W., Hang, J.–L., Kim, P., Lieber, C. M., Nature (1998) 391, 59.
15 Avouris, A., Acc. Chem. Res. (2002) 35, 1026
16 Javey, A., Guo, J., Lundstrom, M., Dai, H., Nature (2003) 424, 654.
17 Heinze, S., Tersoff, J., Avouris, Ph., App. Phys. Lett. (2003) 83, 5038.
18. Shiffler, D. A., Ruebush, M., Zagar, D., LaCour, M., Golby, K., Clark, M. Collins, Haworth, Michael D., and Umstattd, R., IEEE Transactions on Plasma Science (2002) 30, 1592.
19 Akdim, B., Duan, X., Pachter, R., Nano Lett. (2003) 3, 1209.
20 Yakobson, B. I., Avouris, P., "Carbon Nanotubes: Synthesis, Structure, Properties, Applications", (M. S. Dresselhaus, G. Dresselhaus, Ph. Avouris, eds.), Springer-Verlag Berlin, 391, (2000).
21 Du, F., Fischer, J. E., Winey, K. I., (2003) J. Poly. Sci. B: Poly. Phys., 41, 3333.
22 Treacy, M. M. J., Ebbesen, T. W., Gibson, J. M., Nature (1996) 381, 678.
23 Krishman, A., Dujardin, E., Ebbesen, T. W., Yanilos, P. N., Treacy, M. M. J., Phys. Rev. B (1998) 58, 14013.
24 Salvetat, J.-P., Briggs, G. A. D., Bonard, J.-M, Bacsa, R. R., Kulik, A. J., Phys. Rev. Lett. (1999) 82, 944.
25 Demczyk, B. G., Wang, Y. M., Cumings, J., Hetman, M., Han, M., Zettl, A., Ritchie, R. O., Mat. Sci. Eng. A (2002) 334, 173.
26 Delley, B. J., Chem. Phys. (2000) 113, 7756; implemented by Accelyrs, Inc. in DMOL3.
27 Maiti, A., Chem. Phys. Lett. (2000) 331, 21.
28 Dunlap, B. I., Boettger, J. C., J. Phys. B (1996) 29, 4907.
29 Boettger, J. C., Phys. Rev. B (1997) 55, 11202.
30 Yoon, Y. G., Mazzoni, M. S. C., Choi, H. J., Iijima, J., Louie, S. G., Phys. Rev. Lett. (2001) 86, 688.
31 Sanchez-Portal, D., Artacho, E., Soler, J. M., Rubio, A., Ordejon, P., Phys. Rev. B, (1999) 59, 12678.
32 Hernandez, E., Goze, C., Bernier, P., Rubio, A., Phys. Rev. Lett. (1998) 80, 4502.
33 Goze, C., Vaccarini, L., Henrard, L., Bernier, P., Hernandez, E., Rubio, A., Synt. Met., (1999) 103, 2500.
34 Lu, J. P., Phys. Rev. Lett. (1997) 79, 1297.
35 Wang, Y., Perdew, J. P., Phys. Rev. A, (1990) 41, 78.
36 Charlier, J. C., Michenaud, J. P., Phys. Rev. Lett. (1993) 70, 1858
37 Miyamoto, Y., Saito, S., Tomanek, D., Phys. Rev. B. (2001) 65 041402.
38 Cousins, C. S. G., Heggie, M. I., Phys. Rev. B (2003) 67, 024109.
39 Abe, M., Kataura, H., Kira, H., Kodama, T., Suzuki, S., Achiba, Y., Kato, K-I., Takata, M., Fujiwara, A., Matsuda, K., Maniwa, Y., Phys. Rev B (2003) 68, 041405.

40 Vukovic, T., Damnjanovic, M., Milosevic, I. Physica E (2003) 16, 259.
41 Charlier, J. C., Gonze, X., Michenaud, J.-P., Europhys. Lett. (1995) 29, 43.
42 Girifalco, L. A., Miroslav, R. S., Lee, Phys. Rev. (2000) B 62, 13104).
43 Baskin, Y., Mayer, L., Phys. Rev. (1955) 100, 544.
44 Stojkovic, D., Zhang, P., Crespi, V.H., Phys. Rev. Lett. (2001) 87, 125502.
45 Henrard, L., Hernandez, E., Bernier, P., Rubio, A., Phys. Rev. (1999) B 60, R8521.
46 Kuzmany H., Plank W., Hulman M., Kramberger Ch., Gruneis A., Pichler Th., Peterlik H., Kataura H., Achiba Y., Eur. Phys. J. (2001) B 22, 307.
47 Strano, M. R., Hague, R., et al. personal communication, 2002.
48 Rao, A. M., Chen, J., Richter, E., Schlecht, U., Eklund, P. C., Haddon, R. C., Venkateswaran, U. D., Kwon, Y.-K., Tomanek, D., Phys. Rev. Lett. (2001) 86, 3895.
49 Akdim B., Duan X., Adams, W. W., Pachter R., Phys. Rev. B (2003) 67, 245404.

Chapter 20

Quantum Approaches for Nanoscopic Materials and Phenomena

Douglas S. Dudis[1], Alan T. Yeates[1], Guru P. Das[2], and Jean P. Blaudeau[3]

[1]Air Force Research Laboratory, Materials and Manufacturing Directorate, 2941 Hobson Way, Wright-Patterson Air Force Base, OH 45433-7750
[2]GP Das, Inc., Beavercreek, OH 45431
[3]High Performance Technologies, Inc., ASC/HP, 2435 Fifth Street, Building 676, Wright-Patterson Air Force Base, OH 45433–7802

Materials with nanoscopic dimensions exhibit quantized behavior which require quantum mechanical treatments to be understood and modeled. This chapter documents a number of such phenomena of technological importance and then surveys various quantum chemical approaches to treat these materials. Advantages and disadvantages are highlighted. Finally, a new ansatz for treating large systems quantum mechanically, which more naturally connects with classical chemical concepts, is outlined.

Civilian and defense applications for nanomaterials abound as attested by the other chapters in this book, recent reviews and popular literature (1-4). Modeling and simulation are central to understanding nanoscale materials and phenomena as well as providing insight into new avenues for experimental research. In 1993 Merkle (5) boldy stated that many nano-devices and nano-machines could be reasonably modeled by then available computational chemical methods. His central thesis was that many technologically interesting nano-devices and nano-machines can be designed and simulated *years before it becomes feasible to make such entities* with computational methods paving the way toward more productive paths and quickly eliminating dead-ends. This is analogous to the aerospace industry in which aircraft are now 'flown' and optimized computationally prior to being built and the electronics industries in which electronic circuits are optimized by simulation techniques prior to manufacture.

Merkle (5) conveniently divided molecular- and nano-scale simulations into two categories – those which rely on force-field based molecular-mechanics methods (no explicit electronic information) which are utilized for modeling nano-machines, and those that are quantum based, which are suitable for nano-electronic components and devices. Force field methods have been employed to simulate numerous nano-mechanical phenomena and components including Drexler-Merkle gears and neon pumps (6), carbon nanotube based gears (7), atomic-scale friction (8) and semiconductor nanostructures (9).

A defining feature of nanoscopic materials is that they have at least one length of nanometer scale. A consequence of such dimensions is that some properties display quantized behavior whereas other properties may be in the classical limit. This can be easily seen in carbon nanotubes where, in the direction of the tube diameter, quantities must be quantized, but along the tube length, properties should be more classical in nature. It has been argued that classical methods may not even accurately represent mechanical behavior in this regime (10). Hence, an alternative working definition for nanoscopic materials from a modeling and simulation perspective is those materials and phenomena which manifest both quantum and classical behavior due to the nanoscopic dimensionalities.

The focus of the present chapter is on quantum chemical methods to treat phenomena in which electrons must be considered. The emphasis is on broad-summaries of various quantum approaches. Our aim is to summarize for the non-specialist various quantum chemical approaches that are used to treat materials with the emphasis placed on modeling larger molecules and aggregates. Mathematical details are beyond the scope of this work and are well treated elsewhere (11,12). We start by illustrating some nanotechnologically important phenomena which inherently require quantum treatments; we then summarize three forefront quantum methodologies; we next briefly overview approaches to modeling large systems; and finally we outline the advantages of a new quantum chemical formulation particularly suitable to nano-scale materials. One computational area that may be of great interest in the future is in the mesoscale realm. The union of finite element methods with computational chemistry methods in the same calculation holds great promise for nanomaterials but is not discussed further in this chapter.

Nanoscopic Quantum Phenomena

Merkle (5) argued persuasively that computational chemical force-field techniques are already able to address important mechanical nanotechnology problems. Phenomena which are electronic in nature, however, require explicit consideration of the electrons and hence quantum mechanical formulations. Examples of such phenomena with wide ranging civilian and DoD technological applications include:

- Optical Absorption and Emission
- Chemical Reactivity (synthesis, corrosion, degradation etc.)
- Nonlinear Optical Responses
- Electronic Conductivity, Semiconducivity and Superconductivity
- Energy and Charge Transport
- Magnetism
- Radiation Damage

To illustrate some specific areas which require some degree of quantum chemical treatment, we list several biological processes and their corresponding DoD analog.

Bio-process	DoD Interest
Vision	Sensing
Photosynthesis	Energy Production
Nerve responses	Signal transduction/actuation
Green Fluorescent Protein	Sensor Protection
Electron Transport in DNA	Conductive Plastics/Stealth
Cognition	Bio-molecular electronics

The need for practical quantum chemical methods is thus evident and an active area of research. This need was prescient in the early statement by Dirac (13):

"... The underlying physical laws necessary for the mathematical theory of a large part of physics and the whole of chemistry are thus completely known, and the difficulty is only that the exact application of these laws leads to equations much too complicated to be soluble. It therefore becomes desirable that approximate practical methods of applying quantum mechanics should be developed, which can lead to an explanation of the main features of complex atomic systems without too much computation."

Until the advent of methods to implement quantum theory, appropriate algorithms, and modern computers, this remained a vision rather than a reality. Today, many calculations achieve experimental accuracy (14), but the roles of theory, modeling and simulation for materials research and development are broader than simply reproducing or challenging experimental measurements. These roles include:

- Phenomenological Understanding
- Theory and Method Validation
- Property Prediction
- Principle Elucidation
- Materials Design

The decade of the 1990s heralded a very real transformation in the way much of chemistry, physics, materials science and biology are practiced by the integration of modeling and simulation into the mainstream of research and development.

Overview of Computational Quantum Chemistry

Modern solutions of the time independent Schroedinger equation (equation 1) follow two very different theories: wavefunction-based (*ab initio* Hartree-Fock and correlated methods) and electron density-based (density functional theory, DFT) (15). We will outline these approaches and then describe implementations suitable for nanoscopic problems.

$$\left[-\sum_{i=1}^{n}\frac{\hbar^2}{2m_e}\nabla_i^2-\sum_{\alpha=1}^{N}\sum_{i=1}^{n}\frac{Z_\alpha e^2}{\left|\vec{R}_\alpha-\vec{r}_i\right|}+\sum_{i=1}^{n}\sum_{j\neq i}^{n}\frac{e^2}{\left|\vec{r}_i-\vec{r}_j\right|}+V_{nn}\right]\psi\left(\vec{r}_1,\cdots,\vec{r}_n,\left\{\vec{R}_\alpha\right\}\right)$$
$$=\varepsilon_e\left(\left\{\vec{R}_\alpha\right\}\right)\psi\left(\vec{r}_1,\cdots,\vec{r}_n,\left\{\vec{R}_\alpha\right\}\right) \quad (1)$$

Ab Initio Hartree-Fock and Correlated Methods

As its name implies, *ab initio* means the calculations are carried out from first principles. That is, no experimental parameterizations are input into the evaluation of the wavefunction other than fundamental quantities such as Plank's constant, the nuclear charges and masses, and the mass and charge of the electron. These include the most accurate of the quantum chemical methods and can be "competitive with experiment." A calculated result is competitive with experiment when the results are robust enough to be accepted in lieu of an experiment - or - when the theory can challenge an experimental observation. In many respects this level of acceptance of computational results has long existed in several areas of physics, but it is only slowly coming to hold in the chemistry community. Indeed, for his widely celebrated work (16) on methylene, CH_2, Schaefer received the 1992 Centenary Medal of the Royal Society of Chemistry (London) as "the first theoretical chemist successfully to challenge the accepted conclusions of a distinguished experimental group for a polyatomic molecule, namely methylene."

A hallmark of *ab initio* approaches is that they can be systematically improved. By improving both the basis sets used as well as the degree of electron correlation included in the calculations, it is possible to solve the Schroedinger equation to any degree of numerical accuracy desired (17). Such calculations are unquestionably competitive with experiments. Of course, a price must be paid to obtain such accuracy, and this price is a very non-linear scaling in the complexity of the problem – often N^6 or worse (N being the number of electrons).

Traditional *ab initio* methods, especially highly correlated treatments, are not generally (yet) applicable to nanoscopic systems. For smaller molecular systems, many methods such as MultiConfiguration Self-Consistent Field (MCSCF), Configuration Interation (CI), Coupled Cluster (CC) and Many Body Perturbation Theory (MBPT) are available and well documented in the literature (11,12) and hence will be not be discussed further here. These approaches are applicable to nanoscopic problems in providing solid treatments of subcomponents of nanomaterials. Some of their advantages include approaches for ground as well as excited states, and methods to improve a result toward a converged answer. The major disadvantage is scaling (formally $\sim N^4$ for Hartree-Fock, coupled-cluster and configuration interaction methods scale as N^6 or worse). However, such sophistication is often not required.

For many applications, semi-empirical methods are adequate. These are non-first principles methods based on model Hamiltonians which rely on experimentally determined parameterizations as well as some numerical fitting to the parametric data. Foremost among these methods are the AM1, PM3, MNDO and SAM theories and parameterizations (11,12).

Density Functional Theory

Although formulated very early in the development of quantum mechanics – the Thomas-Fermi model (18,19) in the 1920s – modern chemical applications of DFT are based on the Hohenberg and Kohn theorem of 1964 (20). This theorem states that the ground state energy and other properties are uniquely determined if the ground state density is known (this encompasses the ground state of any given symmetry – both spatial and spin). The following year, Kohn and Sham laid the foundations of most modern DFT programs by introducing the Kohn-Sham orbitals, which are defined as the orbitals for a system of noninteracting electrons that reproduce the ground state density (21). This work was rewarded with the award of the 1998 Nobel prize in chemistry to Walter Kohn (shared with John Pople).

The central equation of modern DFT is the Kohn-Sham equation for the total energy (22):

$$E[\rho(\vec{r})] = \sum_{i=1}^{N} \int \psi_i(\vec{r})\left(-\frac{\nabla^2}{2}\right)\psi_i(\vec{r})d\vec{r} - \sum_{A=1}^{M} \int \frac{Z_A}{\left|\vec{r}-\vec{R}_A\right|}\rho(\vec{r})d\vec{r} + \frac{1}{2}\iint \frac{\rho(\vec{r}_1)\rho(\vec{r}_2)}{\left|\vec{r}_1-\vec{r}_2\right|}d\vec{r}_1 d\vec{r}_2$$
$$+ E_{XC}[\rho(\vec{r})] \qquad (2)$$

The first three terms are analogous to those found in the Schrödinger equation: the kinetic energy term, the external potential term due to the interaction of the electrons with the nuclei, and the coulomb term. It should be emphasized that the kinetic energy term is for the kinetic energy of a system of noninteracting electrons and, thus, is different than the kinetic energy term found in the Schrödinger equation. Recall that the density can be defined as:

$$\rho(\vec{r}) = \sum_{i=1}^{N} \left|\psi_i(\vec{r})\right|^2 \qquad (3)$$

The last term is defined as the exchange-correlation term and has three major contributions: the correction to the kinetic energy, the exchange term, and the dynamic correlation term. It is the exchange-correlation term that is the central problem of DFT in that there is no known formulation for it based on first principles, and it is this term which distinguishes various DFT implementations from one another. DFT has the advantage of including some correlation effects in a simple and effective way. On the other hand, there is no systematic way to improve a given functional.

Early exchange-correlation functionals were expressed as functionals of the density only (termed local density approximation, LDA (23,24)), and while useful chemical information resulted from the calculations, there was need for improvement. The incorporation of density gradients leads to the generalized gradient approximation (GGA), which have been broadly accepted and have produced excellent results (25,26,27). Most traditional quantum chemistry programs (e.g. Gaussian (28)) allow DFT calculations with GGAs. There have been many GGA functionals developed, two common ones are PW91 (29) and BLYP (30,31). A further development is that of the hybrid functionals, e.g. B3LYP (32). These functionals are parameterized to express the total exchange as a linear combination of the exact (non-local) Hartree-Fock exchange along with the DFT (local) exchange. Recently Adamo et al. (25) have published a review on DFT where they introduce adiabatic connection methods.

Advantages of DFT include an easy incorporation of correlation effects, an easy incorporation of relativistic effects (e.g. in the Amsterdam Density Functional program, ADF (33)), and much better scaling than traditional wavefunction-based systems. Pure DFT (i.e. non-hybrid functionals) formally scales as N^3 because the coulomb integrals can be expanded in terms of the density (and can be expressed in terms of three body integrals) and there are no formal four-center exchange integrals (22). The disadvantages of DFT include a lack of a theoretically sound treatment of excited states (although the development of time-dependent methods, TDDFT, is addressing this concern), problems with non-dynamical correlation (a weakness of any single determinant method, including Hartree-Fock), and problems with weak interactions, such as van der Waals and hydrogen bonding. One should point out that even in TDDFT methods, the properties calculated (including transition energies) are ground-state properties; no excited state properties are calculated. The problems with non-dynamical correlation relate to the well-known DFT multiplet problem: how to describe states (e.g. open-shell singlets) which are intrinsically multi-determinant, in terms of a single determinant theory. Koch and Holthausen (34) provide an interesting discussion of DFT and hydrogen-bonds. Simplifying matters, as long as the potential along the XH-X axis is single-welled (termed weak hydrogen bonding), DFT performs adequately; it is when the potential is double-welled (termed strong hydrogen-bonding) that DFT fails. This fact should not be surprising, as the latter case is fundamentally a multi-reference system.

In calculations of nanomaterials, the advantages of DFT methods are clear (35-38). Its ease of incorporating correlation effects and its scalability make it an ideal tool for large systems. Furthermore, its extension to periodic boundary conditions have made it a favorite tool of the solid-state physics community.

Tight-Binding Methods

The recent upsurge in the popularity of tight-binding (TB) methods results from the compromise between computational economy and the ability to obtain reasonably accurate electronic and structural properties achieved by these methods. TB methods are not new, first appearing shortly after the development of the Hartree-Fock equations (39,40). Hückel (41,42) proposed the first and most famous TB method. In the Huckel method, only the π-electrons of a conjugated hydrocarbon are treated explicitly via a nearest-neighbor interaction with the remaining electrons and nuclei contributing to a constant additive energy. The 1954 paper by Slater and Koster (43) generalized the approach and is still of great interest and significance to modern TB methods.

The term "tight-binding" encompasses a variety of approaches to electronic structure problems; however, the defining and unifying feature of the TB methods is the use of an atomic-like basis set. They are most often applied to extended periodic systems, such as crystals or polymers, but large aperiodic systems benefit from the simplicity of the methods as well. Typically, a single atomic-like basis set is assigned to each atom or "site" in the system although some systems, such as transition metals, require multiple basis sets with differing angular momenta. By parameterizing the site-site Hamiltonian matrix, an enormous decrease in computational complexity is achieved, while still retaining a considerable amount of electronic information. A TB calculation can frequently be three orders of magnitude faster than a similar *ab initio* computation. The greatest disadvantage of TB methods is that the accuracy of the calculation is strongly dependent on the parameterization. This, in turn, leads to a decrease in the transferability of the methods, i.e., each new problem often requires a new parameterization. An excellent review of the TB approach as applied to the large scale modelling of materials has recently been published by Goringe, et. al. (44).

TB calculations usually begin with the prescription of a model Hamiltonian to describe the relevant physics, typically containing only those terms felt to be important to the problem. The obvious danger with this approach is the possibility that a critical component of the physics might be overlooked. Nevertheless, these simplified Hamiltonians have been quite useful for describing complex problems. For example, the Su-Shrieffer-Heeger (SSH) Hamiltonian (45) was very successful in describing the ground and low-lying electronic and vibrational excited states of polyacetylene (PA).

$$\mathrm{H}_{SSH} = -\sum_{n,\sigma}\left[t_0 - \alpha\left(u_{n+1} - u_n\right)\right]\left(a_{n+1}^{+}a_n + h.c.\right) + \tfrac{1}{2}K\sum_{n=1}^{N-1}\left(u_{n+1} - u_n\right)^2 \qquad (4)$$

In this application, the "site" is a CH unit in the PA chain with N units. Each unit is assumed to host a single basis function for a π-electron. The operators a_n and a_n^+ are the annihilation and creation operators for a π-electron on site "n", and h.c. stands for hermitian conjugate. The variable u_n is the displacement of the site coordinate from its position if all sites were equidistant from each other. The parameter t_0 is called the transfer integral, which corresponds to the intersite Hamiltonian matrix element in the undimerized chain. This transfer integral is modified by the parameter α, which corresponds to the linear electron-phonon coupling constant. Finally K corresponds to the spring constant between nearest-neighbor sites in their undimerized location. This spring constant can be estimated from experimental data or higher-level computations on simpler systems. Although this Hamiltonian is clearly much simpler than the *ab initio* Hamiltonian (equation 1), the SSH Hamiltonian has been very successful at describing the dimerization of the PA lattice as well as the structural and vibrational properties of the lowest electronic excited states. The SSH Hamiltonian can be extended to other polymers as well, for example, poly-p-phenylene vinylene, PPV (46). However, the parameters used to describe PA, cannot be readily extended to the PPV polymer. Recent progress has been made to help alleviate the problem of transferability of TB computations using the self-consistent charge density-functional based approach. A very good review for this approach was published by Frauenheim, et. al. (47). This procedure produces a TB method that requires no input from experiment, or separate *ab initio* calculations on smaller systems.

Finally, TB methods are proving to be useful for studying the dynamic properties of systems in which quantum forces are necessary. Because of the poor scaling properties of the *ab initio* methods, dynamic simulations have generally been limited to systems containing a few atoms at most (48). However, TB methods readily lend themselves to the application of approaches that lead to linear scaling. Thus, quantum molecular dynamic simulations have become feasible for systems in the nanoscopic size range (49,50).

Quantum Treatments of Large Systems

Traditional wavefunction and density functional theories are applicable to large systems, including nanomaterials; however, their implementation often involves different algorithms. These include the various linear scaling methods, hybrid (often referred to as QM/MM) methods, and sparse matrix methods.

Linear scaling methods are dependent on what Kohn (51) has termed the nearsightedness of equilibrium systems, which may be defined as, for large systems, the property that the effects of an external potential decay rapidly (for systems without long-range electric fields) (50). Recently, Goedecker published a review of various linear scaling methods that he defined as being based on the decay properties of the density matrix (52). Practical implementations of the traditional methods scale as N^3 (along with a prefactor, c_3). Linear scaling techniques scale as N, but with a different prefactor, c_1. Since $c_1 > c_3$, for small N, the traditional implementations are faster. For large N, the exponential term dominates and the linear scaling methods are faster. These methods include the Fermi operator method, the Fermi operator projection method, the divide-and-conquer method, the density-matrix minimization approach, the orbital minimization approach, and the optimal basis density-matrix minimization method. Other disadvantages include: no self-consistency for the whole fragment, problematic treatment of excited states, arbitrary fragments, and, for *ab initio* methods, is limited to Hartree-Fock. A couple of representative examples for nanomaterials are the SIESTA project (53), a localized-density-matrix method (54), and a density-minimization approach (55).

Hybrid (usually referred to as QM/MM, but other combinations, such as QM/QM' are possible) methods treat the most important part of a system with a higher level of theory than the remainder of the system (in the ONIOM method up to three layers are allowed) (56, 57, 58). Often the important part is treated with some level of quantum mechanics and the remainder with molecular mechanics. Molecular mechanics treats a molecule in a ball-and-spring fashion (albeit a very sophisticated spring) in classical fields, but does not explicitly treat electrons in the Hamiltonian, and hence is incapable of treating problems where electronic phenomena are key. Disadvantages of these methods include: the proper treatment of the interface between the QM and MM portions, the quantum portion still scales as the parent technique, and an unsatisfactory treatment of electron transport and optical properties. Along these lines, Sauer has introduced a combined quantum mechanics-interatomic potential function approach QM-Pot (59).

A third class of methods is the sparse matrix methods (60), which solve the Hartree-Fock equations for large systems. Disadvantages include that it is limited to the ground state, formulated (much less implemented!) only for semi-empirical methods, and intrinsically incapable of treating electron transport and optical properties such as two-photon absorption.

Fragment Quantum Approaches

Richard Feynman is generally credited with opening the nanoscience and nanotechnology fields in his visionary 1959 lecture entitled "There's Plenty of Room at the Bottom". In his Lectures on Physics, Feynman made the following statement (61): "... And lo and behold! The chemists are almost always correct." The context of this quote is the elucidation of molecular structures, especially a comparison of results derived from organic-chemical experiments against x-ray diffraction. This statement, however, provides a philosophical starting point for viable approaches to treat large systems quantum mechanically distinct from traditional quantum chemical treatments.

Modern quantum chemical methods start with the Linear Combination of Atomic Orbitals (LCAO) ansatz (10) in which the materials under study are constructed from individual *nuclei* and *electrons*. Nevertheless, molecules are *not* atoms. Chemists, biologists, materials scientists, and physicists understand materials in terms encompassing atoms, functional groups, clusters, and/or discrete molecular components. For example, the reality and utility of functional groups – such as phenyl, nitro, and carbonyl groups for organic chemists, and nucleic acids, amino acids, and cofactors for biochemists – is time-honored. When these same materials are treated quantum chemically, all the useful higher level ('chemical intuition') information is discarded. Moreover, once a calculation is performed by building a system from its constituent nuclei and electrons, various transformations, localization schemes, or population analyses can be employed to try to understand the information in more conventional molecular terms. Extracting useful information by transforming from atoms back into groups can be cumbersome, somewhat arbitrary, and often ill defined. A macroscopic analogy might be an electronics engineer resorting to individual transistors to build a complex system rather than starting with predefined and well understood integrated circuit components.

We recently (62) introduced an Ab initio Fragment Orbital (AFOT) method. Rather than starting with the LCAO ansatz, the AFOT model starts by recognizing that the constituent fragments of a large molecule approximately retain their individual orbital description even while interacting with each other within the molecule. Once the fragments are properly identified, the wave function of the total system is generated in terms of what can be called the Linear Combinations of Fragment Molecular Orbitals (LC-FMO) rather than LCAOs. Simplification results because, especially for lower lying excited states, only a small number of higher-lying occupied and lower-lying unoccupied orbitals of each fragment participate as 'active orbitals.' Hence, the global wavefunction is given by equation (8).

$$\phi_i^{GMO} = \sum_j c_{ij} \phi_j^{FMO} \qquad (8)$$

The total single-determinant molecular wave function for the ground state is constructed in the restricted Hartree-Fock form:

$$\Psi_{tot} = \Phi_{core} \Psi_I^F (\phi^L_{I,II})^2 \Psi_{II}^F (\phi^L_{II,III})^2 \ldots\ldots \qquad (9)$$

where Φ_{core} represents the core, Ψ_i^F, i=I,II, ... are products of doubly occupied active orbitals "belonging" to various fragments I, II etc. The ϕ's (to be called link orbitals) are doubly occupied orbitals linking the fragments; for example, the link orbital $\phi^L_{I,II}$ links the fragments I and II.

The details of our AFOT model are beyond the scope of this chapter, so we focus not so much on the merits of the AFOT implementations, but rather on the advantages to be gained from LC-FMO type approaches in general. Starting from molecular fragments as the fundamental units rather than atoms is similar to the way biologists, chemists, materials scientists, and soft-matter physicists think. This gives several advantages over other quantum approaches:

- Systems are constructed in terms familiar to experimentalists
- Various levels of sophistication are possible for different fragments
- No problematic classical/quantum interfaces
- Localization/description/understanding of phenomena (e.g. energy or charge migration) are handled more 'naturally'
- Systematic improvements are possible to validate convergence of results
- Correlated or semi-empirical implementations are possible
- Scaling is outstanding (approaches linear if implemented correctly)
- Essential physics of a problem can be probed through systematic variations

Finally, one of the starting points for the LCAO method is simply that it reproduces one of the right limits. As a molecule dissociates, the picture must become that of independent atomic orbitals, and so constructing a higher system from atomic orbitals makes sense. This analogy can be seen clearly in charge-transfer complexes if the donor molecule is considered one fragment and the acceptor molecule considered the other fragment. Clearly, as the charge-transfer complex dissociates into its consitituent molecules (fragments), the electronic description must become that of the independent fragments (molecules).

Summary

We have overviewed computational chemistry methods applicable to nanomaterials emphasizing that in order to obtain a fundamental understanding of these materials, the electrons need to be explicitly treated. Molecular mechanics can give useful results for their mechanical properties, but many of the DoD interests in these substances are electronic in nature. While there exists a myriad of methods, these can usually be quickly narrowed down, depending on the system and property of interest. Ever increasing capabilities of high performance computing will lead to modification of methodologies and introduction of new ones. As an example, many methods have either been only recently implemented (e.g. true linear scaling and hybrid methods) or have been drastically improved (e.g. the advent of GGAs into DFT). We have introduced a new methodology, AFOT, that has the advantages of being based on a chemical viewpoint, can handle large systems at an *ab initio* level of theory, and is ideal for nanoscale applications.

References

1. Tour, J. M. *Acct. Che. Res.* **2000**; *33,* 791-804
2. *Chemical Reviews*, Vol. 99 (7), 1999. Issue devoted to Nanostructures.
3. *Scientific American*, September 2001, Special Issue: Nanotech: The Science of the Small Gets Down to Business.
4. "Nanotechnology in Biology: The Good of Small Things," *The Economist*, 22 December 2001, 93-94.
5. Merkle, R. C. " *Nanotechnology* **1991**, *2*, 134-141.
6. Cagin, T.; Jaramillo-Botero, A; Gao, G.; Goddard, W. A. III *Nanotechnology*, **1998**, *9*, 143-152.
7. Han, J.; Globus, A.; Jaffe, R.; Deardorff, G. *Nanotechnology*, **1997**, *8*, 95-102.
8. Shimizu, J.; Eda, H.; Yoritsune, M.; Ohmura, E. *Nanotechnology*, **1998**, *9*, 118-123.
9. Nakano, A.; Campbell, T. J. ; Kalia, R. K.; Kodiyalam, S.; Ogata, S.; Shimojo, F.; Vashita, P.; Walsh, P. AIP Conference Proceedings (2001), 583 (Advanced Computing and Analysis Techniques in Physics Research), 57-62.
10. Noid, D. W.; Tuzun, R. E.; Sumpter, B. G. *Nanotechnology*, **1997**, *8*, 119-125.
11. Szabo, A.; Ostlund, N. S. Modern Quantum Chemistry: Introduction to Advanced Electronic Structure Theory, McGraw-Hill, New York, 1989.
12. Encyclopedia of Computational Chemistry, Paul von Rague Schleyer, ed., John Wiley and Sons, Inc., NY, 1998.
13. Dirac, P. A. M. *Proc. Royal Soc. London, Ser. A*, Vol. CXXIII, April 1929, 714.
14. Bartlett, R. J.; Stanton , J. F. "Applications of Post-Hartree-Fock methods: A tutorial" in *Reviews in Computational Chemistry*, Volume 5, p. 65 (D. Boyd and K. Lipkowitz, editors). VCH Publishers, New York, New York (1994).

15. Bartlett, R. J. Quantum Theory Project in "Chemistry for the 21st Century," E. Keinan and I. Schechter (eds), Wiley-VCH, New York, 2001 (pp. 271-286).
16. Bender, C. F.; Schaeffer, H. F. III *J. Am. Chem. Soc.* **1970**, *92*, 984-5.
17. Warren J. Hehre, Leo Radom, Paul v. R. Schleyer and John A. Pople, "Ab Initio Molecular Orbital Theory," John Wiley and Sons, Inc., NY, 1988.
18. Thomas, L. H. *Proc. Camb. Phil. Soc.* **1927**, *23*, 542.
19. Fermi, E. *Rend. Accad. Lincei* **1927**, *6*, 602.
20. Hohenberg, P.; Kohn, W. *Phys. Rev.* **1964**, *136*, B864.
21. Kohn, W.; Sham, L. *J. Phys. Rev.* **1965**, *140*, A1133.
22. Leach, A. R. Molecular Modeling: Principles and Applications 2nd ed. New York, Prentice Hall, 2001.
23. Slater, J. C. Quantum Theory of Molecules and Solids Volume 4: The Self-Consistent Field for Molecules and Solids, New York, McGraw-Hill, 2001.
24. Vosko, S. H.; Wilk, L.; Nusair, M. *Can. J. Phys.* **1980**, *58*, 1200.
25. Adamo, C.; di Matteo, A.; Barone, V. *Adv. Quantum Chem.* **2000**, *36*, 45.
26. Andzelm, J. *Pol. J. Chem.* **1998**, *72*, 1747.
27. Proynov, E. I.; Ruiz, E.; Vela, A.; Salahub, D. R. *Int. J. Quantum Chem. Quantum Chem. Symp.* **1995**, *29* (Atomic, Molecular, and Condensed Matter Theory and Computational Methods, Proceedings of the International Symposium, 1995), 61.
28. Gaussian 03, Revision A.1, M. J. Frisch, G. W. Trucks, H. B. Schlegel, G. E. Scuseria, M. A. Robb, J. R. Cheeseman, J. A. Montgomery, Jr., T. Vreven, K. N. Kudin, J. C. Burant, J. M. Millam, S. S. Iyengar, J. Tomasi, V. Barone, B. Mennucci, M. Cossi, G. Scalmani, N. Rega, G. A. Petersson, H. Nakatsuji, M. Hada, M. Ehara, K. Toyota, R. Fukuda, J. Hasegawa, M. Ishida, T. Nakajima, Y. Honda, O. Kitao, H. Nakai, M. Klene, X. Li, J. E. Knox, H. P. Hratchian, J. B. Cross, C. Adamo, J. Jaramillo, R. Gomperts, R. E. Stratmann, O. Yazyev, A. J. Austin, R. Cammi, C. Pomelli, J. W. Ochterski, P. Y. Ayala, K. Morokuma, G. A. Voth, P. Salvador, J. J. Dannenberg, V. G. Zakrzewski, S. Dapprich, A. D. Daniels, M. C. Strain, O. Farkas, D. K. Malick, A. D. Rabuck, K. Raghavachari, J. B. Foresman, J. V. Ortiz, Q. Cui, A. G. Baboul, S. Clifford, J. Cioslowski, B. B. Stefanov, G. Liu, A. Liashenko, P. Piskorz, I. Komaromi, R. L. Martin, D. J. Fox, T. Keith, M. A. Al-Laham, C. Y. Peng, A. Nanayakkara, M. Challacombe, P. M. W. Gill, B. Johnson, W. Chen, M. W. Wong, C. Gonzalez, and J. A. Pople, Gaussian, Inc., Pittsburgh PA, 2003.
29. Perdew, J. P.; Wang, Y. *Phys. Rev. B* **1992**, *45*, 13244.
30. Becke, A. D. *Phys. Rev. A*, **1988**, *38*, 3098.
31. Lee, C.; Yang, W.; Parr, R. G. *Phys. Rev. B* **1988**, *37*, 785.
32. Stephens, P. J.; Devlin, J. F.; Chabalowski, C. F.; Frisch, M. J. *J. Phys. Chem.* **1994**, *98*, 11623.

33. te Velde, G.; Bickelhaupt, F. M.; Barends, E.J.; van Gisbergen, S. J. A.; Fonseca Guerra, C.; Snijders, J. G.; Ziegler, T. *J. Comput. Chem.* **2001**, *22*, 931.
34. Koch, W.; Holthausen, M. C. A Chemist's Guide to Density Functional Theory, New York, Wiley-VCH, 2001.
35. Chelikowsky, J. R. *J. Phys. D: Applied Phys.* **2000**, *33*, R33.
36. Artacho, E.; Sanchez-Portal, D.; Ordejon, P.; Garcia, A.; Soler, J. M. *Physica Status Solida B* **1999**, *215*, 809.
37. Kohn, W. Rev. Mod. Phys. **1999**, 71, S59.
38. Andreoni, W. Mol. Nanostruct., Proc. Int. Wintersch. Electron. Prop. Novel Mater. **1998**, Meeting Date 1997, 3-13.
39. Fock, V. *Z. Physik*, **1930**, ***61***, 126.
40. Slater, J. C. *Phys. Rev.*, **1930**, *35*, 210.
41. Hückel, E. *Z. Physik*, **1931**, *70*, 204.
42. Hückel, E. *Z. Physik*, **1932**, *76*, 628.
43. Slater, J. C.; Koster, G. F. *Phys. Rev.*, **1954**, *94*, 1498.
44. Goringe, C. M.; Bowler, D. R.; Hernández, E. *Rep. Prog. Phys.*, **1997**, *60*, 1447.
45. Su, W. P.; Schrieffer. J. R.l Heeger, A. J. *Phys. Rev. Lett.*, **1979**, *42*, 1698.
46. Shuai, Z.; Beljonne, D.; Brédas, J. L. *Solid State Commun.*, **1991**, *78*, 477.
47. Frauenheim, T.; Seifert, G.; Elstner, M.; Hajnal, Z.; Jungnickel, G.; Porezag, D.; Suhai. S.; Scholz, R. *Phys. Stat. Sol.* B, **2000**, *217*, 41.
48. Ben-Nun. M.; Martinez, T. J. *Adv. Chem. Phys.* **2002**, *121*, 439-512.
49. Ordejón, P. *Computational Mat. Sci.*, **1998**, *12*, 157.
50. Galli, G. *Phys. Stat. Sol.* B, **2000**, *217*, 231.
51. Kohn, W. *Phys. Rev. Lett.* **1996**, *76*, 3168.
52. Goedecker, S. *Rev. Mod. Phys.* **1999**, *71*, 1085.
53. Ordejón, P. *Phys. Stat. Sol. (b)* **2000**, *217*, 335.
54. Chen, G.; Yokojima, S.; Liang, W.; Wang, X. *Pure Appl. Chem.* **2000**, *72*, 281.
55. Challacombe, M. *J. Chem. Phys.* **1999**, *110*, 2332.
56. Maseras, F.; Morokuma, K. *J. Comput. Chem.* **1995**, *16*, 1170.
57. Dapprich, S.; Komáromi, I.; Byun, K. S.; Morokuma, K.; Frisch, M. J. *J. Molecul. Struc. (Theochem)* **1999**, *461-462*, 1.
58. Shoemaker, J. R.; Burggraf, L. W.; Gordon, M. S. *J. Phys. Chem. A* **1999**, *103*, 3245.
59. Sauer, J.; Sierka, M. *J. Comput. Chem.* **2000**, *21*, 1470.
60. Young, W. *J. Molecul. Struc. (Theochem)*, **1992**, *255*, 461.
61. Feynman, R. P.; Leighton, R. B.; Sands, M., The Feynman Lectures on Physics, Addison-Wesley, Reading, MA, 1977, Vol. 1, p. 8.
62. Das, G. P.; Yeates, A. T.; Dudis, D. S., *Int. J. Quant. Chem.*, submitted.

Chapter 21

Nanotechnology Applications in Space Power Generation

Donna Cowell Senft, Paul Hausgen, and Clay Mayberry

Air Force Research Laboratory, Space Vehicles Directorate, 3550 Aberdeen Avenue, SE, Kirtland Air Force Base, NM 87117–5776

Space power generation needs for generation-after-next satellites have requirements that extend beyond the efficiencies and specific power levels provided by current photovoltaic solar arrays. Nanotechnology and atomic/molecular scale manipulation of material and device properties promise to surpass these limitaions on solar to electric power conversion efficiency and specific power (W/kg) levels. First, nano-engineered luminescent dyes have the capability to absorb and alter the solar spectrum to output frequencies that, when coupled with conventional photovoltaics, increase solar cell efficiencies. Second, the size-dependant bandgap and melting temperatures of nanoparticles may lead to new types of photovoltaic devices. Third, direct conversion of the visible frequencies of the solar electromagnetic spectrum to direct current through arrays of antennas and nanoscale diodes have theoretical efficiencies that exceed those of photovoltaic devices.

Introduction

Photovoltaics have been the primary source of electrical power in space since 1958. The inexhaustibility of the energy source, supplying an energy density in earth orbit of 1367 W/m^2, has resulted in the dominance of solar cells over technologies using consumable energy supplies. The economics of solar radiation to electrical power conversion depend on the cost of the solar cells themselves, the cost of launching the solar array into orbit (>$19,000/kg), and the conversion efficiency of the solar cells. Therefore, device efficiency, cost, and specific power (W/kg) are paramount considerations in solar cell development. A further consideration is materials and device degradation due to exposure to the space environment, in particular electron and proton radiation, ultraviolet light, and atomic oxygen.

Over the past 5-10 years, cell developers have been able to continually increase solar cell efficiencies by increasing the number of junctions in a cell. With III-V triple junction solar cells currently reaching efficiencies of 30% at air mass zero (AM0) and four junction cells under development, crystalline multijunction cells are beginning to approach theoretical efficiency limits (~40% AM0 for a four junction cell). Satellite systems using these high performance solar cells have achieved 70-80 W/kg, 15-20 kW total power, and average cell costs have dropped to $300-$400 per watt produced at the cell level. Beyond four junctions, additional junctions are increasingly difficult to grow and the additional gains from each successive junction decrease proportionately (1). Thin-film solar cells, using amorphous or polycrystalline materials on lightweight, non-crystalline substrates, are expected to result in array-level specific power levels of 200-500 W/kg, meeting the higher specific power needs of next-generation satellites (2). Thin-film arrays are expected to increase performance in terms of $/W, stowed power density, array mass and ultimate cost compared to crystalline multijunction arrays. However, with thin-film efficiencies expected to level off at 15%, a high power application of 100 kW would require a thin-film solar array area of ~700 m^2. Arrays of this size may experience problems with array dynamics and drag, especially at lower altitudes, perhaps necessitating a detached array coupled to the payload through an umbilicle. Concentrator arrays can increase solar cell efficiency by increasing the power density of the solar radiation. Lightweight solar concentrators (3) have been successful in increasing the specific power level of multijunction crystalline arrays, but at the expense of tighter pointing requirements, thereby increasing the necessary propellant mass. There are other approaches such as mechanically stacked solar cells (expensive, suitable for concentrator systems), thermal electric converters (TPV, AMTEC, Stirling, thermionic, etc). These, however, have significant drawbacks such as vibrations, relatively low power density, high cost, and low theoretical efficiencies when compared to the state-of-the-art solar arrays for satellite power systems.

The unique optical and materials properties engendered by nanotechnology have the potential to increase the efficiency and specific power of solar-to-electric conversion, satisfying generation-after-next power generation needs. The full diversity of nanotechnology research is relevant to space power generation. Nanoengineered luminescent organic molecules have the potential to provide a tailored spectral distribution of photons to a photovoltaic device. Large scale patterning at nanoscale resolution can create arrays of energy conversion devices. Nanoparticles with tailored optical and physical properties can increase solar cell efficiency. Current nanotechnology approaches to space power generation include: in-line solar spectrum alteration with organic molecules (possibly enhanced with dendritic molecular attachments), use of the unique optical and melting properties of nanoparticles, and direct energy conversion through nanoscale antenna arrays.

In-line Solar Spectrum Alteration with Nanotechnology

Nano-engineered luminescent dyes (possibly enhanced by dendrimer molecular attachments) that have solar spectrum alteration capabilities have the potential to increase the solar energy conversion efficiency of space photovoltaic devices. The increase in efficiency is obtained by altering the solar spectrum to better match the energy conversion strengths of the photovoltaic device. The dyes alter the solar spectrum by absorbing photons at wavelengths that are not efficiently converted to electricity and emitting them at wavelengths that are efficiently converted to electricity.

These spectrum-altering dyes would be applied in a non-absorbing matrix to the solar incident side of existing photovoltaics. The result would be that the spectral distribution of photons incident on the photovoltaic device would differ from that of the incoming solar spectrum. Thus, the dye would alter the solar spectrum "in-line" (see Figure 1) with no area concentration of energy.

A similar, but sperate, concept to the one presented here is referred to in the literature as a "luminescent solar concentrator" (LSC) (4,5). Like the in-line spectrum alteration concept, the LSC consists of a luminescent material in a solid matrix. However, unlike the in-line concept, the solid matrix of an LSC is designed to channel photons emitted by the luminescent material to the matrix edges. The channeling occurs due to the fact that the matrix has an index of refraction greater than the external medium (air or vacuum), which causes complete internal reflection to occur for photons emitted at angles outside of the escape angle as defined by Snell's law. A photovoltaic material is attached to at least one edge (edges without photovoltaic material are made highly reflective) to convert the photons into electricity (see Figure 1). A prime objective of the LSC concept is the area concentration of solar energy to reduce the amount of photovoltaic material needed, which is not an objective of in-line solar spectrum alteration. While area concentration is not an objective of the in-line solar spectrum alteration concept, increased solar conversion efficiency will reduce the area of photovoltaic material needed to meet a particular power need.

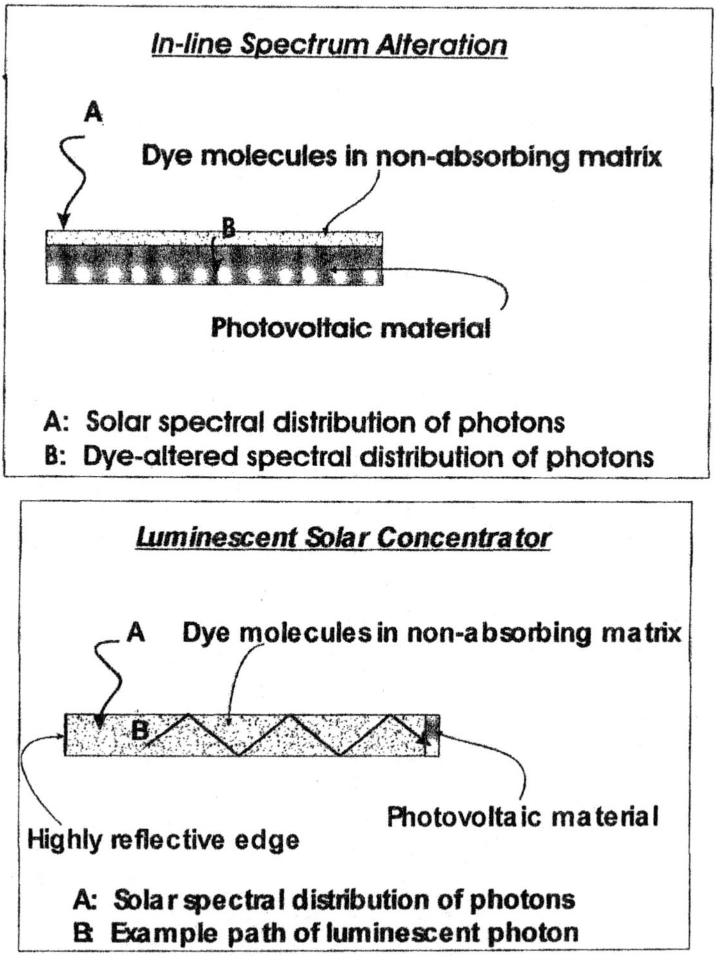

Figure 1. Schematics of the in-line spectrum alteration concept and the luminescent solar concentrator.

Applications of In-Line Solar Spectrum Alteration

In-line spectrum alteration can be used to make a myriad of spectral changes that could result in the increase in conversion efficiency for photovoltaic devices. Regional spectral changes that would cause incremental increases in photovoltaic conversion efficiency, or *partial spectrum shifting*, would involve shifting photons in a wavelength range in which the photovoltaic device has poor quantum efficiency (a measure of current generation capability for photons of a given energy) to another wavelength range with a high quantum efficiency.

A variety of causes can be cited for poor current generation at particular photon energy, and it is often dependent on the device design and type. Therefore, it is beyond the scope of this work to examine them in detail. However, for the purpose of this work, it is important to note that the ability to alter the spectral content of the incident solar energy adds an additional design element that can be used to compensate for deficiencies in the device material.

A representative application of in-line solar spectrum alteration could involve shifting photons with wavelengths less than 400 nm to wavelengths near the photovoltaic device's bandgap. These high-energy photons are typically not converted efficiently or not converted at all, due to shallow absorption in the semiconductor that prevents the resulting electron/hole pair from being separated by the junction. Also, high-energy photons tend to yellow the cover-glass adhesive and are therefore typically rejected with special optical coatings.

A far term application of this technology would be to absorb the entire solar spectrum and emit the energy at wavelengths equal to and slightly lower than a single junction photovoltaic device's bandgap (*full spectrum compression*). The goal would be to transform the polychromatic solar energy into nearly monochromatic light. Full spectrum compression would largely eliminate super-bandgap and sub-bandgap losses. Elimination of these losses could produce efficiencies greater than 45% for single junction devices (6).

Super-bandgap losses are caused by the fact that a photon with energy greater than or equal to the device's bandgap only creates one electron/hole pair at the bandgap energy. This is true no matter how much the photon's energy exceeds that of the bandgap. Photons with energy higher than the bandgap create energetic electron/hole pairs that quickly dissipate their surplus energy (energy greater than the bandgap) in the form of phonons (thermal vibrations) and settle to the edge of the conduction/valence bands. Thus, photon energy in excess of the device bandgap is lost to heat generation and is not harvested as electrical energy. Sub-bandgap losses are due to the fact that a photon cannot create an electron/hole pair unless it possesses energy greater than the device's bandgap energy. Like the super-bandgap losses, the energy loss due to sub-bandgap photons takes the form of heat generation, or depending on the device design, the photons may pass through the device without being absorbed (7).

In-line Solar Spectrum Alteration Challenges

Many issues must be adequately addressed for successful implementation of in-line solar spectrum alteration using organic dyes. Many of these issues for in-line spectrum alteration are shared with the luminescent solar concentrator. Therefore, the wealth of literature that exists for the luminescent solar concentrator is a very useful information source. Among these issues are dye stability, re-absorption, isotropic emission, quantum yield, and matrix absorption. In the following paragraphs, each of these issues will be discussed.

Dye performance stability is a critical concern for application to space photovoltaics where expected lifetimes can reach 15 years. Of course, complete stability is desired, but a limited amount of degradation might be acceptable. For the dyes to be used in space applications, a thorough understanding of the degradation mechanisms and rates must be obtained to adequately design the space power system. Two causes of dye degradation that are discussed in the literature are thermal bleaching (8) and photo bleaching (9,5) (also called optical bleaching).

Thermal bleaching is the deterioration of the dye performance caused by excessive temperature. Experimentation (5) on Rhodamine-6G has shown negligible degradation for temperatures less than 60°C. However, a study by Rico *et al.* (10) indicates that at 80° C Rhodamine-6G (and certain other dyes) do exhibit optical efficiency degradation (~14% in 100 h). Rico speculated that the optical efficiency decrease was caused by aggregation of dye molecules in the polymethyl methacrylate (PMMA) host matrix. Since typical space solar arrays are designed to operate at temperatures below 60° C, thermal bleaching may not be a significant issue for space applications.

Photo bleaching is the breakdown of the dye caused by the absorption of solar energy. The LSC literature is replete with various dyes that do not have adequate stability for space use due to photo bleaching. For example, a commonly used dye, Rhodamine-6G, in a host matrix of PMMA decays to 50% of initial dye population in 0.08 years when subjected to solar illumination (5). Coumarin-6 has better stability, but it is also unacceptable at 0.27 years to reach 50% of initial dye population5. One of the most photo-stable dyes, Lpero, was examined by Bakr *et al.* (11) who found over 95% of the initial dye population to be present after one year. In the end, it may be necessary to use inorganic absorbers as the spectrum-shifting component (5,12) because they would likely be much more robust.

Another issue regarding the use of spectrum shifting dyes is re-absorption of emitted photons before they enter the photovoltaic device (5,13,14). This occurs due to an overlap in the absorption and emission bands of a dye. While this is an enormous problem for luminescent solar concentrators, it should not be as significant for the in-line spectrum alteration concept. This is because the optical path length of the luminescent photon is much shorter for in-line spectrum alteration than for the luminescent solar concentrator, which decreases the probability that the emitted photon will be reabsorbed by another dye molecule. Development of dyes that have minimal overlap of the absorption/emission spectrum will reduce this effect significantly.

Because the dye molecules are typically randomly oriented in their host matrix, they will effectively emit in an isotropic manner. For the in-line spectrum alteration concept, this isotropic emission causes approximately half of the emitted energy to be directed away from the solar cell. However, since the dyes are typically placed in a matrix material with an index of refraction greater than one and the external medium will either be air or vacuum and thus have an index of refraction near one, some of the energy emitted away from the solar cell will be redirected toward the solar cell by internal reflection.

If the dyes are in a matrix with an index of refraction of 1.5, the internal reflection reduces the isotropic emission loss to approximately 13% (for higher matrix indices, this amount is reduced even further). Since the index of refraction of the solar cell material is greater than the typical matrix index of 1.5, total internal reflection will not occur at the solar cell/matrix interface. However, some reflection will occur based on the solar cell surface reflectance that can be reduced by an appropriately designed anti-reflection coating (7) (see Figure 2). Complete internal reflection may be increased further by geometrically designing the matrix/external medium interface to decrease the effective escape cone angle (12). In addition, preferential orientation of the emitting molecules may reduce the directional emission loss (5). It should be noted that the isotropic emission loss for the LSC concept is twice the amount for the in-line spectrum alteration concept (~25% for a matrix index of refraction of 1.5) because it possess two interfaces with the external medium (5,15).

High quantum yield of the spectrum shifting dye is essential for success of the in-line spectrum alteration concept. The quantum yield of a dye is the ratio of the number of emitted photons to the number of absorbed photons. This parameter is important because each photon at or above the photovoltaic material's bandgap that enters the device will generate an electron/hole pair (the collection of this electron/hole pair is device dependant). Thus, the higher the quantum yield, the higher the potential increase in short circuit current of the photovoltaic device and correspondingly, the efficiency. Many of the dyes discussed in the LSC literature had high quantum yields (approaching 90%) (14), but had low stability. Therefore, the challenge is to develop a dye that has high quantum yield, high stability, and appropriate spectrum shifting capability. A potential method of increasing quantum yield and stability is to dope the matrix with Ag ions in addition to the organic dye (16, 17).

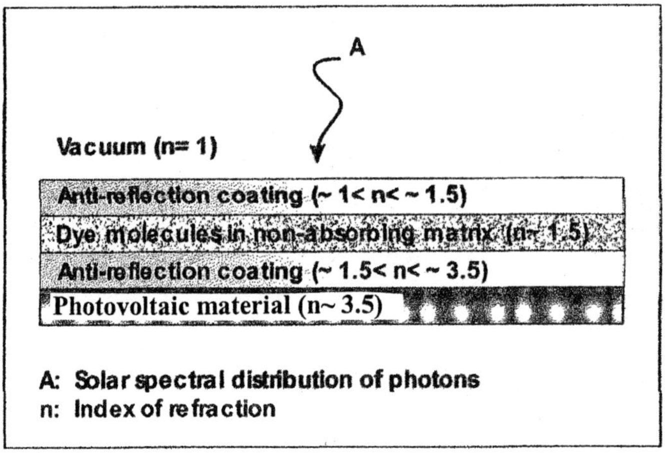

Figure 2. In-line spectrum alteration with anti-reflection coatings.

High quantum yield of the spectrum shifting dye is essential for success of the in-line spectrum alteration concept. The quantum yield of a dye is the ratio of the number of emitted photons to the number of absorbed photons. This parameter is important because each photon at or above the photovoltaic material's bandgap that enters the device will generate an electron/hole pair (the collection of this electron/hole pair is device dependant). Thus, the higher the quantum yield, the higher the potential increase in short circuit current of the photovoltaic device and correspondingly, the efficiency. Many of the dyes discussed in the LSC literature had high quantum yields (approaching 90%) (14), but had low stability. Therefore, the challenge is to develop a dye that has high quantum yield, high stability, and appropriate spectrum shifting capability. A potential method of increasing quantum yield and stability is to dope the matrix with Ag ions in addition to the organic dye (18, 19).

The host matrix choice for the spectrum altering dye is an important consideration. Dye matrix absorption must be very low for wavelengths that are absorbed by the dye and wavelengths that are converted to electricity by the solar cell. Therefore, the matrix material must either have a very low extinction coefficient in the usable wavelength range or it must be made very thin. If the matrix is very thin, the dye must be made very absorbent. The use of dendritic molecular attachments to the base molecule can potentially increase the absorbency of the dye. The choice of matrix affects the effective quantum yield of the dye and its stability.

While difficult to incorporate in a host matrix of conventional glasses or crystals due to the high process temperatures required, dyes have been incorporated in sol-gel SiO_2 (16,20). An inorganic matrix such as SiO_2 has an advantage over a polymeric matrix (typical for a LSC) in that it is expected to be much more resistant to optical transmission changes caused by very high-energy solar photons. An added benefit of SiO_2 is that it is typically a candidate for space protection of solar cells due to its high resistance to atomic oxygen and its excellent optical transmission.

Estimate of Performance Impact

Due to the inherent flexibility of partial spectrum shifting, a general statement regarding its impact on device performance is difficult to make. Therefore, to illustrate its potential impact, a particular spectral shift was chosen and its effect on three representative solar cell types was examined in simple calculation. The three solar cell types included: dual junction crystalline cell (GaAs based), thin-film triple junction amorphous silicon cell (a:Si), and a thin-film single junction CIGS ($CuInGaSe_2$) cell. The analysis was subject to the following assumptions: (1) spectrum shifting dye is applied in a non-absorbing matrix on the solar incident side of the solar cell, (2) matrix and dye do not absorb photons with wavelengths greater than 400 nm, (3) the uncoated solar cell does not convert any photons less than 400 nm into electricity, (4) all photons in the 240-400 nm range are absorbed by the dye, (5) a fraction of these absorbed photons are then emitted at prescribed wavelength(s), and (6) reflectance properties of the photovoltaic device are equal for both the cells coated with the dye and the uncoated cells. The analysis used measured electrical performance data when the three cells were subjected to an AM0 spectrum with photons less than 400 nm filtered out (these values are designated in the equations below with a superscript of "measured"). Other assumptions were made that were particular to each solar cell type and will be presented in the description of the results.

Throughout this derivation the following superscript definitions apply: "new" is the total solar cell parameter with spectrum shifting, "measured" is the actual experimentally measured value of a real device without spectrum shifting, and "add" is the increase in the parameter due to spectrum shifting.

First, the number of solar photons is calculated at each wavelength from 240 nm to 400 nm:

$$N_{ph}(\lambda) = \frac{P(\lambda)}{E(\lambda)} \qquad \left\{ \frac{\#\ of\ photons}{m^2} \right\}$$

where $P(\lambda)$ is the solar energy flux at wavelength λ with units of $[W/m^2]$[21] and $E(\lambda)$ is the energy of a photon at wavelength of λ with units of [J]. The additional electrons collected due to the spectral shift are then calculated using the following relationship:

$$\delta = \left[\sum_{\lambda=240\,nm}^{400\,nm} N_{ph}(\lambda) \right] * QY * IP * EQE(\lambda_e) \qquad \left\{ \frac{\#\ of\ e^-}{m^2 s} \right\}$$

where $N_{ph}(\lambda)$ is defined as the number of solar photons at wavelength λ per area [#/m^2]; QY is the quantum yield, which is defined as the number of photons emitted divided by the number of photons absorbed; IP is fraction of photons emitted that are incident on the surface of the solar cell; and $EQE(\lambda_e)$ is the external quantum efficiency of the device at a wavelength of λ_e (EQE is the number of collected electrons divided by the number of incident photons at a particular wavelength at short circuit). Using the calculated number of additional collected electrons, the additional short circuit current is:

$$I_{sc}^{add} = \delta * q * Area \qquad \{A\}$$

where q is the electron charge [Coulombs] and $Area$ is the active device area [m^2].

The short-circuit current of the device subjected to the altered solar spectrum is then obtained by adding the additional short circuit current to the measured short circuit current of a device with out the spectrum shifting :

$$I_{sc}^{new} = I_{sc}^{measured} + I_{sc}^{add} \qquad \{A\}$$

The new maximum cell power output when subjected to the modified solar spectrum is calculated by assuming the fill factor and the open circuit voltage are not changed significantly by the change in short circuit current:

$$P_{max}^{new} = FF^{measured} * I_{sc}^{new} * V_{oc}^{measured} \qquad \{W\}$$

where FF is the fill factor of the device and V_{oc} is the open circuit voltage. The new efficiency is subsequently calculated as:

$$\eta_{new} = \frac{P_{max}^{new}}{P_{in}} * 100 \qquad \{\%\} \qquad (1)$$

where P_{in} is the solar power incident on the device.

Using this procedure, efficiencies for the three devices subjected to an altered solar spectrum were calculated assuming various values of *QY*, *IP*, and wavelength of emission. The results are tabulated in Tables I, II, and III. The table headings include *QY* (quantum yield), *IP* (percent photons emitted that are incident on the solar cell), η_{wo} (measured conversion efficiency without any spectrum shifting), η_w (calculated efficiency with spectrum shifting – Equation 1), and η_w-η_{wo} (percentage point difference in conversion efficiency between the device with spectrum shifting and the device without spectrum shifting).

Table I shows that measurable gains in efficiency are possible using a high performance spectrum shifting dye (high quantum yield and low emission loss) for a dual-junction GaAs based crystalline solar cell. Since the two junctions were likely well current-matched, the additional photons were assumed to be emitted at two different wavelengths, with each wavelength being absorbed by a different junction. This was necessary because the two junctions are connected in series and therefore an increase in current production by one junction will be limited by the current production of the other junction. Therefore, the photons where divided such that the addition current added to each junction was equivalent. Quantum efficiency measurements showed that the top junction was active with minimum bottom junction absorption at 490 nm and that the bottom junction was active at 800 nm with minimal top junction absorption. The shifted photons were therefore split between 490 nm and 800 nm. The external quantum efficiency of the device was 0.8 and 0.88 at 490 nm and 800 nm, respectively.

For the triple junction a:Si device, the top cell was assumed to be current limiting. It was assumed that the other two junctions (middle and bottom) had sufficient current producing capacity to match the increase in the top cell current caused by the additional photons. External quantum efficiency measurements of the device showed that the top junction absorbs at 470 nm with an external quantum efficiency of 0.78 without significant absorption from the middle and bottom junctions. Therefore, a wavelength of 470 nm was assumed for emission of the shifted photons. The calculated results based on these assumptions are presented in Table II.

Gains obtained for the single junction CIGS cell were not very significant, as can be seen in Table III. For this cell, the shifted photons were assumed to be emitted at 650 nm. At this wavelength, the cell exhibited an external quantum efficiency of 0.58. The poor performance increase of the single junction CIGS cell compared to the multi-junction a:Si and GaAs cells can be explained by the fact that its short circuit current density was already relatively high (~40 mA/cm^2) and the additional current supplied by the shifted photons (~2 mA/cm^2 for QY=IP=1) did not increase it by a significant percentage. The short circuit current for the a:Si and the dual-junction GaAs based cells were ~9 mA/cm^2 and 16 mA/cm^2, respectively.

Table I. Calculated increase in efficiency for crystalline dual-junction solar cell.

QY	*IP*	η_{wo} *[%]*	η_w *[%]*	$\eta_w - \eta_{wo}$
0.33	0.5	22.37	22.81	0.44
0.33	0.75	22.37	22.98	0.61
0.66	0.75	22.37	23.48	1.11
0.9	0.75	22.37	23.84	1.47
1	0.75	22.37	23.99	1.62
1	1	22.37	24.49	2.12

Table II. Calculated increase in efficiency for a:Si triple junction solar cell.

QY	*IP*	η_{wo} *[%]*	η_w *[%]*	$\eta_w - \eta_{wo}$
0.33	0.5	9.29	9.72	0.43
0.33	0.75	9.29	9.94	0.65
0.66	0.75	9.29	10.61	1.32
0.9	0.75	9.29	11.1	1.81
1	0.75	9.29	11.3	2.01
1	1	9.29	11.98	2.69

Table III: Calculated increase in efficiency for CIGS single junction solar cell.

QY	*IP*	η_{wo} *[%]*	η_w *[%]*	$\eta_w - \eta_{wo}$
0.33	0.5	7.97	8.04	0.07
0.33	0.75	7.97	8.08	0.11
0.66	0.75	7.97	8.17	0.2
0.9	0.75	7.97	8.24	0.27
1	0.75	7.97	8.27	0.3
1	1	7.97	8.37	0.4

It is important to reiterate that the above analysis is top-level and it is meant only to demonstrate the potential for increased performance for a particular example of spectrum alteration. Removal of the assumptions may cause the actual performance to differ markedly from that predicted. Also, this particular example is not meant to exhaust the potential applications of partial spectrum shifting. High performance spectrum shifting dyes would add an additional design element that could be used in the overall design of a photovoltaic device, which could result in significant performance gains.

From the above discussion, it is evident that in-line spectrum alteration for increasing conversion efficiency of space solar cells has potential, but it is yet to be proven and there are significant hurdles to cross. Future research should focus on developing spectrum-shifting dyes that have high quantum yield, good stability, and greater absorption at desired wavelengths while maintaining minimal absorption at undesired wavelengths. Dye carriers should be developed that have minimal absorption across the spectral range of the photovoltaic device and the spectral range of dye absorption/emission. Dye carriers must also maintain their transmission characteristics when subjected to the space environment.

Solar Cells Fabricated with Nanoparticles

Semiconductor clusters with sizes of the order of 1 nm have size-dependent physical properties distinct from the behavior found in either bulk materials or isolated atoms (22). Some of these unusual properties may eventually be exploited to overcome intransigent barriers for increased photovoltaic performance. Among the unusual properties of nanoparticles are variations in melting temperatures, charge transport behavior, and optical excitations that differ significantly from the bulk properties. Quantum confinement in nanoscale clusters increases the density of states, which increases optical absorptivity, an important parameter determining the optimal thickness of an absorber layer in a solar cell. In addition, the energy required to add extra charge to a nanoparticle increases as the size of the cluster decreases. This increased energy, due to coulombic repulsion, leads to efficient charge transport from the cluster to its surroundings, necessary for efficient current collection. These quantum size effects become important if the size of the cluster is on the order of or less than the Bohr exciton radius, a_B, given by

$$a_B = \frac{\varepsilon_0 \varepsilon h^2}{\pi \mu e^2}$$

where ε is the relative permittivity of the semiconductor, μ is the reduced effective mass of the exciton, and e is the electron charge (23). The value of the Bohr exiton radius in CdS is about 2.4 nm, but can be as large as 18 nm for PbS (21).

Anomalous bright-red photoluminescence (PL) was first observed in porous silicon (24), which does not normally fluoresce in the visible region. Although some have ascribed the blue-shift in photoluminescence to emission from surface states, several experimental studies (25,26) on nanoparticles have established that the PL shift is due to quantum confinement and that the magnitude of the shift is a function of the size of the nanoparticle. The bandgap of CdSe nanoparticles can be varied from 2.0 to 2.6 eV by varying the particle diameter from 6 to 2 nm. The same variation in CdS nanoparticle sizes will produce bandgaps from 2.6 to 3.1 eV (27,28).

In nanoparticles, the ratio of surface to bulk atoms is much larger than for macroscopic structures. Atoms in the interior of a nanocluster may exhibit a bulk-like crystal structure, but the large surface area results in an increased influence of surface forces on the nanoparticle. The relative instability arising from the surface forces results in a size-dependent melting temperature that decreases substantially as the size of the nanoparticle decreases. The melting temperature of nanoparticles can be observed by monitoring diffraction patterns during transmission electron microscopy (TEM), and entire nanoparticles, including the interior atoms of the cluster, have been observed to melt within a narrow band of temperature. In semiconductors, the melting temperature has been measured as function of size for CdS, GaAs, and Si nanoparticles (29,30). A silicon nanoparticle with a diameter of 7 nm, for example, melts at half the bulk melting temperature of 1687 K. CdS nanoparticles melt at half the bulk value of 1678 K at a size of ~3.5 nm. Thermodynamic models based on the size of the particle and the difference in surface tension between the liquid and solid phases have been fairly successful at predicting the observed variation in melting temperature (20).

Applications of Optical Properties of Nanoparticles

Since the electronic bandgaps of nanoparticles can be tuned over a large portion of the solar spectrum, they could be incorporated into a solar cell to more efficiently absorb light. A device that incorporates a high density of nanoparticles of an appropriate size distribution could be imbedded in a semiconducting matrix to absorb and convert to electrical power the majority of the solar spectrum. The nanoparticle solar cell would essentially function in the manner a crystalline multijunction solar cell increases the efficiency of conversion by allowing high energy photons to be absorbed at a junction of relatively large bandgap and lower energy photons to be absorbed at the lower energy bandgap junction. Since the photon energy in excess of the bandgap is converted to waste heat, multi-bandgap solar cells increase the efficiency of solar conversion efficiency. Since the theoretical multijunction solar cell efficiency increases with the number of junctions, increases in efficiency could be seen by increasing the range and number of nanoparticle bandgaps. Nanoparticle solar cells could approximate a multijunction system with a large number of junctions. The maximum theoretical efficiency of a three bandgap multijunction cell is 56% and for 36 junctions, the theoretical efficiency increases to 72%. To approach the theoretical multijunction efficiencies, a method to collect the current produced by the nanoparticles is needed. In traditional crystalline multijunction cells, current is maximized by minimizing carrier recombination and maximizing charge mobility through a defect-free crystal lattice. The nanoparticle device will also require that the photon flux is incident on the highest bandgap junction succeeded by junctions of decreasing bandgap energy.

The first attempts to realize this type of nanoparticle device have been inorganic/organic composite solar cells produced by embedding inorganic semiconductor nanoparticles in a polymer matrix (20). These types of solar cells can be produced inexpensively on lightweight substrates and under non-vacuum conditions. In a traditional solar cell, the p-n junction is responsible for the charge separation of the electron-hole pair produced by the excitation of the absorbed photon. In the nanoparticle/matrix cell, the nanoparticle-polymer interface is responsible for charge separation after excitation in the inorganic nanoparticle. Holes are transferred to the polymer matrix, which transports them to the electrode. The nanoparticles act as electron acceptors and can transport the these charges to the opposite electrical contact, provided that the density of the nanoparticles is high enough to form a percolation network. The open circuit voltage of this type of cell is determined by the difference in work functions of the opposite electrodes, which are typically an ITO ($\phi = 4.7 - 4.9\,\text{eV}$) and a metal layer ($\phi = 4.3\,\text{eV}$ for Al). Incorporation of CdSe nanoparticles into a conjugated polymer matrix has been shown to improve efficiencies over that of polymer devices (31,32). However, top efficiencies for conjugated polymer devices have not exceeded ~2% due to low charge mobilities. Another disadvantage of polymer-based solar cells is susceptibility to damage in the space environment by UV and radiation exposure. The efficiency of the nanoparticle-to-polymer charge transfer is sensitive to the chemical termination of the nanoparticle surfaces (33). In particular, oxidation of the nanoparticles can hinder charge transfer.

For space applications, a more long-lived cell may be made of nanoparticles imbedded in an inorganic semiconductor matrix (34). One method for growing nanoparticles on crystalline wafers in vacuo is through Stranski-Krastanov growth regimes (35). Growth of this type of nanoparticle and the surrounding semiconductor matrix will most likely require traditional vacuum CVD and crystalline wafer substrates, increasing production and array costs. However, charge transfer in the inorganic matrix will not suffer from the low mobilites found in a conjugated polymer matrix, and both the nanoparticles and the matrix may be grown uninterrupted via CVD. Results on investigations into this type of photovoltaic device have not yet been published.

Nanoparticles are also employed in dye-sensitized or Grätzel-type solar cells (36). In these cells, light is absorbed by dye molecules (usually Ru complexes) bound to the surface of wide-bandgap nanoparticles, normally TiO_2. When the dye absorbs a photon and an exiton is generated, the electron is injected into the TiO_2 nanoparticle, while the hole is transferred to the surrounding material. The nanoparticles are immersed or embedded in an electrolyte, which can be a liquid, iodine-based solution employing a redox reaction or a solid conjugated polymer matrix. Liquid electrolytes have been the more successful of the two, producing cells with efficiencies of up to ~10% (AM 1.5). Both the cell materials and production methods for dye-sensitized cells are inexpensive, requiring only a sintering step to ensure good mobility of the electrons through the TiO_2 matrix. Cells have been fabricated on metal foil substrates in addition to the glass superstrates favored for terrestrial applications. However, liquid electrolytes are ill-suited for space applications, and cells with solid polymer electrolytes have much lower efficiencies. Solid electrolyte materials may suffer from degradation under space conditions. However, if solid electrolyte cell efficiencies can be improved to equal that of the liquid electrolyte cells, protective coatings may enable these cells to protect the cells from significant levels of degradation. Finally, the functioning of dye molecules may degrade over time through photobleaching and radiation damage. Optimization of the stability and sensitivity of dyes is an ongoing area of research.

Application of Melting Properties of Nanoparticles

An intriguing application of nanoparticles is the use of their reduced melting temperatures to form thin films of tailored crystallinity. Current processes for recrystallization occur at temperatures at or near the bulk melting point. Atomic mobilities at these temperatures allow incorporation of atoms into sites that continue the single crystal structure of the seed. At lower temperatures, the kinematics of crystal growth favor polycrystalline growth and the incorporation of grain boundaries into the structure. The reduced melting temperatures of nanoparticles results from a relative instability due to the higher energy surface states. Therefore, atoms comprising a nanoparticle have increased mobilities at lower temperatures than the builk material. The coalescence of two nanoparticles results in a larger particle with a smaller proportion of surface atoms and an increase in the original melting temperature.

Surface atoms of nanoparticles are mobile at a temperature slightly below the melting temperature, resulting in a sintering temperature about 2/3 of the reduced melting temperature. Sintering of semiconductor nanoparticle precursors has been shown to form a densified polycrystalline thin film (7). The sintering process temperature is compatible with polymer substrates, and a thin-film transistor device has been produced on a polymer substrate from nanoparticle precursors (37). These thin films were formed from colloidal CdSe nanoparticles deposited from a liquid solution suspension onto the substrate. Annealing produced crystallites slightly larger than the original nanoparticle sizes. Yang *et al.* has observed nanoparticles fusing into long parallel strips when annealed. When annealed further, the strips fused to form a textured crystalline layer (24). Ginley *et al.* has recrystallized metal nanoparticle precursors for application as metal contacts on solar cells (38).

For space-based solar cells, the ultimate goal is the inexpensive deposition of thin, crystalline photovoltaic films onto lightweight, flexible substrates, producing substantially higher specific power levels than current solar cell technologies. Single crystal, multijunction solar cells have produced record-high efficiencies, currently at 30.1%. These cells, grown on single crystal wafers are brittle and fragile and require a rigid support structure. The result is a specific power limited to less than 100 W/kg. Thin-film crystalline solar cells on lightweight substrates are capable of specific power levels of >1000 W/kg because of the elimination of the rigid support structure. Several methods have been developed to produce crystalline thin-film silicon cells for terrestrial uses (39,40). High efficiency single crystal or large-grain crystalline cells on foreign substrates are produced either by transfer of single crystal layers onto foreign substrates or by the growth of thin crystalline layers in high temperature processes. The latter methods, such as Si ribbon technology and zone melting recrystallization take advantage of the large equilibrium crystals grown at temperatures near the melting point of the material. Unfortunately, these high temperature substrates are too massive and fragile for space applications. The high temperatures require a large thermal budget, increasing the cost of these solar cells. The substantially reduced melting temperature of nanoparticles provides an avenue for novel semiconductor processing techniques. These techniques may lead to the production of crystalline thin-film solar cells on low-temperature, lightweight, and non-epitaxial substrates. The high atomic mobilities within nanoparticles at relatively low temperatures allow for the coalescence of nanoparticles with solid surfaces.

The growth of large-grain solar cells from nanoparticle precursors faces several hurdles. Polycrystalline solar cells have lower efficiencies than single crystal cells due to minority carrier recombination at the unpassivated grain boundaries. In previous work that demonstrates the efficiencies possible with large-grain photovoltaic devices, Venkatasubramanian *et al.* have produced 20% efficient (AM1.5) GaAs cells using cast Ge wafers with sub-mm grain size as substrates (41,42). Using a model, Kurtz and McConnell (43) showed that a polycrystalline solar cell with grain sizes greater than 20 μm and sufficiently passivated grain boundaries is capable of performance on par with single crystal devices.

Understanding the dynamics of nanoparticle atoms at temperatures near their melting temperature will be imperative to obtaining a tailored crystalline thin-film structure from nanoparticle precursors. Owing to the size of nanoparticles, their properties and kinetics are accessible to being explored through molecular dynamics simulations. Several simulations have been conducted that shed light on the melting, coalescence, and stability of nanoparticles. A series of molecular dynamics experiments on Au nanoparticles by Lewis (44) using the embedded-atom method, melt and coalesce particles at a number of temperatures. Lewis has found that initially premelting occurs for the surface atoms, but when the particle is heated to slightly higher temperatures, the center of the nanoparticle also becomes liquid-like. Two liquid nanoparticles will coalesce into a solid, larger cluster with a melting temperature appropriate for the size of the coalesced particle. More interesting, the simulations show that a liquid nanoparticle will coalesce with a larger, solid nanoparticle. Semiconductor nanoparticles are more difficult to simulate than Au nanoparticles owing to their covalent bonding, but the structure and stability of Ge nanoparticles have been examined using *ab initio* potentials (45). The binding energy of a nanoparticle composed of a Group IV element, Pb, has also been examined via Hartree Fock calculations (46). However, molecular dynamics simulations using classical potentials are more practicable and may be adequate if compared to experimental observations.

With the insight provided by molecular dynamics simulations and experimental studies on the melting, coalescence, and freezing behavior of semiconductor nanoparticles, design of a thermal processing technique may allow the growth of thin films of tailored crystallinity on arbitrary substrates. This technique may involve the suppression of coalescence among the individual nanoparticles for the preferential coalescence onto a larger, "seed" crystal.

Synthesis of Nanoparticles for Solar Cell Applications

Unlike many semiconductor devices, solar cells require production in large areas with high uniformity. The synthesis of nanoparticles with controlled size distributions, chemical composition, and chemical termination is critical to produce large area, high efficiency photovoltaics. There are several methods of nanoparticle production that are currently under investigation. For solar cell applications, it will be practicable to work with colloidal nanoparticles deposited from solution. The nanoparticles must be passivated in solution to retard their growth and to cap the size distribution. Solution methods of synthesis have been more successful in avoiding the oxidation that results from methods such as pyrolysis. However, the terminating species must not hinder the formation of the solar cell device and should not contribute organic contamination to the final device. Colloidal Si and Ge nanoparticles have successfully been synthesized by metathesis reaction with several species of passivating layers (47,48). High volume production of colloidal nanoparticles will be important for future applications.

Rectifying Antennas Applied to Solar Power Conversion

Of all the types and configurations of power conversion devices, the most practical for earth orbiting space applications in terms of efficiency, mass, cost and stowed power density has been an array of crystalline solar cells. The performance of systems using these cells has been ever improving, and recent advances in monolithic crystalline solar cell technologies have pushed triple-junction solar cells to 30% efficiencies on prototypical sized cells (4 cm^2) and the minimum average efficiency of commercially available cells to 27.5%. Satellite systems using these high performance solar cells have achieved 70-80 W/kg, 15-20 kW total power, and average cell costs have dropped to \$300-\$400 per watt produced at the cell level. Programs are in place to push the efficiency of monolithic crystalline multijunction solar cells ever higher with goals as high as 35% for near term progress, this in turn will increase satellite total available power, reduce array mass and cost. Reaching beyond 32-35% represents a significant challenge however, as new material systems must be developed that are a very close match to existing material systems in terms of lattice match and current match at selected bandgaps. This poses a significant hurdle to the industry as these material systems become increasingly complex (quaternary systems are now being explored) and difficult to fabricate with high quality. An alternative approach to increasing the performance of satellite energy conversion systems is to utilize thin-film photovoltaics with low density or very thin substrates. The efficiency of thin-film photovoltaics at the present time is approximately 10% for devices with large areas, and as high as 18% for very small area devices at the research stage of development. The reduced efficiency requires a significantly increased area for array sizes, the final performance is projected to be higher than a monolithic crystalline based array in terms of \$/W, stowed power density, array mass and ultimate cost. The increased size of the array does have significant implications associated with attitude control and power transfer from the array to the payload since the system may use the free flyer concept with umbilical cord for very high power satellites (100 kW). There are other approaches such as mechanically stacked solar cells (expensive, suitable for concentrator systems), thermal electric converters (TPV, AMTEC, Stirling, thermionic, etc). These, however, have significant drawbacks such as vibrations, relatively low power density, and high cost when compared to the state of the art for satellite power systems, and low theoretical efficiencies. There is a technology that offers very high theoretical efficiencies but at the cost of enormous development, namely rectifying antennas (rectennas) with feature size small enough to utilize the wave nature of light coupled to rectifying diodes fast enough to convert this alternating current to direct current power. This technology is well understood, has long history of development and has performed well at high powers and GHz frequencies. The history, performance, theoretical limits and present state of the art and direction of the technology will be discussed as well as the Air Force involvement in the technology.

History of Rectennas

The history of rectifying antennas (rectennas) is directly linked to the history of wireless power transmission, and this history has been documented in sufficient detail in several articles by William C. Brown (47,48,49). This history begins with the spark gap experiments and discoveries by Heinrich Hertz, in which he demonstrated a complete system to generate and transmit high frequency RF energy, to the efforts to develop a solar power satellite (beaming power from a satellite to a receiver on earth), a time period from the late 1800s to about 1980 (50). During this time systems were developed to transfer significant amounts of power over lengthy distances with enough efficiency to begin to consider them for practical applications. The devices of merit that were developed were the klystron tube, the magnetron, the phased array antenna system, fairly high speed recitifiers, and a rectenna system that reduced harmonic radiation and increased efficiency. These systems were linearly polarized systems that simplified the transmission and receiving antenna configurations. The frequency for which systems were designed stagnated around the 2.4-2.5 GHz bandwidth that were reserved for industrial, scientific and medical (ISM) applications. This band of frequencies also presents a window for minimum attenuation through the earth's atmosphere.

Work on the rectenna portion of a power beaming system produced significant improvements to power systems, and a number of milestones that make the concept interesting from a satellite power system point of view. Development work during this time period resulted in an 85% efficient rectifying antenna utilizing linearly polarized radiation at 2.45 GHz with the efficiency defined at the ratio of output dc power from the rectenna terminals and input power as the incident radiation at the surface of the rectenna. Rectifying antenna grids were developed that reached over 500 W/kg and this compares very well with present SOA solar cell arrays. Thin-film rectennas were developed and shown to be highly tolerant to space environments. In fact, during this time period there was much development work on a 'Satellite Power System' where power beaming from the to satellites and from satellites to earth were explored. Here power levels would reach gigawatt levels and distances on the order of a hundred miles were examined. Towards that goal, a demonstration of beamed power over a distance of one mile was conducted at the JPL Goldstone facility in the Mojave Desert. Of the microwave power intercepted by the rectenna array, 84% was converted into a dc power level of over 30 kW. The frequency was 2.39 GHz and the dc power was expended in a resistive load.

There has been increasing interest in raising the frequency of operation of microwave power beaming systems as a result of space based power beaming satellite systems and the advantage of reduced array size, and because the harmonics exist at much higher frequencies reducing the electromagnetic interference problem. The difficulty associated with increasing the frequency is the deterioration of device properties with increased frequency and at extremely high frequencies the device sizes themselves become challenging as well replicating a component across the expanse of an array large enough to intercept appreciable power levels is also a major challenge. Results of rectennas at 35 GHz have shown promise but not without significant reduction in efficiency as a result of rectifier losses.

To fully appreciate the significance of the problem with utilizing the wave nature of light and rectennas, it is important to note that the solar spectrum consists of frequencies in the 100 – 1000 THz range, or about 10 to 100 times the highest frequency being considered for use with power beaming systems. This poses significant problems with rectifying diodes operating in this regime, these frequencies are beyond the capacity of even fast Schottky barriers to operate efficiently or at all. Typical diodes are shunted at these frequencies and a tunnel diode with very small device area is thought to be the path to success in this field. Further, the size and geometry of the antennas must be carefully considered. The feature size of antennas is on the order of the wavelength of the frequencies of interest, and as shown in Figure 3 the feature sizes would be approximately 250 nm to 1500 nm. A further complication with a rectenna system employed to convert optical power is that light is randomly or elliptically polarized, complicating antenna design. Once a rectenna element has been successfully proven, the final problem to overcome is replicating this extremely small device over large areas. Therefore the problems associated with fabricating halfwave antennas that are randomly polarized with efficient rectifiers can be

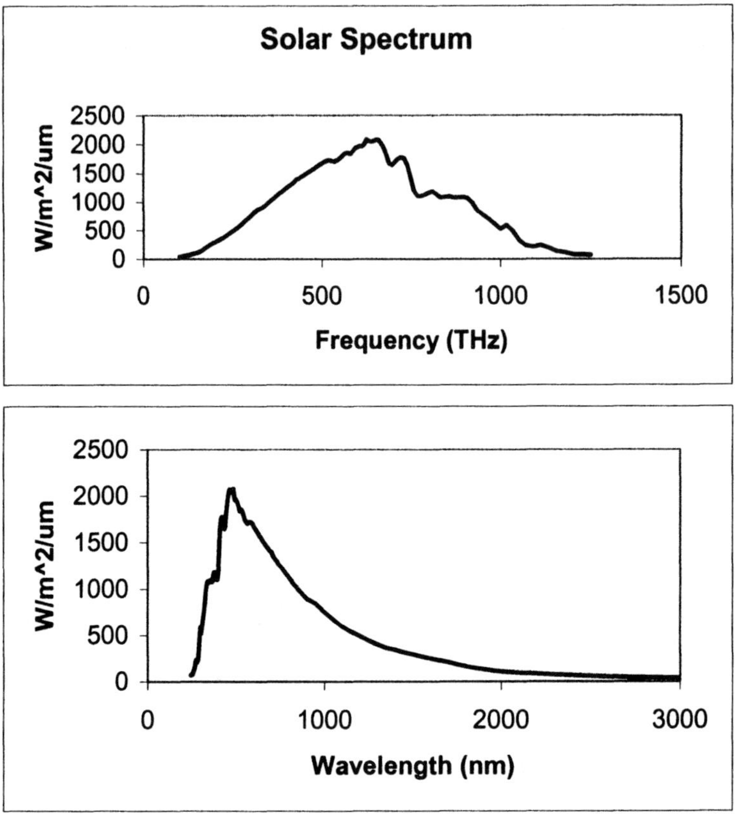

Figure. 3. Solar spectrum plotted versus frequency and wavelength of light.

appreciated. These are small devices and structures that would benefit enormously from nanotechnology development.

Component Development and Applications

Research and development associated with rectenna systems suitable for use as detectors and power converters centers around the development of broadband, broadband polarized antennas and rectifiers for use at 100s of terahertz, as well as replication techniques. While the development of antennas for this application seems to be making progress, the development of rectifiers will require much more development and innovation. There are two components that have been developed that show promise for the ultimate challenge of converting the sun's power into direct current electrical power using the wave nature of light rather than the particle (photon) nature of light; the several designs for an elliptically polarized antenna system and the metal-oxide-metal (MOM) junction.

Beginning with the elliptically polarized antenna, new approaches have been developed to efficiently intercept arbitrarily polarized incident waves with broad spectral content. The work done to date has been in the ISM band (2.4-2.5 GHz), a band of frequencies reserved for industrial, scientific and medical applications, and one that exhibits a good window for propagation through the atmosphere. The emphasis to date has been on power beaming, transferring power over great distances at high efficiencies. These applications have, in most cases, utilized non-time varying, continuous wave, well defined RF power whose characteristics were known *a priori*. New applications for rectenna systems include infrared detectors, infrared power converters, and power recycling. The power recycling concept is an interesting one since RF power is continuously radiated into space at very high power levels with a variety of carrier frequencies and sidebands. Most of this power is wasted and can contribute to health and safety concerns to operating personnel. There may be situations where it may be profitable to intercept this power and convert it back into usable electric power, for instance utilizing the side of a building that is situated in a location where it naturally intercepts a goodly amount of RF power. The major difficulty here is that this power has various polarizations, frequencies, and uncorrelated phases. This problem has been solved to some extent with the development of broadband rectenna arrays such as the grid and spiral arrays (51). These arrays were operated from 4.5 to 15 GHz and achieved 35% - 40% efficiency. The antennas are operated not as a resonant element but with flat response to frequency over as large a band as possible. The rectifiers used in any rectenna are more efficient when operated in the very nonlinear portion of the I-V curve, and in some applications this will require biasing, an inherent loss. In this application the rectenna was shown to operate with self biasing as a result of the circuit layout. Both the grid and spiral rectenna designs were shown to have good response over a broad fractional bandwidth, fairly good efficiency as a function of input RF Power, and fairly good efficiency as a function of input wave polarization. Certainly the antenna designs are on a good trajectory towards eventual use in infrared and visible wavelengths.

The detectors, on the other hand, have not faired as well in development towards good efficiencies at frequencies at much higher than the ISM band. While the efficiencies exhibited by rectenna systems in the ISM have been very good, as high as 85% for 2.4 GHz systems, projected efficiencies in frequencies a decade or so higher have been much lower. Even with well defined radiation, the rectenna characteristics suffered a 30% - 53% reduction from the efficiency measured at 2.4 GHz when the frequency was raised to 35 GHz, based on predictions and experimental evidence provided by Tae-Whan Yoo of Texas A&M University (52). The efficiency was reduced mainly as a result of the capacitance and the shunt and series resistances of the rectifying diode. There is hope even here however, and success may be achieved with the development of metal-oxide-metal (MOM) devices. The earliest MOM device reported in the literature utilized a cat whisker tungsten wire in contact with a polished metal plate for detection of submillimeter waves (53). The contact was not ohmic due to some trapped material at the contact and conduction was assumed to happen by means of electrons that tunnel through this uncontrolled insulator. The device has a very small capacitance (10^{-2} pF) and a spreading resistance of a few ohms which led to a limiting RC time constant of 10-13 s. The device was very unstable mechanically and a clever advance in the state of the art was reported in 1978 (54) and named the edge-MOM device.

One of the key difficulties in achieving the required performance from a rectifying device is to minimize the device capacitance (as well as resistance) and this relates directly to the area of the device and inversely to the distance between the plates. These devices were able to achieve areas as small as 10^{-10} cm^2 that responded to the electric field to at least 0.6 μm, and distances in between the plates of ~100 Å. The devices were manufactured by overlapping of a 1 μm metal strip and an oxidized edge of a metal strip, ~100 Å thick, that achieved an active area of 100 cm^2. The device was fabricated using conventional fabrication techniques. To reduce lead resistance two of these devices are fabricated from a cental pad, and this arrangement results in a convenient half-wave dipole configuration. Currents from the devices were developed as a result of three different mechanisms, photo-assisted tunneling, thermal-assisted tunneling, and field-assisted tunneling (Fermi level modulation). Photo-assisted tunneling is a fast process whereby electrons in the vicinity of the junction surmount the barrier with larger probabilities resulting from gaining an energy $h\omega$ from incoming photons. The quantum yield from this process is low (less than 0.1 percent) and since the devices have an inherently small area this current is small. Thermal assisted tunneling results from incoming photons coupling to phonons and raising the temperature of the device. As a result of the increase in temperature Fermi electrons will gain ~kdT additional energy and the tunneling current will increase. This effect is slow and is not directly related to the junction area. The open circuit thermal assisted signals were observed to increase monotonically with bias. Finally, field assisted tunneling results from a coupling mechanism between the incoming photon and surface plasmons (or other mechanisms) in the visible range. This creates an optical voltage that will alternate at the applied frequency around the quiescent point. Due to the nonlinearity of the I-V characteristic of the junction, harmonics of the driving frequency result. The direct current voltage delivered from the device is related to these harmonically driven currents. This optical voltage is a function of temperature and is greater at lower temperatures, due to an enhancement in the sensitivity of the device. The results from testing these devices show that they are indeed sensitive to radiation in the visible range to 600 THz encompassing a good part of the solar spectrum, although efficiencies, operating voltage and overall performance is unclear (55).

The results of the development of broadband elliptically polarized antenna arrays and rectifying junctions places the rectenna arrays suitable for the AM0 spectrum merely on the first rung of development. The real attraction of this technology are the high theoretical efficiencies and the performance of linearly polarized rectennas in ISM bands to date.

Road to Further Development and Challenges

The road to further development of rectennas and their use as power converting arrays or even as detectors for IR frequencies hinges on the development of efficient rectifiers as there seems to be a number of proven techniques for fielding broadband elliptically polarized antennas that offer substantial efficiencies. Data for the rectifiers mentioned above were published in 1978, improvements are slow and it will require much innovation on the part of researchers to achieve useable performance. There are, however, a number of interested parties seeking to develop optical rectenna technology for applications ranging from detection of intelligence signals, extending the frequency range of present communications systems, and sensing very weak signals. An envisioned optical rectenna system may prove to be very radiation hard. Certainly the excitement for such a system is present, and the significant difficulties that must be surmounted are clearly understood.

References

1. Sze, S.M. *Physics of Semiconductor Devices*; Wiley & Sons: New York, 1981; p. 798.
2. Tringe, J.; Merrill, J.; and Reinhardt, K. *28th IEEE Photovoltaic Specialists Conference*, Anchorage, AK, USA, **2000**, 1242-5.
3. Murphy, D. M.; Allen, D. M. *26th IEEE Photovoltaic Specialists Conference,* Anaheim, CA, USA, **2000**, 1027-30.
4. Weber, W. H.; Lambe, J. *Applied Optics* **1976,** *15,* 2299-2300.
5. Batchelder, J. S.; Zewail, A. H.; Cole, T. *Appl. Optics* **1979,** *18,* 3090-3110.
6. Green, M. A.; Zhao, J.; Wang, A.; Wenham, S. R. *IEEE Electron Device Lett.* **1992,** *13,* 317-318.
7. Green, M. A. *Solar Cells,* Dept of Registrar/Publications Division, University of New South Wales, Kensington, New South Wales, **1992.**
8. Weber, J. *Physics Lett.* **1976,** *57A*, 465-466.

9. Beer, D.; Weber, J. *Optics Comm.* **1972,** *5*,307-309.
10. Rico, F. J. M.; Jaque, F. *J. Power Sources*, **1981,** *6*, 383-388.
11. Bakr, N. A.; Mansour, A. F.; Hamman, M. *J. Appl. Polymer Sci.* **1999,** *74*, 3316-3323.
12. Zewail, A. H.; Batchelder, J. S. US Patent 4,227,939, 1980.
13. Kondepudi, R.; Srinivasan, S. *Solar Energy Mater.*, **1990,** *20*, 257-263.
14. Olson, R. W.; Loring, R. F.; Fayer, M. D. *Appl. Optics* **1981,** *20*, 2934-2940.
15. Reisfeld, R.; Eyal, M.; Valeri, C.; Zusman, R. *Solar Energy Mater.* **1988,** *17*, 439-455.
16. Hinseh, A.; Zastrow, A.; Wittwer, V. *Solar Energy Mater.* **1990,** *21*, 151-164.
17. Wilson, H. R. *Solar Energy Mater.* **1987,** *16,* 223-234.
18. Reisfeld, R. *J. Non-Crystalline Solids*, **1990,** *121*, 254-266.
19. Tada, H. Y.; Carter, J. R.; Anspaugh, B. E.; Downing, R. G. *Solar Cell Radiation Handbook, Third Edition*, JPL Publication 82-69, 1982.
20. Alivisatos, A.P. *Science,* **1996,** *271*, 933-937.
21. Zhang, J. Z. *J. Phys. Chem. B* **2000,** *104*, 7239-7253.
22. Canham, L. T. *Appl. Phys. Lett.* **1990,** *57,* 1046.
23. Ledoux, G.; Guillois, O.; Porterat, D.; *et al. Phys. Rev B* **2000,** *62,* 15942.
24. Yang, C.-S.; Kauzlarich, S. M.; Wang, Y. C.; Lee, H. W. H. *J. Cluster Sci.* **2000,** *11,* 423.
25. Bawendi, M. G.; Steigerwald, M. L.; Brus, L. E. *Annu. Rev. Phys. Chem.* **1990,** *41,* 477.
26. Vossmeyer, T.; Katsikas, L.; Giersig, M.; Popovic, I. G.; Diesner, K.; Chemseddine, A.; Eychmuller, A.; Weller, H. *J. Phys. Chem.* **1994,** *98,* 7665.
27. Goldstein, A. N.; Echer, C. M.; Alivisatos, A. P. *Science,* **1992,** *256,* 1425-1427.
28. Goldstein, A. N. *App. Phys. A* **1996,** *62*, 33-37.
29. Greenwood, N. C.; Peng, X. G.; Alivisatos, A. P. *Phys. Rev. B* **1996,** *54,* 17628.
30. Erwin, M. M.; Kadavanich, A. V.; McBride, J.; *et al. Euro. Phys. J. D* **2001,** *16,* 275.
31. Greenham, N. C.; Peng, X.; Alivaisatos, A. P. *Phys. Rev. B* **1996,** *54*, 17628-17637.
32. Bawendi, M. G. *Sol. Stat. Comm.* **1998,** *107,* 1998.
33. Seifert, W.; Carlsson, N.; Miller, M.; Pistol, M.-E.; Samuelson, L.; Wallenberg, L. R *Prog. Crystal Growth Charach.* **1996,** *33,* 423-471.
34. O'Regan, B.; Grätzel, M. *Nature,* **1991,** *353*, 737.
35. Ridley, B. A.; Nivi, B.; Jacobson, J. M. *Science,* **1999,** *286*, 746-749.
36. Ginley, D. S. *Photovoltaics for the 21st Century Proceedings* **1999,** *99-11,* 103.

37. Slaoui, A.; Monna, R.; Poortmans, J.; Vermeulen, T.; Evrard, O.; Said, K.; Nijs, J. *J. Mater. Res.* **1998,** *13*, 2763-2774.
38. Bergmann, R. B.; Rinke, T. J. *Prog. In Photovoltaics* **2000,** *8*, 451-464.
39. Venkatasubramanian, R.; O'Quinn, B. C.; Hills, J. S.; *et al.* *25th Photovoltaics Specialists Conference* **1996,** Washington, D.C., 31-36.
40. Venkatasubramanian, R.; O'Quinn, B. C.; Silvola, E. *et al.* *26th Photovoltaics Specialists Conference* **1997,** Anaheim, CA, 811-814.
41. Kurtz, S. R.; McConnell, R. *Future Generation PV Technologies: 1st NREL Conference,* AIP, **1997,** *404*, 191.
42. Lewis, L. J.; Jensen, P.; Barrat, J.-L. *Phys. Rev. B* **1997,** *56*, 2248-2257.
43. Pizzagalli, L.; Galli, G.; Klepeis, J. E.; Gygi, F. *Phys. Rev. B* **2001,** *63*, 165324/1-5.
44. Mazzone, A. M. *Computational Materials Sci.* **2001,** *18*, 185-192.
45. Taylor, B. R.; Kauzlarich, S. M.; Lee, H. W. H.; Delgado, G. R. *Chem. Mater.* **1998,** *10*, 22-24.
46. Taylor, B. R.; Kauzlarich, S. M.; Delgado, G. R.; Lee, H. W. H. *Chem. Mater.* **1999,** *11*, 2493-2500.
47. Brown, W. C. IRE Int. Conv. Rec. **1961,** *9*, 93-105.
48. Brown, W. C. *IEEE Trans. Microwave Theory Techniques* **1984,** *MTT-32*, .
49. Brown, W. C. *IEEE Trans. Microwave Theory Techniques* **1992,** *40*, .
50. Hagerty, J. A.; Lopez, N. D.; Popovic, B.; Popovic, Z. *European Microwave Conf. Digest* **2000,** *2*, 13.
51. Dickinson, R. NASA Tech Briefs, Winter 1978.
52. Yoo, T. Ph.D. Thesis, Texas A&M University, TX, 1993.
53. Does, J. W. *Microwave J.* **1966,** *9,* 48-55.
54. Heiblum, M.;Wang, S.; Whinnery, J. R.; Gustafson, T.K. *IEEE J. Quantum Electronics* **1978,** *QE-14*, .
55. Fumeaux, C.; Alda, J.; and Boreman, G. D. *Optics Lett.* **1999,** *24*.

Indexes

Author Index

Subject Index

D

G

H

I

O

Q

T

U

V

W

X

Y

Z